AF594927

The Era of Assisted Reproductive Technologies Tailored to the Specific Necessities of Species, Industry and Case Reports

The Era of Assisted Reproductive Technologies Tailored to the Specific Necessities of Species, Industry and Case Reports

Editor

David Martín Hidalgo

MDPI • Basel • Beijing • Wuhan • Barcelona • Belgrade • Manchester • Tokyo • Cluj • Tianjin

Editor
David Martín Hidalgo
Universidad de Extremadura
Spain

Editorial Office
MDPI
St. Alban-Anlage 66
4052 Basel, Switzerland

This is a reprint of articles from the Special Issue published online in the open access journal *Animals* (ISSN 2076-2615) (available at: https://www.mdpi.com/journal/animals/special_issues/The_era_Assisted_Reproductive_Technologies_tailored_the_specific_necessities_species_industry_case_reports).

For citation purposes, cite each article independently as indicated on the article page online and as indicated below:

LastName, A.A.; LastName, B.B.; LastName, C.C. Article Title. *Journal Name* **Year**, *Volume Number*, Page Range.

ISBN 978-3-0365-0828-3 (Hbk)
ISBN 978-3-0365-0829-0 (PDF)

Cover image courtesy of David Martín Hidalgo.

Contents

About the Editor

David Martín Hidalgo DVM, Ph.D. After finished his veterinary degree in Cáceres (Spain), David Martín was enrolled in the Ph.D. program of the University of Extremadura (Spain). It was during this period (2009–2013) that his interest in the improvement of current ART outcomes grew. Dr. Martín's Ph.D. topic was the improvement of seminal dose conservation at low temperatures using, as the principal strategy, the use of antioxidants. In addition, he participated in the description of AMPK's roles, a master energy sensor, in sperm features of several mammals. Later, looking for a period of postdoctoral specialization, he moved to the University of Massachusetts (USA) where he studied how spermatozoa manage their metabolism during capacitation (2015–2017). Thus, based on sperm energy substrates, he designed a capacitation medium that increases the percentage of hyperactivated spermatozoa, bypassing male subfertility. Because of the importance and experience and knowledge gained during his first postdoc, he performed a second postdoc at the University of Porto (Portugal) (2018–2020) with the final outcome of a patented medium that enhances human sperm capacitation. Currently, he is working at the University of Extremadura (Spain) with the aim of spreading the enhanced capacitation media to other species with issues in achieving fertilization using classic ART procedures.

David Martín also participates in scientific life as a reviewer of different journals, such as: Scientific Reports, Reproduction Fertility Development, Toxicology of Reproduction, Theriogenology, Reproduction and Animal Reproduction Science.

Preface to "The Era of Assisted Reproductive Technologies Tailored to the Specific Necessities of Species, Industry and Case Reports"

This book covers the manuscripts of the Special Issue of Animals titled: *"The Era of Assisted Reproductive Technologies Tailored to the Specific Necessities of Species, Industry and Case Reports"* with a total of 12 original manuscripts and three reviews. All of them are focused on different strategies to enhance the reproductive outcome of our main domestic farm animals (pig, cattle, horse, rabbit and sheep) by the use of novel assisted reproductive technologies (ARTs).

Fertilization is a complex event to orchestrate that can be affected by countless factors, for instance, toxicants, diseases or genetic issues. Thus, for example, it has been demonstrated that the ovine gene BMP15 presents a mutation named FecX. Interestingly, in heterozygosis, ewes exhibit an increase in the ovulation rate, producing more lambs per birth. Unfortunately, homozygosis induces ewe sterility. It has been described that rams carrying the FecX mutation exhibited a better sperm motility parameters and a higher pregnancy rate than non-carrier rams after performing artificial inseminations (AIs) (Abecia et al., 2020). Therefore, when looking to increase profit associated with a better farrowing rate, the selection of males carrying FecX is highly recommended. Nevertheless, before performing AI with a male carrying FecX, the ewe population should test as FecX wild-type or non-carrier. In the same line of thought, in the improvement of farm industry profits, it is a common strategy to select parents that transmit a desired phenotype. For instance, in rabbits, selection can be based on post-weaning daily weight gain. Nevertheless, this selected trait seems to be exhausted after 37 generations of paternal selection (Juarez et al., 2020). At this point, other inherited phenotype selection criteria must be considered (Juarez et al., 2020), for example, feed efficiency, body fat content, litter size, etc.

One of the first steps of the fertilization process begins with ejaculation into the vagina and the female tract undergoes dynamic modifications after it interacts with the ejaculate. Thus, using the rabbit as an animal model, a species where ovulation is induced after mating, it was shown that seminal plasma triggers the differential expression of the glucocorticoid receptor (NR3C1/GR) in the female rabbit reproductive tract (Ruiz-Conca et al., 2020). This exhaustive work analyzed the expression levels of the glucocorticoid receptor changes along seven different subsections of the female rabbit reproductive tract. In addition, differential NR3C1/GR expression throughout early embryo development suggested a relevant role of GR action, mediated by NR3C1 in oviductal and uterine embryo transport (Ruiz-Conca et al., 2020). Further investigation should confirm this observation and be kept in mind when seeking protocols with the final aim of enhancing ART outcome in rabbits.

Sperm, oocyte and embryo cryopreservation is one of the most common ARTs used worldwide that allows long-term preservation of genetic diversity. Nevertheless, it is not a harmless process and cells show declines after the freezing/thawing process. Although a great improvement has been achieved in the cryopreservation medium, cryobiology extenders are still under constant development. The bovine industry is by far one of the sectors where sperm cryopreservation has been increasingly developed in recent decades and where cryopreserved sperm are used routinely to perform AI. Recently, it has been shown that bovine extenders can be successfully used to cryopreserve stallion sperm (Nikitkina et al., 2020). The comparison of commercial versus homemade cryopreservation extenders showed that both can be used interchangeably, with the only advantage being that commercial ones already contain egg yolk within their components, saving time in the process of extender elaboration (Nikitkina et al., 2020).

In their travel along the reproductive tract, sperm must undergo several events, named sperm capacitation, that allow them to have the ability to fertilize an oocyte. Hypermotility and an increase in phospho-tyrosine (PY) levels have been historically used as hallmarks of sperm capacitation status. Recently, it has been shown that two pollutants, perfluorooctane sulfonate (PFOS) and perfluorohexane sulfonate (PFHxS), impair boar sperm capacitation (Oseguera-López et al., 2020). The harmful effects were associated with the induction of higher sperm mortality associated with a higher level of reactive oxygen species (ROS) production that led to an increase in the percentage of damaged DNA and lower sperm PY levels (Oseguera-López et al., 2020).

The importance of capacitation has been highlighted in equines, where it has been suggested that in vitro fertilization (IVF) fails due to sperm capacitation defects. As a consequence, a lot of effort have been made lately to improve stallion sperm capacitation medium. For instance, the addition of D-penicillamine (PEN) to the capacitation medium improved stallion sperm hypermotility and PY levels (Ruiz-Díaz et al., 2020). In addition, PEN allowed for the selection of sperm in vitro by thermotaxis. Thus, those sperm selected by a temperature gradient had better levels of PY and lower levels of DNA fragmentation when compared to the unselected fraction (Ruiz-Díaz et al., 2020). The authors proposed to improve equine IVF and intracytoplasmic sperm injection (ICSI) outcomes by selecting sperm by thermotaxis after capacitating them in a medium containing PEN (Ruiz-Díaz et al., 2020). Others proposed a protocol to select cryopreserved stallion spermatozoa with higher fertilizing capacity by incubating them in an in vitro model of oviduct epithelial cells (OECs) (Gimeno et al., 2021). The sperm that detached from OECs after capacitation presented higher potential fertilizing capacity, described by an increase in living sperm with progressive motility and with the acrosome intact (Gimeno et al., 2021).

Following from stallion sperm cryopreserved as a model of study, the irradiation of these cells with red light has been proposed as a tool to enhance sperm quality after the process of cryopreservation. Red light induces changes in sperm mitochondrial membrane potential, and intracellular ROS, without affecting the integrity of the plasma membrane and acrosome (Catalán et al., 2021). Interestingly, because different results can be obtained, the turbidity of the cryopreservation extender and the color of the straw chosen (Catalán et al., 2021) must be taken into consideration. Recently, extracellular vesicles have emerged in the field of reproduction due to their roles in reproductive processes. Moreover, extracellular vesicles have been used as a novel tool to ameliorate the harmful effects of sperm cryopreservation (Saadeldin et al., 2020) as well as to improve the outcome of other ARTs (Gervasi et al., 2020).

Female reproductive tract peristaltic movement helps sperm to advance in their trip seeking an oocyte. Once sperm are close to the fertilization place (ampulla), they are guided to the oocyte by chemotaxis (mainly by substances secreted by the oocyte, cumulus cells and others poured into the oviduct). The development of chemotactic chambers specially designed for sperm selection has allowed the discovery that boar sperm selected by chemotaxis using follicular fluid provide better IVF outcomes in comparison to other chemoattractants tested, such as periovulatory oviductal fluid (pOF), conditioned medium (CM) from the in vitro maturation of oocytes and progesterone (P4) (Vieira et al., 2020).

The success of IVF is determined, among other factors, by the quality of both sperm and oocytes. To increase the efficiency of embryo production by in vitro procedures, different non-invasive selection strategies have been summarized by Aguila et al. (2020) to help to predict bovine oocyte competence. In addition, before performing IVF, oocytes need to be maturated in vitro. Nevertheless, ovaries of farm animals are usually obtained in distant locations from the assisted reproductive laboratory. Consequently, the development of appropriate media is needed to transport ovaries and to keep oocyte developmental competence in the best condition. In this regard, it has been shown that transport of up to 5 h in a saline solution protected ovine oocytes better than TCM199 (Martín-Maestro et al., 2020). In addition, IVF outcomes were improved when using an oocyte obtained from ovaries transported in saline solution in comparison to TCM199 (Martín-Maestro et al., 2020). In brief, when ovary transport is needed, a saline solution is the recommended medium to preserve oocyte developmental competence and to produce more and better quality embryos (Martín-Maestro et al., 2020).

The final goal of IVF is the obtention of in vitro embryo production (IVP) that eventually will transfer back to a receptor female. As happens with IVF medium, IVP medium production is under constant optimization to enhance its final yield. For instance, the successful equine pregnancy and foaling rates obtained after ICSI are only 10% of the oocytes matured in vitro. These poor results underline that conditions used for oocyte in vitro maturation are not optimized for equine oocytes. Following the rationale that before ovulation, oocytes are surrounded by follicular fluid (FF), Fernández-Hernández et al. (2020) studied the metabolome of periovulatory FF. Their final aim was to improve our insight into the in vivo conditions to eventually translate this knowledge to optimize equine in vitro maturation protocols. The authors found nine metabolites in the FF not included in traditional IVP medium, confirming that bigger efforts have to be made to adapt current ART media to every species' specific needs (Fernández-Hernández et al., 2020). Finally, also intrigued by a better IVP medium composition, no differences were found when a comparison of bovine embryo transference results of classic IVP medium containing BSA versus IVP medium containing female reproductive fluids was made (Lopes et al., 2020). Interestingly, some hormone levels (anti-Müllerian hormone and estrogen:progesterone ratio) were different in the receptor of embryos depending on the IVP medium chosen in comparison with control (pregnant by AI) (Lopes et al., 2020).

In summary, this Special Issue has shown an overview of the new alternatives and protocols to enhance domestic animals' ART outcomes. The editor expects that this group of manuscripts will be used as an essential tool to help veterinarians to increase farm profits and offer them new options to bypass cases of subfertility. At the same time, it is expected to serve as a source of inspiration for researchers of this amazing branch of the biology of reproduction interested in the enhancement of fertility results by the use of novel ARTs, and new ideas may emerge after the reading this Special Issue.

David Martín Hidalgo
Editor

Article

Semen Quality of Rasa Aragonesa Rams Carrying the *FecXR* Allele of the *BMP15* Gene

José Alfonso Abecia [1,*], Ángel Macías [2], Adriana Casao [1], Clara Burillo [2], Elena Martín [2], Rosaura Pérez-Pé [1] and Adolfo Laviña [2]

[1] University Institute of Research in Environmental Sciences of Aragon (IUCA), University of Zaragoza, Miguel Servet, 177, 50013 Zaragoza, Spain; adriana@unizar.es (A.C.); rosperez@unizar.es (R.P.-P.)

[2] National Association of Rasa Aragonesa Breeders (ANGRA), Cabañera Real, s/n, 50800 Zuera, Spain; angel@rasaaragonesa.com (Á.M.); aabecia0@gmail.com (C.B.); jabecia1@alumno.uned.es (E.M.); adolfo@rasaaragonesa.com (A.L.)

* Correspondence: alf@unizar.es

Received: 16 August 2020; Accepted: 9 September 2020; Published: 11 September 2020

Simple Summary: It has been demonstrated that the ovine gene *BMP15* presents a mutation in the Rasa Aragonesa Spanish sheep breed, which has been called *FecXR*. In heterozygosis, ewes exhibit a variable increase in the ovulation rate, producing 0.35 additional lambs per birth and, in homozygosis, sterility. Since the importance of carrying this polymorphism in rams has not been studied, sperm quality and fertility of rams carrying the *FecXR* mutation of the ovine gene *BMP15* has been determined, comparing semen quality, testicle characteristics, and fertility rate of rams presenting or not the allele. FecXR rams exhibited a higher masal motility and a higher proportion of rapid sperm than did non-carrier rams; however, no differences in scrotal circumference or testicular length and diameter were found, although FecXR rams produced a higher proportion of pregnant ewes after artificial insemination. Thus, it seems that the *FecXR* allele creates high-quality semen and improves some sperm parameters in this breed, making these males especially valuable for artificial insemination to produce prolific ewes, when wild-type, non-carrier ewes, are inseminated.

Abstract: The *FecXR* mutation is a variant of the ovine gene *BMP15* in the Rasa Aragonesa breed. Information on the physiological importance of carrying the *FecX* polymorphism in rams is limited. The aim of this study was to compare semen quality, testicle characteristics, and fertility rate of rams that carry the *FecXR* allele. Rams (n = 15) were either FecXR allele carriers (n = 10) or non-carriers, wild type (++) (n = 5). FecXR rams exhibited higher mass motility ($p < 0.05$), proportion of rapid sperm ($p < 0.05$), and a lower proportion of slow sperm ($p < 0.0001$) than did ++ rams. The presence of the *FecXR* allele was not associated with mean scrotal circumference or testicular length and diameter, although season had a significant ($p < 0.05$) effect on these traits. Genotype ($p < 0.05$) and season ($p < 0.01$) had a significant effect on mean fertility rate, FecXR rams had a higher proportion of pregnant ewes than did ++ rams ($p < 0.05$). In conclusion, the *FecXR* allele produced high-quality semen throughout the year, and corresponded with an improvement in some sperm parameters, particularly, mass motility and the proportion of rapid sperm.

Keywords: *BMP15*; ram; semen

1. Introduction

Several mutations in genes of the transforming growth factor-beta (TGF-β) superfamily have positive effects on ovulation rate and litter size; e.g., FecB or Bone Morphogenetic Protein (BMP) R1B, FecX or BMP15, and FecG or GDF9 (for a review [1]). Galloway et al. [2] identified a mutation in the *BMP15* gene, that introduced a stop codon on the X chromosome, which prevents the normal translation of the protein encoded by this gene and, subsequently, demonstrated its effect on the ovulation rate in a population of the Inverdale (*FecXI*) sheep breed. The mutation is sex-linked because it is located in the non-recombinant region of the X chromosome and, therefore, males can have one copy of the gene, only, but females can be hetero or homozygous for the mutation (review [3]). A male carrier transmits the mutation to all of his daughters but to none of his sons, and heterozygous females pass on the mutation to, on average, half of their offspring. However, homozygous females are sterile because they do not develop ovarian follicles correctly.

Since the discovery of the mutation in the Inverdale breed, mutations in *BMP15* have been identified in other prolific breeds including Belclare and Cambridge (*FecXB*), Hanna (*FecXH*), Galway (*FecXG*), Lacaune (*FecXL*), Rasa Aragonesa (*FecXR*), Grivette (*FecXGr*), and Olkuska (*FecXO*) [2,4–9]. In all, the mechanism of action is similar (amino acid substitutions, deletions, or stop codons), and all have received the same name (*FecX*) and the first initial of the breed in which it was discovered because the phenotypic effects are similar; i.e., in heterozygosis, a variable increase in ovulation rate and, in homozygosis, sterility [10].

The *FecXR* mutation is a variant of the ovine gene *BMP15* in the Rasa Aragonesa breed. Rasa Aragonesa is one of the most important meat sheep breeds in Spain, where there are about 1.1 million head, and 360,000 are registered in the Stud Book of the National Association of Rasa Aragonesa Breeders (ANGRA) [11]. Mean litter size is 1.2–1.5 lambs/birth [12], and ANGRA is developing a genetic improvement program that includes prolificacy as one of the important objectives. *FecXR* has been included in the selection scheme under the commercial denomination "Gen ANGRA Santa Eulalia", and there are >5000 ewes that carry this mutation. The allele has been used to increase prolificacy in Rasa Aragonesa sheep through artificial insemination (AI) of wild type, non-carrier ewes, which are used to disseminate the allele across those farms interested in improving litter size. The positive effect of *FecXR* on prolificacy is well known; viz., 0.35 additional lambs per birth [11], which has increased cost effectiveness and profits.

Most of the studies on the expression of *FecXR* have involved female sheep and information on the physiological importance of the *FecXR* polymorphism in males, particularly, rams, is limited. Studies on *BMP15* in rams have investigated tissue expression pattern in rams that differ in fecundity [13], fertility rate [14], and the influence of the *FecB* genotype on semen attributes [15]. The aim of this study was to compare the semen quality, testicle characteristics, and fertility rate, through AI, of Rasa Aragonesa rams that carry the *FecXR* allele during different seasons, so that it is hypothesized that the efficiency of AI using these particular rams may be improved.

2. Materials and Methods

2.1. Animals

Rams were housed at CERSYRA (Regional Centre for Animal Selection and Reproduction) in Zaragoza, Spain (41°N), and were breeding males for AI in the stud book of ANGRA. Inseminations were performed on commercial farms by veterinarians of ANGRA. Approval from the Ethics Committee of the University of Zaragoza was not a prerequisite for this study. The study met the Spanish Policy for Animal Protection RD1201/05, which meets the European Union Directive 2010/63 on the protection of animals used for experimental and other scientific purposes.

Fifteen adult Rasa Aragonesa rams (age: 5.7 ± 2.8 yr) used in the study were either FecXR allele carriers (n = 10) or non-carriers, wild type (++) (n = 5). The laboratory procedures (DNA extraction, polymerase chain reaction (PCR) amplification prior to sequencing, DNA sequencing and analysis) after they had been exposed the localization of the allele are described by Monteagudo et al. [8].

2.2. Semen Collection and Analyses

Rams were housed together and were fed to meet their maintenance requirements. Throughout the year, semen samples (96 per ram) were collected twice a week [16], starting at 9:00 am, in an artificial vagina at 35–40 °C lubricated with petroleum jelly. Each collection day, a routine, simplified semen analysis was performed that included concentration measured by spectrophotometry (AccRead,

IMV Technologies, L'Aigle, France) (1:400 dilution in saline solution plus 0.2% glutaraldehyde), volume (ml), measured in a graduated collection tube, and mass motility estimated by optical microscopy at 100× magnification and scored from 0 to 5. Once per month, the proportion of static sperm, total motile (TM) sperm, non-progressive (NPM) and progressive (PM), and motile sperm subpopulations (rapid, medium, or slow sperm) were measured in a computer-assisted sperm analysis (CASA) using ISAS software (Integrated Semen Analysis System, Proiser, Paterna, Valencia, Spain). Semen sample processing and motility and viability assessment followed the method of Palacín et al. [17]. Briefly, before motility or viability analysis, 200×10^6 sperm/mL semen samples were mixed and re-diluted to a final concentration of 50×10^6 sperm/mL using INRA 96 (IMV Technologies, L'Aigle, France) extender. An Olympus BX40 microscope under 100× magnifications, provided with heated stage set at 37 °C, was used to estimate sperm motility. The grade of the forward progression (fast progressive, slow progressive and motile but not progressive) determined on the TM sperm were recorded. Sperm with curvilinear velocity (VCL) ≥ 75 m/sg and straightness (STR) ≥ 80% were considered rapid progressive and with VCL < 5 m/sg and STR ≥ 80% slow progressive.

Thereafter, the semen was diluted (INRA 96) and put into French mini-straws for AI (0.25 mL, 300 × 106 spermatozoa/mL).

2.3. Testicular Measurements

Once every month, scrotal circumference (SC) (pulling the testes firmly down into the lower part of the scrotum and placing a measuring tape into a loop around the greatest diameter over the scrotum), length (placing the fixed arm of a caliper at the proximal end and the sliding arm at the distal end of the testes) and diameter (placing one arm of a caliper at the medial aspect and the other at the lateral aspect of the testes, at the point of maximum width) of each testicle, were determined. Testicular length (TL) and diameter (TD) were calculated as the mean of both testicles.

2.4. Artificial Inseminations (AI)

In the 12 months of the study, 1412 AI were performed on 29 farms. To synchronize estrus, vaginal sponges containing 30 mg of fluorogestone were applied for 12 d. At pessary withdrawal, ewes received 480 IU of eCG. Cervical AI [18] was performed 54 ± 1 h after sponge withdrawal (14:00 p.m.), using an ovine AI gun (IMV, Instruments de Medicine Veterinaire, L'Aigle, France) and 0.25 mL French mini-straws. All of the inseminated ewes were *FecXR* allele non-carriers.

Births from AI were recorded on the farms throughout the year of the study. Fertility rate was the proportion of ewes lambing after AI, prolificacy was the number of lambs born per lambing, and fecundity rate was the number of lambs born per inseminated ewe.

2.5. Statistical Analyses

Semen quality, testicular dimensions, and reproductive performance after AI were evaluated statistically based on a multifactorial model that included the presence/absence of the *FecXR* allele (FecXR or ++ wild rams) and season as fixed effects in the Least-Squares Method of the GLM procedure in SPSS v.26 (IBM Corp., Released 2019). The seasons were defined based on the Northern Hemisphere Meteorological Season Division [19]. An ANOVA identified significant differences between genotypes and between seasons. The general representation of the model is as follows: $y = xb + e$, where y is $N \times 1$ vector of records, b denotes the fixed effect in the model within the association matrix x, and e is the vector of residual effects. To test for significant differences between effect combinations, a post-hoc Fisher's Least Significant Difference (LSD) test was used.

3. Results

3.1. Semen Quality

Mean (±S.E.M.) sperm count (3762 ± 1060), ejaculate volume (0.93 ± 0.04 cm^3) and semen concentration (4055 ± 100 × 10^6) did not differ between the two genotypes, but concentration was significantly ($p < 0.05$) higher in summer than it was in autumn and winter ($p < 0.05$). Mass motility (4.26 ± 0.19) was significantly ($p < 0.05$) affected by the presence of the allele and season, with a significant ($p = 0.01$) interaction between effects. FecXR rams exhibited a higher mass motility ($p < 0.05$), a higher proportion of rapid sperm ($p < 0.05$), and a lower proportion of slow sperm ($p < 0.0001$) than did ++ rams (Figure 1). Mean proportion of NPM, PM, TM, and medium-speed sperm did not differ significantly between genotypes or among seasons.

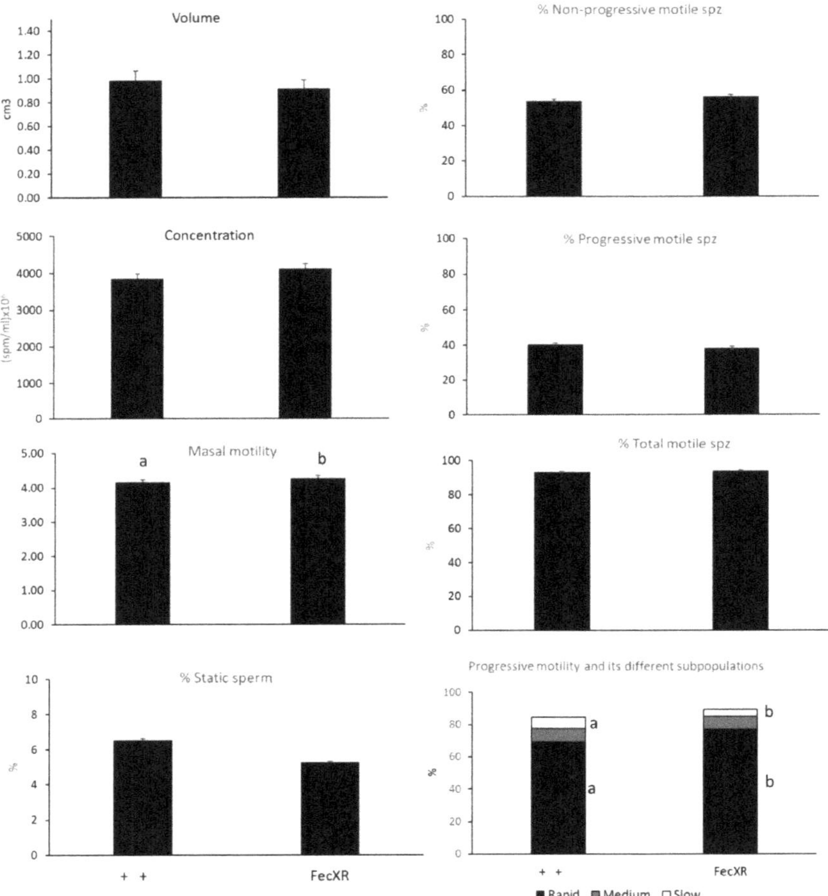

Figure 1. Seminal traits (mean ± S.E.M.) of Rasa Aragonesa rams of the wild genotype (+ +; $n = 5$) or those carrying the *FecXR* allele of the BMP15 gene ($n = 10$) (a,b indicate $p < 0.05$) (spz: spermatozoon). Values calculated from semen samples collected twice a week for one year.

The proportion of slow sperm was significantly ($p < 0.05$) lower in summer (Figure S1). In winter, FecXR rams tended to present a higher mass motility and a lower proportion of static sperm than did ++ rams ($p < 0.10$) (Table 1). Furthermore, FecXR rams had higher proportions of rapid sperm in spring ($p < 0.10$) and winter ($p < 0.001$), and lower proportions of slow sperm in spring ($p < 0.05$), summer ($p < 0.05$), and winter ($p < 0.01$) than did ++ rams (Table 1).

Table 1. Seminal traits (mean ± S.E.M.) of Rasa Aragonesa rams of the wild genotype (++; $n = 5$) or carrying the *FecXR* allele of the *BMP15* gene ($n = 10$) (* indicate differences $p < 0.10$ within season) (** indicate significant differences at $p < 0.05$ within season). Values calculated from semen samples collected twice a week for one year. NPM: non-progressive motile sperm; PM: progressive motile sperm; TM: total motile sperm.

	Spring		Summer		Autumn		Winter	
	++	FecXR	++	FecXR	++	FecXR	++	FecXR
Sperm count (×10^6^)	4020 ± 311	3665 ± 533	4381 ± 91	4456 ± 364	3465 ± 553	3297 ± 375	3352 ± 338	3525 ± 458
Volume (cm^3^)	0.91 ± 0.04	0.86 ± 0.10	1.10 ± 0.02	0.99 ± 0.10	0.92 ± 0.09	0.84 ± 0.09	0.97 ± 0.03	0.97 ± 0.11
Concentration (×10^6^)	4420 ± 140	4163 ± 256	3987 ± 146	4572 ± 209	3713 ± 252	4002 ± 242	3452 ± 241	3609 ± 143
Mass motility (0–5)	4.30 ± 0.00	4.28 ± 0.01	4.27 ± 0.03	4.29 ± 0.01	4.29 ± 0.01	4.28 ± 0.02	3.83 ± 0.42 *	4.30 ± 0.00
% Static spz	3.43 ± 0.88	5.65 ± 2.24	8.33 ± 3.34	4.93 ± 0.90	5.89 ± 0.95	6.79 ± 1.80	7.44 ± 0.29	5.48 ± 0.61
% NPM spz	57.93 ± 0.88	54.99 ± 1.84	52.78 ± 3.58	53.93 ± 2.32	54.78 ± 3.16	58.55 ± 2.45	50.67 ± 4.10	57.57 ± 1.19
% PM spz	41.66 ± 3.01	40.71 ± 2.76	38.89 ± 4.08	41.13 ± 2.09	39.33 ± 2.22	34.67 ± 1.72	41.89 ± 4.37	36.95 ± 1.41
% TM spz	96.41 ± 1.04	93.87 ± 2.21	91.67 ± 3.34	95.07 ± 0.90	94.11 ± 0.95	93.21 ± 1.80	92.56 ± 0.29	94.38 ± 0.65
% Rapid spz	69.97 ± 5.27	75.50 ± 3.83	69.78 ± 8.42 *	80.23 ± 2.21	76.33 ± 2.96	75.09 ± 3.68	60.89 ± 2.63 **	76.52 ± 2.69
% Medium spz	6.31 ± 0.94	8.43 ± 1.16	8.33 ± 1.20	7.17 ± 1.17	8.11 ± 1.46	8.33 ± 1.14	10.33 ± 2.19	8.52 ± 0.89
% Slow spz	18.69 ± 5.26 **	8.95 ± 1.79	11.78 ± 4.11 **	6.23 ± 0.54	9.00 ± 3.02	8.94 ± 1.83	20.67 ± 1.86 **	8.48 ± 1.58

3.2. Testicular Measurements

SC (FecXR: 32.9 ± 0.6; ++: 31.8 ± 1.7 cm), TD (FecXR: 6.5 ± 0.2; ++: 6.2 ± 0.5 cm), and TL (FecXR: 9.3 ± 0.2; ++: 8.6 ± 0.7 cm) did not differ significantly between carriers and non-carriers of the *FecXR* allele; however, SC was highest in summer and winter ($p < 0.05$), TD was highest in summer and autumn ($p < 0.01$), TL was lowest in spring and winter ($p < 0.05$) (Table S1).

3.3. Reproductive Parameters

FecXR rams impregnated a significantly higher proportion ($p < 0.05$) of ewes (62.5 ± 2.5%) than did ++ rams (56.7 ± 2.9%), and fertility rates were lowest in spring and winter inseminations (Table S1). FecXR rams had significantly ($p < 0.05$) higher fertility rates than did ++ rams in winter inseminations, only (0.66 ± 0.08 vs. 0.40 ± 0.06%).

Prolificacy (FecXR: 1.73 ± 0.06; ++: 1.71 ± 0.06 lambs/lambing) and fecundity (FecXR: 1.10 ± 0.06; ++: 0.98 ± 0.06 lambs/ewe) did not differ significantly between FecXR and ++ rams, but differed significantly ($p < 0.001$) among seasons (Table S1).

4. Discussion

To our knowledge, this is the first study of the semen quality of rams carrying the *FecXR* allele. Ejaculate volume and sperm concentration did not differ significantly between the FecXR and wild rams in any season, which suggests that the production of seminal plasma or spermatogenesis are not affected by the *BMP15* gene. These results are similar to the observations of Kumar et al. [20] in Garole x Malpura rams carrying the *FecB* allele, and parallel the absence of differences in testicular size. Mass motility and the proportion of rapid sperm were significantly higher, and the proportion of slow sperm lower in the FecXR than they were in the wild-type rams.

Furthermore, ewes that had been inseminated with semen collected from FecXR rams had the highest mean annual fertility rate. Sperm motility and velocity are two of the most important aspects of semen quality because they are correlated with fertility [21]. In a study of Rasa Aragonesa breed at the same latitude as in our study, it has been reported that high-fertility rams produced a higher proportion of fast and linear spermatozoa than did low-fertility rams [22]. In Iberian deer, mean and maximum straight-line velocity of sperm and fertility are significantly correlated, and it appears that sperm swimming velocity is a main determinant of fertility in mammals [23]. Thus, it is likely that high mass motility and high proportion of rapid sperm contributed to the high fertility rates in FecXR rams. On the other hand, Lahoz et al. [24] did not detect significant differences between genotypes in a program that involved cervical insemination. Given the number of external factors that can affect the proportion of ewes that become pregnant after AI (year, farm, technician) [25] including weather [26] and climate [27], differences in the conditions at the time of experiments involving AI might have contributed to the presence or absence of differences between genotypes. Further study is needed to determine how external factors might influence the effect of *FecX* on reproductive parameters.

The finding that the highest fertility rate occurred in summer is similar to previous observations [28] in the same breed and at the same latitude as in our study, where the lowest AI fertility was between March and June, and the highest was in the first months of increasing daylength (July and August). Rasa Aragonesa is a reduced-seasonal anestrous breed [29], in which females exhibit an onset of the breeding season in July and a peak in ovulation rate in late August. Thus, our study confirms that summer is the peak breeding season for rams and ewes of this breed.

Differences in the pregnancy rates related to polymorphisms of the *BMP15* gene have been reported by Sun et al. [30], who found that Chinese Holstein bulls of the CT genotype had a significantly lower sperm motility than did bulls of the CC or TT genotypes. In sheep, Chen et al. [13] reported the expression of *BMP15* in the epididymis of rams, which was significantly higher in a less-fecund breed (Sunite) than it was in a high-fecundity breed (Small Tail Han). Possibly, the expression level of *BMP15* and fecundity in rams are negatively correlated. Garole × Malpura rams that carry the *FecB* genotype had a significantly higher proportion of rapid motile sperm and with higher linearity, and a higher FSH concentration than did the wild type [15].

In our study, testicular dimensions did not differ significantly between rams that carried the *FecXR* allele and those that did not. Rasa Aragonesa light lambs that did or did not carry the *FecXR* allele did not exhibit significant differences in birth weight, growth rate, or carcass quality [31]; moreover, it appears that *FecXR* allele may not influence testicular morphology or fleece weight at 13 months of age in carrier Romney rams [32]. The absence of differences in testicular measurements between genotypes parallels the lack of differences in sperm volume and concentration, which are highly correlated to testicle size [33].

Season had a significant effect on testicular measurements, which was similar to the effects reported by Avdi et al. [34] in Chios and Serres rams. Similarly, Chios and Friesian rams had semen characteristics that were generally better in summer and peaked in quality in autumn [35]. Although seasonal variations in reproductive traits in sheep are less marked in rams than they are in ewes, the consequences of the non-reproductive season are smaller testicular volume and diameter, lower semen quality, and hormone profiles that differ from those in the breeding season [36]. Photoperiod is the key environmental signal that dictates the timing of the reproductive cycle of the ram [37], which is synchronized through changes in daily melatonin secretion [38]. Rams exhibit a seasonal decrease in sexual behavior and spermatogenesis at about the time that ewes are in sexual rest, but with a 1- to 2-month advance in phase [39].

5. Conclusions

In conclusion, this study demonstrated that carriers of the *FecXR* allele produce good-quality semen throughout the year, and corresponded with an improvement in some sperm characteristics—particularly mass motility and the proportion of rapid sperm—along with an interaction effect with season. In addition, the ability to pass the allele to their female offspring, through the insemination of wild type, non-carriers ewes, makes these males especially valuable for AI to produce prolific ewes.

Supplementary Materials: The following are available online at http://www.mdpi.com/2076-2615/10/9/1628/s1, Figure S1: Annual seminal traits (mean ± S.E.M.) of Rasa Aragonesa rams (a,b indicate $p < 0.05$) (spz: spermatozoon). Values calculated from semen samples collected twice a week for one year, Table S1: Mean (± S.E.M.) testicular measurements and reproductive traits of Rasa Aragonesa rams ($n = 15$) (a,b,c indicate significant differences $p < 0.05$). Values calculated from semen samples collected twice a week for one year.

Author Contributions: Conceptualization, J.A.A., Á.M., R.P.-P. and A.L.; methodology, J.A.A., Á.M., C.B., E.M.; formal analysis, A.C., R.P.-P., Á.M.; investigation, Á.M., C.B. and E.M.; writing—original draft preparation, J.A.A.; writing—review and editing, Á.M., A.C. and R.P.-P.; supervision, J.A.A., Á.M. and A.L.; project administration, J.A.A.; funding acquisition, J.A.A. All authors have read and agreed to the published version of the manuscript.

Funding: This research was partially funded by Gobierno de Aragón, grant number A07_20R.

Acknowledgments: The authors thank the farmers involved in the study and the CERSYRA staff for their collaboration in the collection of semen samples and the preparation of doses for AI. We thank Bruce MacWhirter for the English revision of the manuscript. Partially funded by Gobierno de Aragón, group A07_20R.

Conflicts of Interest: The authors declare no conflict of interest.

References

1. Vadhana, E.; Santhosh, A.; Pooja, G.S.; Vani, A.; Kumar, S. FecB: A major gene governing fecundity in sheep. *J. Entomol. Zool. Stud.* **2019**, *7*, 270–274.
2. Galloway, S.M.; McNatty, K.P.; Cambridge, L.M.; Laitinen, M.P.E.; Juengel, J.L.; Jokiranta, T.S.; McLaren, R.J.; Luiro, K.; Dodds, K.G.; Montgomery, G.W.; et al. Mutations in an oocyte-derived growth factor gene (BMP15) cause increased ovulation rate and infertility in a dosage-sensitive manner. *Nat. Genet.* **2000**, *25*, 279–283. [CrossRef] [PubMed]
3. Moore, R.K.; Shimasaki, S. Molecular biology and physiological role of the oocyte factor, BMP-15. *Mol. Cell. Endocrinol.* **2005**, *29*, 67–73. [CrossRef] [PubMed]
4. Montgomery, G.W.; Galloway, S.M.; Davis, G.H.; McNatty, K.P. Genes controlling ovulation rate in sheep. *Reproduction* **2001**, *121*, 843–852. [CrossRef]

5. Hanrahan, J.P.; Gregan, S.M.; Mulsant, P.; Mullen, M.; Davis, G.H.; Powell, R.; Galloway, S.M. Mutations in the genes for oocyte-derived growth factors GDF9 and BMP15 are associated with both increased ovulation rate and sterility in Cambridge and Belclare sheep (Ovis aries). *Biol. Reprod.* **2004**, *70*, 900–909. [CrossRef]
6. Davis, G.H. Major genes affecting ovulation rate in sheep. *Genet. Sel. Evol.* **2005**, *37*, 11–23. [CrossRef]
7. Bodin, L.; Di Pasquale, E.; Fabre, S.; Bontoux, M.; Monget, P.; Persani, L.; Mulsant, P.A. A novel mutation in the bone morphogenetic protein 15 gene causing defective protein secretion is associated with both increased ovulation rate and sterility in Lacaune sheep. *Endocrinology* **2007**, *148*, 393–400. [CrossRef]
8. Monteagudo, L.V.; Ponz, R.; Tejedor, M.T.; Laviña, A.; Sierra, I. A 17 bp deletion in the Bone Morphogenetic Protein 15 (BMP15) gene is associated to increased prolificacy in the Rasa Aragonesa sheep breed. *Anim. Reprod. Sci.* **2009**, *110*, 139–146. [CrossRef]
9. Demars, J.; Fabre, S.; Sarry, J.; Rossetti, R.; Gilbert, H.; Persani, L.; Tosser-Klopp, G.; Mulsant, P.; Nowak, Z.; Drobik, W.; et al. Genome-wide association studies identify two novel BMP15 mutations responsible for an atypical hyperprolificacy phenotype in sheep. *PLoS Genet.* **2013**, *9*, e1003482. [CrossRef]
10. Otsuka, F.; McTavish, K.J.; Shimasaki, S. Integral role of GDF-9 and BMP-15 in ovarian function. *Mol. Reprod. Dev.* **2011**, *78*, 9–21. [CrossRef]
11. ANGRA. Available online: https://www.rasaaragonesa.com (accessed on 15 August 2020).
12. Sierra, I. La raza ovina Rasa Aragonesa: Caracteres morfológicos y productivos. *Anim. Genet. Resour. Inf.* **1992**, *10*, 65–74.
13. Chen, W.; Tian, Z.; Ma, L.; Gan, S.; Sun, W.; Chu, M. Expression analysis of BMPR1B, BMP15, GDF9, Smad1, Smad5, and Smad9 in rams with different fecundity. *Pak. J. Zool.* **2020**, *52*, 1665–1674. [CrossRef]
14. Lahoz, B.; Blasco, M.E.; Sevilla, E.; Folch, J.; Roche, A.; Quintin, F.J.; Martínez-Royo, A.; Galeote, A.I.; Calvo, J.H.; Fantova, E.; et al. Fertility of Rasa Aragonesa rams carrying or not the FecXR allele of BMP15 gene when used in artificial insemination. In Proceedings of the European Association for Animal Production (EAAP), Barcelona, Spain, 24–27 August 2009.
15. Kumar, D.; Joshi, A.; Naqvi, S.M.K.; Kumar, S.; Mishrac, A.K.; Maurya, V.P.; Arora, A.L.; Mittal, J.P.; Singh, V.K. Sperm motion characteristics of Garole × Malpura sheep evolved in a semi-arid tropical environment through introgression of FecB gene. *Anim. Reprod. Sci.* **2007**, *100*, 51–60. [CrossRef] [PubMed]
16. Evans, G.; Maxwell, W.M.C. Handling and examination semen. In *Salamon'S Artificial Insemination of Sheep and Goats*; Salamon, S., Ed.; Butterworths: Sydney, Australia, 1987; pp. 93–106.
17. Palacín, I.; Vicente-Fiel, S.; Santolaria, P.; Yániz, J.L. Standardization of CASA sperm motility assessment in the ram. *Small Rumin. Res.* **2013**, *112*, 128–135. [CrossRef]
18. Macías, A.; Ferrer, L.M.; Ramos, J.J.; Lidón, I.; Rebollar, R.; Lacasta, D.; Tejedor, M.J. Technical Note: A new device for cervical insemination of sheep—Design and field test. *J. Anim. Sci.* **2017**, *95*, 5263–5269. [CrossRef]
19. Trenberth, K.E. What are the Seasons? *Bull. Am. Meteorol. Soc.* **1983**, *64*, 11. [CrossRef]
20. Kumar, D.; Naqvi, S.M.K.; Kumar, S. Sperm motion characteristics of FecBBB and FecBB + Garole × Malpura rams during the non-breeding season under hot semi-arid environment. *Livest. Sci.* **2012**, *150*, 337–341. [CrossRef]
21. Aitkin, R.J. Motility parameters and fertility. In *Control of Sperm Motility: Biological and Clinical Aspects*; Gagnon, C., Ed.; CRS Press: Boca Raton, FL, USA, 1990; pp. 285–302.
22. Yániz, J.L.; Palacín, I.; Vicente-Fiel, S.; Sánchez-Nadal, J.A.; Santolaria, P. Sperm population structure in high and low field fertility rams. *Anim. Reprod. Sci.* **2015**, *156*, 128–134. [CrossRef]
23. Gomendio, M.; Roldán, E.R.S. Implications of diversity in sperm size and function for sperm competition and fertility. *Int. J. Dev. Biol.* **2008**, *52*, 439–447. [CrossRef]
24. Lahoz, B.; Alabart, J.L.; Jurado, J.J.; Calvo, J.H.; Martínez-Royo, A.; Fantova, E.; Folch, J. Effect of the FecXR polymorphism in the bone morphogenetic protein 15 gene on natural or equine chorionic gonadotropin-induced ovulation rate and litter size in Rasa Aragonesa ewes and implications for on-farm application. *J. Anim. Sci.* **2011**, *89*, 3522–3530. [CrossRef]
25. Anel, L.; Kaabi, M.; Abroug, B.; Alvarez, M.; Anel, E.; Boixo, J.C.; de la Fuente, L.F.; Paz, P. Factors influencing the success of vaginal and laparoscopic artificial insemination in Churra ewes: A field assay. *Theriogenology* **2005**, *63*, 1235–1247. [CrossRef] [PubMed]
26. Palacios, C.; Abecia, J.A. Meteorological variables affect fertility rate afterintrauterine artificial insemination in sheep in a seasonal-dependent manner: A 7-year study. *Int. J. Biometeorol.* **2015**, *59*, 585–592. [CrossRef]

27. Abecia, J.A.; Máñez, J.; Macias, A.; Laviña, A.; Palacios, C. Climate zone influences the effect of temperature on the day of artificial insemination on fertility in two Iberian sheep breeds. *J. Anim. Behav. Biometeorol.* **2017**, *5*, 124–131. [CrossRef]
28. Palacín, I.; Yániz, J.L.; Fantova, E.; Blasco, M.E.; Quintín-Casorrán, F.J.; Sevilla-Mur, E.; Santolaria, P. Factors affecting fertility after cervical insemination with cooled semen in meat sheep. *Anim. Reprod. Sci.* **2012**, *132*, 139–144. [CrossRef] [PubMed]
29. Forcada, F.; Abecia, J.A.; Sierra, I. Seasonal changes in oestrous activity and ovulation rate in Rasa Aragonesa ewes maintained at two different body condition levels. *Small Rumin. Res.* **1992**, *8*, 313–324. [CrossRef]
30. Sun, L.P.; Song, Y.P.; Du, Q.Z.; Song, L.W.; Tian, Y.Z.; Zhang, S.L.; Hua, G.H.; Yang, L.G. Polymorphisms in the bone morphogenetic protein 15 gene and their effect on sperm quality traits in Chinese Holstein bulls. *Genet. Mol. Res.* **2014**, *13*, 1805–1812. [CrossRef] [PubMed]
31. Roche, A.; Ripoll, G.; Joy, M.; Folch, J.; Panea, B.; Calvo, J.H.; Alabart, J.L. Effects of the FecX R allele of BMP15 gene on the birth weight, growth rate and carcass quality of Rasa Aragonesa light lambs. *Small Rumin. Res.* **2012**, *108*, 45–53. [CrossRef]
32. Davis, G.H.; Dodds, K.G.; McEwan, J.C.; Fennessy, P.F. Liveweight, fleece weight and prolificacy of Romney ewes carrying the Inverdale prolificacy gene (FecXI) located on the X-chromosome. *Livest. Prod. Sci.* **1993**, *34*, 83–91. [CrossRef]
33. Kheradmand, A.; Babaei, H.; Abshenas, J. Comparative evaluation of the effect of antioxidants on the chilled-stored ram semen. *Iran. J. Vet. Res.* **2006**, *7*, 40–45.
34. Avdi, M.; Banos, G.; Stefos, K.; Chemineau, P. Seasonal variation in testicular volume and sexual behavior of Chios and Serres rams. *Theriogenology* **2004**, *62*, 275–282. [CrossRef]
35. Karagiannidis, A.; Varsakeli, S.; Alexopoulos, C.; Amarantidis, I. Seasonal variation in semen characteristics of Chios and Friesian rams in Greece. *Small Rumin. Res.* **2000**, *37*, 125–130. [CrossRef]
36. Casao, A.; Vega, S.; Palacín, I.; Pérez-Pe, R.; Laviña, A.; Quintín, F.J.; Sevilla, E.; Abecia, J.A.; Cebrián-Pérez, J.A.; Forcada, F.; et al. Effects of Melatonin Implants During Non-Breeding Season on Sperm Motility and Reproductive Parameters in Rasa Aragonesa Rams. *Reprod. Dom. Anim.* **2010**, *45*, 425–432. [CrossRef]
37. Lincoln, G.A.; Short, V. Seasonal Breeding: Nature's Contraceptive. *Recent Prog. Horm. Res.* **1980**, *36*, 1–52. [PubMed]
38. Malpaux, B.; Viguié, C.; Skinner, D.C.; Thiéry, J.C.; Pelletier, J.; Chemineau, P. Seasonal breeding in sheep: Mechanism of action of melatonin. *Anim. Reprod. Sci.* **1996**, *42*, 109–117. [CrossRef]
39. Lincoln, G.A. Significance of seasonal cycles in prolactin secretion in male mammals. In *Perspectives in Andrology*; Serio, M., Ed.; Raven Press: New York, NY, USA, 1989; pp. 299–306.

Article

Rederivation by Cryopreservation of a Paternal Line of Rabbits Suggests Exhaustion of Selection for Post-Weaning Daily Weight Gain after 37 Generations

Jorge Daniel Juarez [1], Francisco Marco-Jiménez [2], Raquel Lavara [2] and José Salvador Vicente [2,*]

1 Facultad de Zootecnia, Universidad Nacional Agraria de la Selva, Tingo María 10131, Peru; jorjua@posgrado.upv.es

2 Instituto de Ciencia y Tecnología Animal, Universitat Politècnica de València, 46022 Valencia, Spain; fmarco@dca.upv.es (F.M.-J.); lavara.raquel@gmail.com (R.L.)

* Correspondence: jvicent@dca.upv.es

Received: 16 July 2020; Accepted: 13 August 2020; Published: 17 August 2020

Simple Summary: This study was conducted to evaluate the effect of a long-term selection for post-weaning daily weight gain after 37 generations, using vitrified embryos with 18 generational intervals to rederive two coetaneous populations, reducing or avoiding genetic drift, environmental and cryopreservation effects. This study reports that the selection programme had improved average daily weight gain without variations in adult body weight but, after 37 generations of selection, this trait seems exhausted.

Abstract: Rabbit selection programmes have mainly been evaluated using unselected or divergently selected populations, or populations rederived from cryopreserved embryos after a reduced number of generations. Nevertheless, unselected and divergent populations do not avoid genetic drift, while rederived animals seem to influence phenotypic traits such as birth and adult weights or prolificacy. The study aimed to evaluate the effect of a long-term selection for post-weaning average daily weight gain (ADG) over 37 generations with two rederived populations. Specifically, two coetaneous populations were derived from vitrified embryos with 18 generational intervals (R19 and R37), reducing or avoiding genetic drift and environmental and cryopreservation effects. After two generations of both rederived populations (R21 vs. R39 generations), all evaluated traits showed some progress as a result of the selection, the response being 0.113 g/day by generation. This response does not seem to affect the estimated Gompertz growth curve parameters in terms of the day, the weight at the inflexion point or the adult weight. Moreover, a sexual dimorphism favouring females was observed in this paternal line. Results demonstrated that the selection programme had improved ADG without variations in adult body weight but, after 37 generations of selection, this trait seems exhausted. Given the reduction in the cumulative reproductive performance and as a consequence in the selection pressure, or possibly/perhaps due to an unexpected effect, rederivation could be the cause of this weak selection response observed from generation 18 onwards.

Keywords: selection programme; embryo vitrification; Gompertz growth curve; biobanking; reproductive performance

---

1. Introduction

Traits related to prolificacy for maternal lines, and feed conversion rate and carcass and meat quality for paternal lines, are commonly used in selection programmes in rabbits [1–8]. Some of them are difficult or expensive to measure (for instance, feed conversion rate trait), so correlated traits such as growth rate have been successfully used in selection [2,7,9–14]. Accurately, post-weaning daily weight gain has been estimated from 0.45 to 1.73 g/day (18 to 68 g) per generation in rabbit [2,8,11,12,15–18]. Body weight at slaughter time ranged in heritability from 0.12 to 0.67 as a consequence of environmental variability, the improvement of facilities, nutrition and feed systems, making it difficult to assess during the selection programme [7].

The success of rabbit selection programmes has been evaluated using a control population, in some cases from rederived cryopreserved embryos or by divergent selection [13,17–21]. Nevertheless, unselected control or divergent populations do not avoid genetic drift, while rederivation of animals by cryopreservation seems to affect phenotypic traits such as birth and adult weights or prolificacy, not only in animals born after transfer, but also their offspring [22,23]. Whether the phenotypic changes are only intragenerational, as a consequence of the embryonic stress response to cryopreservation and transfer, or transgenerational, as a consequence of heritable changes introduced at the epigenome level, is yet to be assessed [24,25]. However, cryopreservation offers the great advantage of being able to measure the genetic progress through generations, making evaluation possible in the same environment, facilities and feed diets of individuals separated by many generations [26].

A control population obtained by cryopreservation has rarely been used in selection experiments. In rabbits, it has been used to evaluate both the response to selection for litter size in maternal lines and for average daily weight gain in paternal lines [18,27]. García and Baselga [27] observed that the estimated responses obtained when using a rederived population of cryopreserved embryos or a mixed model that took into account the kinship matrix (genetic relationship) were different, suggesting that this latter model is less appropriate. Piles and Blasco [18] demonstrated that the selection for average daily gain between the 3rd–4th and 10th generations was successful and the response obtained was similar, using a rederived control population or a model with genetic relationships (0.62 to 0.65 g/day by generation). The latter study was performed with the paternal line used in this work. This study aimed to evaluate the effect of a long-term selection for average daily gain (37 generations) on commercial growth traits and Gompertz parameters, using two population rederived from vitrified embryos with 18 generational intervals to reduce or avoid genetic drift, environmental and cryopreservation effects.

2. Materials and Methods

All the experimental procedures used in this study were performed following Directive 2010/63/EU EEC for animal experiments and reviewed and approved by the Ethical Committee for Experimentation with Animals of the Universitat Politeècnica de València, Spain (research code: 2015/VSC/PEA/00061).

2.1. Animals

A rabbit paternal line (R) selected at the Universitat Politècnica de València was used. This line was founded in 1989 from two closed paternal lines selected according to individual weight gain from weaning to end of fattening period (77 days old) during 12 and 9 generations [11]. Since then, the line has been selected for individual daily weight gain from 28 days (weaning) to 63 days of age (end of fattening phase). Animals from two different generations of selection were used. R19V population was rederived from 256 embryos of 25 donors belonging to ten different sire families of 18th generation vitrified in 2000. R37V population was rederived from 301 embryos of 28 donors belonging to 15 different sire families of 36th generation and they were vitrified in 2015. Both animal groups were transferred at the same time in 2015 (see details in Marco-Jiménez et al. [26]). Offspring were bred in non-overlapping generations during 2 generations (R20–R21 and R38–R39 from R19V and R37V, respectively). Figure 1 and Table 1 show the experimental design and the parents used to generate the offspring analysed in this study.

The animals were housed at the Universitat Politècnica de València experimental farm in flat deck indoor cages (75 × 50 × 30 cm), with free access to water and commercial pelleted diets (minimum of 15 g of crude protein per kg of dry matter (DM), 15 g of crude fibre per kg of DM and 10.2 MJ of digestible energy (DE) per kg of DM). The photoperiod was 16 h of light and 8 h of dark, with a regulated room temperature between 14 °C and 28 °C.

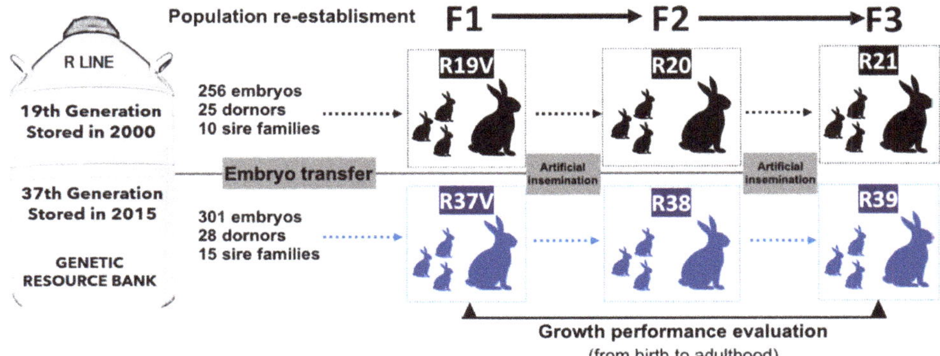

Figure 1. Experimental design: Two experimental progenies were developed from vitrified embryos stored in 2000 (19th generation of selection) and 2015 (37th generation of selection). All rabbits were identified and weighted at weaning and end of the fattening period to calculate the average daily gain. A sample of males and females were weighted weekly from birth to 20 weeks age in F1 y F3 generation to estimate Gompertz curve parameters.

2.2. *Embryo Vitrification and Transfer to Population Rederivation*

Five-hundred-and-fifty-seven embryos were used from donors belonging to R18 and R36 generations. Vitrification and transfer were described elsewhere [26,28,29]. Briefly, the vitrification was carried out in two steps at room temperature (approximately 20–22 °C). In the first step, embryos from each donor doe were placed for 2 min in an equilibrium solution consisting of 12.5% dimethyl sulfoxide (DMSO) and 12.5% of ethylene glycol (EG) in Dulbecco's phosphate-buffered serum (DPBS) supplemented with 0.1% (*w*/*v*) of bovine serum albumin (BSA). In the second step, embryos were suspended for 1 min in the vitrification solution containing 20% DMSO and 20% EG in DPBS supplemented with 0.1% of BSA. Then, embryos suspended in vitrification medium were loaded into 0.125 mL plastic straws (ministraws, L'Aigle, France) and plunged directly into liquid nitrogen. After storage in liquid nitrogen, embryos were warmed and vitrification solution was removed, loading the embryos into a solution containing DPBS and 0.33 M sucrose for 5 min, followed by one bath in a solution of DPBS for another 5 min before transfer. Immediately after warming, the embryos were evaluated morphologically, and only embryos without damage in mucin coat or pellucid zone were transferred by laparoscopy into the oviduct of synchronised recipient females from a maternal line (Line A [29]) following the procedure described by Besenfelder and Brem [30] and García-Domínguez et al. [29]. The re-establishing of both populations to generate the 19th (R19V) and 37th (R37V) generation was described by Marco-Jiménez et al. [26].

2.3. *Reproduction Management in Rederived and Filial Generations*

Rederived rabbits were conducted in non-overlapping generations and the generation interval was approximately 9–10 months. The first reproductive cycle took place at ~5 months of age, and after kindling the new insemination was tried 10–12 days later. Insemination between relatives sharing a grandparent was avoided. Briefly, two ejaculates per male were collected in each replica using an artificial vagina. Ejaculates were diluted 1:5 with Tris-Citrate-Glucose extender and both motility and abnormal spermatozoa were assessed under phase optic at 200×. Only ejaculates with total motility higher than 70% and less than 30% of abnormal sperm were used. After semen evaluation, optimal ejaculates from each male were pooled and extended to 40 million/mL. All females were synchronised with 15UI eCG injected intramuscularly 48h before being inseminated with 0.5 mL of extended semen using a plastic curved pipette. Females were induced to ovulate by intramuscular injection of 1 μg of buserelin acetate at insemination time. Pregnancy was checked at 14 days from insemination and non-pregnant does were inseminated again at 21 days after the previous insemination. In addition, it was noted whether rabbits underwent a lactation–gestation overlap, classifying the reproductive status of their dams in four levels: offspring from primiparous does without overlapping (PD), offspring from primiparous lactating does (females that were pregnant while suckling their first litter, PLD), multiparous lactating does (females with more than one birth that were pregnant while suckling their litter, MLD) and multiparous non-lactating does (females from more than one parturition that were pregnant after lactation, MD).

2.4. Growing Traits Analysis

The offspring born of vitrified embryos were named R19V and R37V, the first filial generation (R20 and R38) and the second filial generation (R21 and R39). Individual weaning weight (WW, 30 days old), individual weight at end of the fattening period (EFW, 63 days old) and average daily weight gain (weight gained from day 28 to 63 divided by 33, ADG) during the fattening period were noted for all generations. To determine differences in growth curve, 76 and 113 rabbits from the rederived population (18 and 21 females and 16 and 21 males from R19V and R37V, respectively) and second filial generation (29 and 39 females and 18 and 27 males from R21 and R39, respectively) were identified at birth with chip and weighed weekly from born to 20 weeks old.

2.5. Statistical Analysis

A first analysis for growth traits was performed, attending to the generation interval of rederived population,

$$Y_{ijklm} = \mu + P_i + R_j + MY_k + PR_{ij} + CO_l + Cov\ X_m + e_{ijklm} \quad (1)$$

where Y_{ijkl} was the trait to analyse, μ was the general mean, P_i was the fixed effect of selection generation of rederived population R19 (R19V, R20, R21) and R37 (R37V, R38, R39), R_j was the fixed effect of reproductive status of the doe used in the WW analysis (PD, PLD, MLD and MD), M_{Yk} was the fixed effect of month-year in which the fattening period ended (39 levels), PR_{ij} was the effect of interaction between rederived population and reproductive status of the mothers used for WW analysis, CO_l was the random effect of common litter, Cov X_m was the covariate of the number born alive (BA) used for the WW trait or the covariate of WW used for weight at the end of the fattening period (EFW) and ADG traits and e_{ijklm} was the error term of the model.

A second analysis of growth traits to evaluate each filial generation (F1-R19V vs. R37V, F2-R20 vs. R38- and F3-R21 vs. R39) was performed using the mixed linear model described above:

$$Y_{ijklm} = \mu + P_i + R_j + MY_k + PR_{ij} + CO_l + Cov\ X_m + e_{ijklm} \quad (2)$$

where Pi was the fixed effect of filial generation (R19V and R37V or R20 and R38 or R21 and R38). In the WW analysis of the first filial generation, the effect of reproduction status of the mothers (R_j) and the interaction (PR_{ij}) were not included, all hosted females were multiparous non-lactating (MD) and the fixed effect month-year had 4 levels. In the analysis of filial generation 2, the fixed effect month-year (MY_k) had 19 levels. For filial generation 3, MY_k fixed effect had 25 levels.

Finally, Gompertz curve parameters were estimated for each rabbit from R19V and R37V, and R21 and R39 for each rabbit:

$$X_{jm} = a_m \exp[-b_m \exp(-k_m t)] + e_{jm} \quad (3)$$

where X_{jm} is the body weight (BW) of m animal at t age (days); am, bm and km are the Gompertz growth curve parameters of m animal; and e_{jm} is the residual. As to the meaning of parameters, a can be interpreted as the mature BW maintained independently of short-term fluctuations, b is a timescale parameter related to the initial BW and k is a parameter related to the rate of maturing. The inflexion time (t_I) is the moment when growth rate is maximum and is determined by $t_I = (\log e^b)/k$ and inflexion weight (w_I) were fixed at 0.37 of adult weight (a) by Gompertz curves.

A mixed linear model was fitted for the analysis of Gompertz growth curve parameters as follows:

$$Y_{ijkl} = \mu + P_i + S_j + PS_{ij} + CO_k + CovB_l + e_{ijkl} \quad (4)$$

where Y_{ijkl} was the Gompertz parameter, P_i is the fixed-effect population (R19V and R37V or R21V and R39V), S_j the fixed effect of the sex, PS_{ij} was the effect of interaction between population and sex, CO_k is the random effect of common litter, $CovB_l$ was the weight at birth as a covariate and e_{ijkl} was the error term of the model.

3. Results

3.1. Descriptive Growing Traits

A total of 2025 (2991 liveborn) animals from the three filial generations that finished fattening were obtained from 437 parturitions (27, 134 and 276 parturitions in the first, second and third generations respectively, Table 1).

Table 1. The total number of parents used to generate the offspring analysed.

Rederived Population	Filial Generation	Generation	Females	Males	Total
19th	F1	R19V	14	11	**25**
	F2	R20	22	7	**29**
	F3	R21	50	16	**66**
		Total	**86**	**34**	**120**
37th	F1	R37V	11	10	**21**
	F2	R38	26	13	**39**
	F3	R39	78	18	**96**
		Total	**122**	**41**	**163**

V: Rederived from vitrified embryos.

Average body weight at the end of fattening was 2.240 kg (EFW, 63 days old) and ADG was 43.95 g/day (Table 2).

Table 2. Number of young rabbits weighted from weaning to end fattening period and average and standard deviation to birth alive (BA), weaning weight (WW), weight at end fattening (EFW) and average daily weight gain (ADG).

Trait	Generation						
	R19V	**R37V**	**R20**	**R38**	**R21**	**R39**	**Total**
BA	3.4 ± 1.40	5.8 ± 3.30	6.7 ± 2.75	6.6 ± 2.95	6.8 ± 2.41	7.4 ± 3.16	6.8 ± 2.94
WW(Kg)	0.75 ± 0.236	0.64 ± 0.139	0.72 ± 0.162	0.72 ± 0.171	0.69 ± 0.165	0.71 ± 0.173	0.71 ± 0.171
EFW(Kg)	2.20 ± 0.302	2.17 ± 0.249	2.24 ± 0.279	2.33 ± 0.309	2.17 ± 0.323	2.25 ± 0.314	2.24 ± 0.313
ADG(g/day)	43.8 ± 5.72	46.3 ± 5.57	44.2 ± 5.98	46.9 ± 6.81	42.0 ± 7.08	43.6 ± 6.37	44.0 ± 6.72
Animals	42	52	285	347	489	810	2025

V: Rederived from vitrified embryos.

Mortality by rederived generations (R19V-R20-R21 vs. R37V-R38-R39) was 17.5% and 20.5% in the lactation period and 15.2% and 16.6% in the fattening period, respectively. For each filial generation, the mortality at the end of the lactation and fattening period reached levels of 3.3% and 19.6% for F1, 21.3 and 10.2 for F2 and 19.3% and 18.4% for F3, respectively. Data not shown in tables.

3.2. Growing Traits by Generation Interval of Rederived Populations. Selection Effect

Selection had a significant effect on studied traits and the estimated effects on WW, EFW and ADG were 0.031 ± 0.014 kg, 0.058 ± 0.013 kg and 1.55 ± 0.392 g/day (R37-R19, $p < 0.05$, Table 3). Moreover, reproductive status of does affected the WW, observing that the young mothers without lactation-gestation overlap (ND) had the lowest WW (0.632 ± 0.016 vs. 0.734 ± 0.019, 0.746 ± 0.015 and 0.718 ± 0.016 kg to PD, PLD, MLD and MD, respectively, $p < 0.05$, data not shown in tables). No interaction between selection generation of rederived population and reproductive status was found. MY fixed and common litter random effects were significant for all analyses (data not shown in tables). As expected, the warm months of July, August and September from each year had adverse effects on the traits studied.

Table 3. Effect of selection for growth rate on weaning weight (WW, kg), weight at end of fattening period (EFW, kg) and average daily weight gain (ADG, g/day) from two rederived population after 18 generations.

Rederived Population	n	WW (kg)	EFW (kg)	ADG (g/day)
19th (R19V+R20+R21)	816	0.692 ± 0.014 ^{b}	2.219 ± 0.013 ^{b}	43.85 ± 0.375 ^{b}
37th (R37V+R38+R38)	1209	0.723 ± 0.012 ^{a}	2.277 ± 0.011 ^{a}	45.41 ± 0.324 ^{a}
Covariate effect		−0.028 ± 0.0024 ***	1.35 ± 0.029 ***	10.58 ± 0.847 ***

n: number of young rabbits. V: Rederived from vitrified embryos. Data are expressed as least squared mean ± standard error of means. ^{a,b} Values with different superscripts in column differ significantly ($p < 0.05$). Significance of estimated value of covariates (*** $p < 0.001$).

3.3. Evolution of Growth Traits by Filial Generation

Weaning weight (WW) was not different in the first two filial generations (F1 and F2), and it was different and favourable to R39 at third filial generation (0.053 ± 0.022 and 0.066 ± 0.019, respectively, $p < 0.05$). On the contrary, the estimated effects on weight at the end of fattening (EFW) and ADG were always significant and favourable in each comparison to the last generations (Table 4).

Table 4. Differences of least square means of weaning weight (WW, Kg), weight at end of fattening period (EFW, kg) and average daily weight gain (ADG, g/day) from filial generations. V: Rederived from vitrified embryos.

Trait	Generation		
	F1 (R37V-R19V)	F2 (R38-R20)	F3 (R39-R21)
WW (Kg)	0.043 ± 0.066	−0.040 ± 0.025	0.066 ± 0.019 **
EFW (Kg)	0.135 ± 0.056 *	0.062 ± 0.020 **	0.056 ± 0.018 **
ADG (g/day)	4.270 ± 1.578 *	1.658 ± 0.606 **	1.468 ± 0.544 **
Covariate effect			
BA(WW)	−0.028 ± 0.0122 *	−0.030 ± 0.0038 ***	−0.027 ± 0.0030 ***
WW(EFW)	1.20 ± 0.120 ***	1.33 ± 0.046 ***	1.36 ± 0.039 ***
WW(DG)	5.60 ± 3.581	10.4 ± 1.35 ***	11.0 ± 1.13 ***

Significance of estimated effects and covariates (* $p < 0.05$, ** $p < 0.01$, *** $p < 0.001$).

Reproductive status of does in F2 and F3 affected the WW. It was observed that the young mothers without lactation–gestation overlap had the lowest WW. No interaction between selection generation of rederived population and reproductive status was found. MY fixed were significant for all analyses, while common litter effect was not significant for EFW and ADG in F1 (R37V-R19V).

3.4. Estimated Growth Curves: F1 Rederived Populations (R19V and R37V)

No differences in Gompertz parameters were found for R19V and R37V population and sex (Table 5).

Estimated growth patterns showed that both population and sex reached a maximum growth rate between 59 and 61 days old and with a weight about 1800 g. In addition, the estimated adult weights were between 4800 and 5000 g (Figure 2).

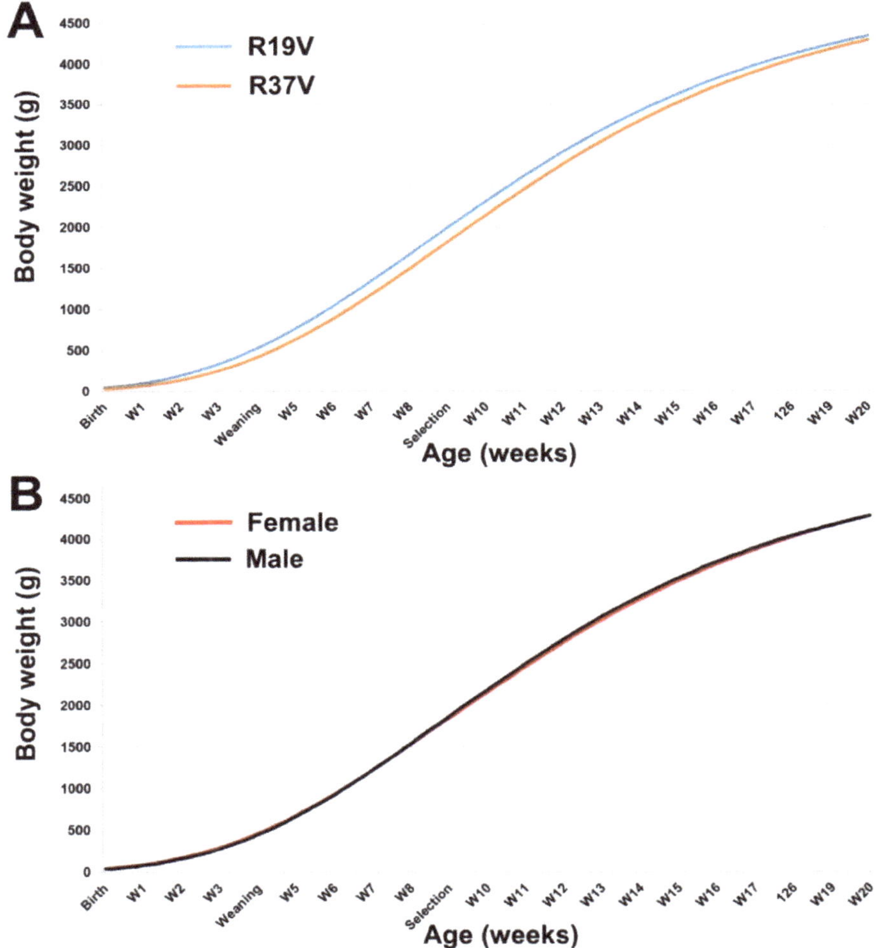

Figure 2. Growth curves from birth to 20-week-old between F1 rederived populations from R19V and R37V generations. V: Rederived from vitrified embryos. Growth curves were fitted using the Gompertz equation for (**A**) R19V and R37V generations and (**B**) sex.

Table 5. Gompertz curve parameters from R19V and R37V generations.

Group	Gompertz Parameters				
	a	b	k	t (days)	W (g)
Rederived Population					
R19V	4906 ± 133	4.62 ± 0.269	0.026 ± 0.001	60.6 ± 2.33	1815 ± 49
R37V	4905 ± 128	5.08 ± 0.288	0.026 ± 0.001	62.5 ± 2.46	1815 ± 47
Sex					
Female (F)	4939 ± 113	4.75 ± 0.211	0.025 ± 0.001	62.1 ± 1.84	1828 ± 42
Male (M)	4872 ± 114	4.93 ± 0.210	0.026 ± 0.001	61.0 ± 1.84	1803 ± 42
Rederived Population × Sex					
R19V × F	4887 ± 170	4.64 ± 0.297	0.030 ± 0.002	60.6 ± 2.63	1808 ± 63
R19V × M	4925 ± 168	4.57 ± 0.292	0.030 ± 0.002	60.6 ± 2.58	1822 ± 62
R37V × F	4992 ± 159	4.85 ± 0.306	0.030 ± 0.002	63.6 ± 2.66	1847 ± 59
R37V × M	4818 ± 155	5.29 ± 0.303	0.030 ± 0.002	61.4 ± 2.63	1783 ± 57

V: Rederived from vitrified embryos. ^{a,b}: Different superscript between columns indicate statistical differences ($p < 0.05$). a: mature body weight; b: a timescale parameter related to the initial body weight; k: growth velocity; t: Inflexion age; w: weight at inflexion. LSM ± SEM: least square mean ± standard error of means.

3.5. Estimated Growth Curves: F3 Rederived Population (R21 and R39)

Gompertz parameters for R21 and R39 populations suggest no relevant differences between both populations (Table 6, Figure 3A). However, Gompertz parameters showed sexually dimorphic growth. Thus, males reached the inflexion point sooner (~3 days) and had a lower adult weight (Table 6, Figure 3B). Moreover, an interaction between population and sex was observed, suggesting a significant dimorphic pattern in R21 population as a consequence of a minor growth rate of R21 females (Figure S1).

Table 6. Gompertz curve parameters from R21 and R39 generations.

Group	Gompertz Parameters				
	a	b	k	t (days)	W (g)
Population					
R21	5518 ± 154	4.77 ± 0.095 ^{a}	0.026 ± 0.001	60.1 ± 1.22	2042 ± 56.9
R39	5400 ± 135	4.49 ± 0.083 ^{b}	0.027 ± 0.001	56.9 ± 1.07	1998 ± 50.1
Sex					
Female (F)	5815 ± 113 ^{a}	4.63 ± 0.070	0.026 ± 0.000 ^{a}	60.1 ± 0.89 ^{a}	2151 ± 41.7 ^{a}
Male (M)	5102 ± 132 ^{b}	4.64 ± 0.082	0.027 ± 0.001 ^{b}	56.9 ± 1.05 ^{b}	1888 ± 48.8 ^{b}
Rederived Population × Sex					
R21 × F	5925 ± 172	4.68 ± 0.107	0.025 ± 0.001 ^{a}	62.7 ± 1.37	2192 ± 63.6
R21 × M	5110 ± 195	4.86 ± 0.122	0.028 ± 0.001 ^{b}	57.4 ± 1.56	1891 ± 72.1
R39 × F	5704 ± 153	4.57 ± 0.095	0.027 ± 0.001 ^{b}	57.5 ± 1.22	2111 ± 56.5
R39 × M	5095 ± 171	4.42 ± 0.106	0.027 ± 0.001 ^{b}	56.4 ± 1.37	1885 ± 63.3

^{a,b} Different superscript between columns indicate statistical differences ($p < 0.05$). a: mature body weight; b: a timescale parameter related to the initial body weight; k: growth velocity; t: inflexion age; w: weight at inflexion. LSM ± SEM: least square mean ± standard error of means.

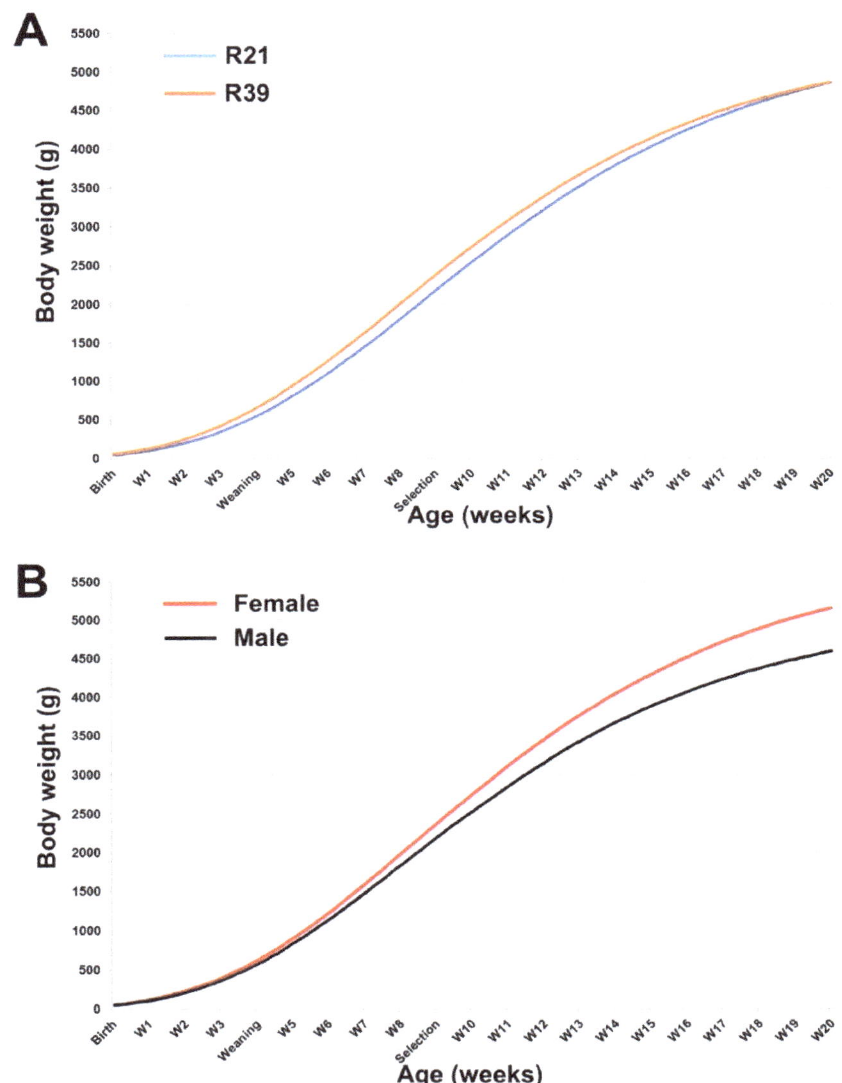

Figure 3. Growth curves from birth to 20-week-old between F3 filial generations from R21 and R39 generations. V: Rederived from vitrified embryos. Growth curves were fitted using the Gompertz equation for (**A**) R21 and R39 generations and (**B**) sex.

4. Discussion

Rabbit meat programmes, like other animal breeding programmes, have based selection on the improvement of target criteria in controlled environments (e.g., litter size, BW daily gain, meat quality or feed efficiency [1,3,5,6,8,11,14,15,18,20,31–34]). Individual selection by ADG has been a common criterion for selecting rabbits in paternal lines, as it is effortless to record and has a moderate heritability (0.15–0.18), and favourable genetic correlation with the conversion index [12,14]. Related to this, response to selection for ADG has been estimated by comparison with a control population [2,8,17], or by divergent selection [12]. The response obtained by generation ranged between 0.56 g/day to 6 g/day [12,18]. Nevertheless, Magheni and Christensen [9] demonstrated an asymmetrical response to divergent selection, the response being greatest in the downward rather than in the upward lines, and therefore higher realised heritabilities were recorded in the downward than in upward lines. It is worth mentioning that in all studies, the response was lower than expected, attending to heritability estimates [7]. In our study, the average daily gain was compared between two rederived populations separated by 18 generations (approximately 15 years). This experimental approach reduces the influence of environmental effects, management, nutrition and even the rederivation procedure, as both populations were cryopreserved, transferred in the same host maternal line and reproductively conducted in the same way under the same environmental and feeding conditions. In this study, the response to the selection for ADG between the 19th and 37th generations was lowest compared to the results obtained in the same line between the 3–4th versus 11th generation [18]. These findings suggest that the response for ADG has been reduced across generations. Despite further research being required, we hypothesise that in part it would be as a consequence of an unfavourable effect on reproductive performance [35,36] and high mortality rates both during lactation (26% in Lavara et al. [37] and 17.5 and 20.5% in the present study to all R19 and R37 rederived generations, respectively) and the fattening period (15.2% and 16.6% to all R19 y R37 rederived generations, respectively), reducing the applied selection pressure and variability. On the other hand, rederivation from cryopreserved embryos could be a noneffective way to generate a control population [24,25]. All this evidence contributed to the response to selection being lower in the last 18 generations from a line with a total of 37 generations (0.09 g/day, Table 1). To the best of our knowledge, this is the first study carried out in rabbit after more than 37 generations of selection for ADG, if we take into account the lines from which it was founded in 1989. On the contrary, other studies have reported better results for genetic progression achieved after a few generations of selection [2,11,17,18,38].

In recent years, different studies have shown that assisted reproduction technologies and specifically embryo cryopreservation are not neutral [24,25]. Embryos are subjected to extreme environments in either recovery, cryopreservation and transfer that alter the gene expression, prenatal and postnatal development, and even modify reproductive performance (for example, focused on rabbit [22–25,39–42]). These effects not only modify the phenotype of those born from cryopreserved embryos, but their effects might also be affected and transferred to subsequent generations through heritable and non-heritable epigenetic changes [24,25]. First, in the first generation, newborns could be affected both by the vitrification transfer procedure and by maternal effects when developing in a different uterine and lactation environment, due to the mother in this experiment and the resulting litter size [26,43–46]. The second generation can be affected by the conditions in which parents developed. Finally, in the third generation, the heritable epigenetic effects generated by the rederived technique could emerge and become present. Lavara et al. [23] demonstrated that vitrification and transfer procedures involved in a rederivation programme for rabbit embryos have long-term consequences on rabbit growth patterns and might affect some growth-related traits in rabbits. Therefore, if the genetic artefacts introduced by the rederivation process were similar in both generations, evaluating their phenotypic differences from the third generation post-rederivation should be the most appropriate method with a minor cumulative genetic drift variance [47]. It has been observed that rederived populations showed differences in WW, end fattening weight and ADG as a consequence of selection. However, these differences were not constant after rederivation. Thus, the growth pattern was similar between the rabbits born from vitrified embryos (rederived populations, R19V and R37V) and the subsequent filial generations (R20 and R38). At weaning and end of fattening, the body weights were not different between generations. Only the first filial generation had a relevant difference for ADG (4.27 g/day), with a response of 0.237 g/day by generation. In the third filial generation (R21 and R39), all traits showed some progress as a result of the selection and the response was 0.113 g/day by generation. These deviations can be appreciated in the growth curve pattern, although they did not reach a significant difference either for the day or the weight at which the inflexion point is reached or the estimated adult weight. It should be noted that the inflexion point was slightly above the rabbit marketing age in Spain (56–58 days old) and near the selection age (63 days old). Blasco et al. [13], comparing a cryopreserved rederived population from 3rd and 4th generations from this paternal line with a 10th generation, concluded that estimated adult BW was increased by 7% after six generations, while other parameters of the Gompertz curve were scarcely affected by selection. In this context, the reduced sample size of rederived population advises caution regarding the conclusions of this study.

The parameters and the resulting growth curve revealed a significant female sexual dimorphism pattern at the inflection point and estimated adult BW. Favourable female dimorphism was already being reported by several authors [48–50] in different breeds or synthetic lines of this species. Rall [48] described that females were larger than males by a proportion of 1.3:1, while Fuente and Rosell [50] reduce this ratio to 1.03. In this line, Blasco and Gómez [38] found no effect of sex in Gompertz parameters after 12 generations of ADG selection. Our data showed that sexual dimorphism was affected by ADG, going from a ratio of 1.0:1 to 1.14:1 in 18 generations.

An undesirable consequence of selection for ADG could be the increment in adult BW, as it augments the costs of maintaining a parent population. Blasco et al. [13] found that Gompertz estimated adult BW increased with selection, whereas the parameters related to the slope of the curve practically did not change. In this circumstance, these latter authors predicted that male lines will become giant lines and the reproduction management will be more difficult, unless artificial insemination is used. An interesting finding of the study on long-term selection in these populations separated by 18 generations was that the selection for ADG did not change the estimated adult weight, although it seems to have reduced the adult female body weight (221 g, approximately −4% between R21 and R39 females). It is necessary to emphasise that from the parameters obtained in the populations directly rederived by vitrified embryo transfer (R19V and R37V), it was not possible to observe differences. This is probably due to the associated procedure and maternal effects derived from the transfer (smaller litter size and mother of the maternal line). Recent studies that demonstrated the incorporation of inheritable epigenetic marks [25] and the lack of knowledge as to whether or not they selectively affect certain genetic loci, could question the model used. Therefore, although rederived populations from cryopreserved embryos have some advantages, by optimising experimental facilities and reducing genetic drift this rederivation procedure could also contribute to distorting the control population [24,25]. Recently, García-Domínguez et al. [51] compared naturally conceived animals with progeny generated after embryo cryopreservation, observing transgenerational effects in differentially expressed transcripts and metabolites in hepatic tissue that could be associated with a lower adult weight of the rederived population. Another effect to take into account would be the differential storage time, although these studies were performed to analyse the effect on post-thaw embryo survival and pregnancy outcomes, but not on the liveborn development [52–57]. No studies have been performed to evaluate if the pattern of omics and phenotypic alteration was influenced by storage time.

As expected, the effect of the rabbit's reproductive status (number of parturitions and overlap or not with lactation) on the WW was significant and unfavourable in the rabbits at first parity. Although the effect is restricted in young mothers without lactation and gestation overlap, our results were similar to those reported in earlier generations of R line [58], and in line with other paternal and maternal lines between primiparous and multiparous does [59–61]. This effect was related to the capacity of the doe to produce milk, which depends on the maturity of the female and the number of suckling kits [62–65]. In this regard, we observed that young females that became pregnant again after 10–12 days post-partum achieved a litter size weight at weaning similar to that of multiparous rabbits. Likewise, they showed a better health and physiological status than young mothers that were not pregnant after being inseminated.

5. Conclusions

In conclusion, the present study describes the selection programme to improve post-weaning daily weight gain without increasing adult body weight in R line, but after 37 generation of selection this trait seems exhausted. A low accumulative reproductive performance or an unexpected rederived effect on growth traits might be the cause of this little selection response. A refoundation of line and selection for other criteria such as feed efficiency and maternal traits will be necessary.

Supplementary Materials: The following are available online at http://www.mdpi.com/2076-2615/10/8/1436/s1, Figure S1: Gompertz growth curve from birth to 20-week-old for interaction between populations and sex of animals.

Author Contributions: Conceptualisation F.M.-J. and J.S.V.; Methodology and data curation F.M.-J., J.D.J., and J.S.V.; writing original draft preparation, J.D.J.; writing and analysis review and editing, F.M.-J., R.L., and J.S.V. All authors have read and agreed to the published version of the manuscript.

Funding: Funding from the Ministry of Economy, Industry and Competitiveness (Research project: AGL2014-53405-C2-1-P) is acknowledged. English text version was revised by N. Macowan English Language Service.

Acknowledgments: The authors would to thank retired professor Baselga for his dedication to genetic improvement and diffusion of selected rabbit lines during these forty years.

Conflicts of Interest: The authors declare no conflict of interest.

Abbreviations

ADG	Daily weight gain
R	Rabbit paternal line
DM	Dry matter
DE	Digestible energy
DMSO	Dimethyl sulfoxide
EG	Ethylene glycol
DPBS	Dulbecco´s phosphate buffered serum
PD	Primiparous does
PLD	Primiparous lactating does
MLD	Multiparous lactating does
MD	Multiparous non-lactating does
WW	Weaning weight
EFW	End of the fattening period
BW	Body weight
R19V	Rederived animals from generation 19
R37V	Rederived animals from generation 37
R20	First filial generation from R19V generation
R21	Second filial generation from R19V generation
R38	First filial generation from R37V generation
R39	Second filial generation from R37V generation

References

1. Estany, J.; Baselga, M.; Blasco, A.; Camacho, J. Mixed model methodology for the estimation of genetic response to selection in litter size of rabbits. *Livest. Prod. Sci.* **1989**, *21*, 67–75. [CrossRef]
2. Lukefahr, S.D.; Odi, H.B.; Atakora, J.K.A. Mass Selection for 70-Day Body Weight in Rabbits. *J. Anim. Sci.* **1996**, *74*, 1481–1489. [CrossRef] [PubMed]
3. Baselga M Genetic improvement of meat rabbits. programmes and diffusion. In *Proceedings of the 8th World Rabbit Congress*; World Rabbit Science Association: Puebla, Mexico, 2004; pp. 1–13.
4. Nagy, I.; Ibáñez, N.; Romvári, R.; Mekkawy, W.; Metzger, S.; Horn, P.; Szendro, Z. Genetic parameters of growth and in vivo computerized tomography based carcass traits in Pannon White rabbits. *Livest. Sci.* **2006**, *104*, 46–52. [CrossRef]
5. Khalil, M.H.; Al-Saef, A.M. Methods, criteria, techniques and genetic responses for rabbit selection: A review. In *Proceedings of the 9th World Rabbit Congress*; World Rabbit Science Association: Verona, Italy, 2008; pp. 1–22.
6. Martínez-Álvaro, M.; Hernández, P.; Blasco, A. Divergent selection on intramuscular fat in rabbits: Responses to selection and genetic parameters. *J. Anim. Sci.* **2016**, *94*, 4993–5003. [CrossRef] [PubMed]

7. Blasco, A.; Nagy, I.; Hernández, P. Genetics of growth, carcass and meat quality in rabbits. *Meat Sci.* **2018**, *145*, 178–185. [CrossRef] [PubMed]
8. de Rochambeau, H.; de la Fuente, L.; Rouvier, R.; Ouhayoun, J. Sélection sur la vitesse de croissance post-sevrage chez le lapin. *Genet. Sel. Evol.* **1989**, *21*, 527. [CrossRef]
9. Mgheni, M.; Christensen, K. Selection Experiment on Growth and Litter Size in Rabbits. *Acta Agric. Scand* **1985**, *35*, 287–294. [CrossRef]
10. Torres, C.; Baselga, M.; Gómez, E.A. Effect of weight daily gain selection on gross feed efficiency in rabbits. *J. Appl. Rabbit Res.* **1992**, *15*, 872–878.
11. Estany, J.; Camacho, J.; Baselga, M.; Blasco, A. Selection response of growth rate in rabbits for meat. *Genet. Sel. Evol.* **1992**, *24*, 527–537. [CrossRef]
12. Moura, A.S.A.M.T.; Kaps, M.; Vogt, D.W.; Lamberson, W.R. Two-Way Selection for Daily Gain and Feed Conversion in a Composite Rabbit Population. *J. Anim. Sci.* **1997**, *75*, 2344–2349. [CrossRef]
13. Blasco, A.; Piles, M.; Varona, L. A Bayesian analysis of the effect of selection for growth rate on growth curves in rabbits. *Genet. Sel. Evol.* **2003**, *35*, 21. [CrossRef] [PubMed]
14. Piles, M.; Gomez, E.A.; Rafel, O.; Ramon, J.; Blasco, A. Elliptical selection experiment for the estimation of genetic parameters of the growth rate and feed conversion ratio in rabbits1. *J. Anim. Sci.* **2004**, *82*, 654–660. [CrossRef] [PubMed]
15. Drouilhet, L.; Gilbert, H.; Balmisse, E.; Ruesche, J.; Tircazes, A.; Larzul, C.; Garreau, H. Genetic parameters for two selection criteria for feed efficiency in rabbits. *J. Anim. Sci.* **2013**, *91*, 3121–3128. [CrossRef] [PubMed]
16. Garreau, H.; Szendro, Z.; Larzul, C.; Rochambeau, H. Genetic parameters and genetic trends of growth and litter size. In *Proceedings of the 7th World Rabbit Congress*; World Rabbit Science Association: Valencia, Spain, 2000; pp. 403–406.
17. Larzul, C.; Gondret, F.; Combes, S.; de Rochambeau, H. Divergent selection on 63-day body weight in the rabbit: Response on growth, carcass and muscle traits. *Genet. Sel. Evol.* **2005**, *37*, 105. [CrossRef] [PubMed]
18. Piles, M.; Blasco, A. Response to selection for growth rate in rabbits estimated by using a control cryopreserved population. *World Rabbit Sci.* **2003**, *11*, 53–62. [CrossRef]
19. Gondret, F.; Combes, S.; Larzul, C.; De Rochambeau, H. Effects of divergent selection for body weight at a fixed age on histological, chemical and rheological characteristics of rabbit muscles. *Livest. Prod. Sci.* **2002**, *76*, 81–89. [CrossRef]
20. García, M.L.; Baselga, M. Estimation of correlated response on growth traits to selection in litter size of rabbits using a cryopreserved control population and genetic trends. *Livest. Prod. Sci.* **2002**, *78*, 91–98. [CrossRef]
21. Piles, M.; David, I.; Ramon, J.; Canario, L.; Rafel, O.; Pascual, M.; Ragab, M.; Sánchez, J.P. Interaction of direct and social genetic effects with feeding regime in growing rabbits. *Genet. Sel. Evol.* **2017**, *49*. [CrossRef]
22. Lavara, R.; Baselga, M.; Marco-Jiménez, F.; Vicente, J.S. Long-term and transgenerational effects of cryopreservation on rabbit embryos. *Theriogenology* **2014**, *81*, 988–992. [CrossRef]
23. Lavara, R.; Baselga, M.; Marco-Jiménez, F.; Vicente, J.S. Embryo vitrification in rabbits: Consequences for progeny growth. *Theriogenology* **2015**, *84*, 674–680. [CrossRef]
24. Garcia-Dominguez, X.; Marco-Jiménez, F.; Peñaranda, D.S.; Vicente, J.S. Long-Term Phenotypic and Proteomic Changes Following Vitrified Embryo Transfer in the Rabbit Model. *Animals* **2020**, *10*, 1043. [CrossRef] [PubMed]
25. Garcia-Dominguez, X.; Vicente, J.S.; Marco-Jiménez, F. Developmental plasticity in response to embryo cryopreservation: The importance of the vitrification device in rabbits. *Animals* **2020**, *10*, 804. [CrossRef] [PubMed]
26. Marco-Jiménez, F.; Baselga, M.; Vicente, J.S. Successful re-establishment of a rabbit population from embryos vitrified 15 years ago: The importance of biobanks in livestock conservation. *PLoS ONE* **2018**, *13*. [CrossRef]
27. García, M.L.; Baselga, M. Genetic response to selection for reproductive performance in a maternal line of rabbits. *World Rabbit Sci.* **2002**, *10*, 71–76. [CrossRef]
28. Vicente, J.-S.; Viudes-de-Castro, M.-P.; García, M.-L. In vivo survival rate of rabbit morulae after vitrification in a medium without serum protein. *Reprod. Nutr. Dev.* **1999**, *39*, 657–662. [CrossRef] [PubMed]
29. Garcia-Dominguez, X.; Marco-Jimenez, F.; Viudes-de-Castro, M.P.; Vicente, J.S. Minimally invasive embryo transfer and embryo vitrification at the optimal embryo stage in rabbit model. *J. Vis. Exp.* **2019**, *16*, 147. [CrossRef]
30. Besenfelder, U.; Brem, G. Laparoscopic embryo transfer in rabbits. *J. Reprod. Fertil.* **1993**, *99*, 53–56. [CrossRef]

31. Larzul, C.; De Rochambeau, H. Selection for residual feed consumption in the rabbit. *Livest. Prod. Sci.* **2005**, *95*, 67–72. [CrossRef]
32. Garreau, H.; Gilbert, H.; Molette, C.; Larzul, C.; Balmisse, E.; Ruesche, J.; Drouilhet, L. Réponses à la sélection pour deux critères d'efficacité alimentaire chez le lapin. 1. Croissance, ingéré et efficacité alimentaire. In *Proceedings of the 16èmes Journées de la Recherche Cunicole*; ITAVI: Le Mans, France, 2015; pp. 161–164.
33. Garreau, H.; Gilbert, H.; Molette, C.; Larzul, C.; Balmisse, E.; Ruesche, J.; Secula-Tircazes, A.; Gidenne, T.; Drouilhet, L. Direct and correlated responses to selection in two lines of rabbits selected for feed efficiency under ad libitum and restricted feeding. In *Proceedings of the 11 the World Rabbit Congress*; World Rabbit Science Association: Qingdao, China, 2016; pp. 43–46.
34. Martínez-Álvaro, M.; Hernández, P.; Agha, S.; Blasco, A. Correlated responses to selection for intramuscular fat in several muscles in rabbits. *Meat Sci.* **2018**, *139*, 187–191. [CrossRef]
35. Vicente, J.S.; Llobat, L.; Viudes-de-Castro, M.P.; Lavara, R.; Baselga, M.; Marco-Jiménez, F. Gestational losses in a rabbit line selected for growth rate. *Theriogenology* **2012**, *77*, 81–88. [CrossRef]
36. Naturil-Alfonso, C.; Lavara, R.; Millán, P.; Rebollar, P.G.; Vicente, J.S.; Marco-Jiménez, F. Study of failures in a rabbit line selected for growth rate. *World Rabbit Sci.* **2016**, *24*, 47–53. [CrossRef]
37. Lavara, R.; Viudes-De-Castro, M.P.; Vicente, J.S.; Mocé, E. Effect of synthetic prostaglandin F2α analogue (cloprostenol) on litter size and weight in a rabbit line selected by growth rate | Lavara | World Rabbit Science. *World Rabbit Sci.* **2002**, *10*, 1–5.
38. Blasco, A.; Gómez, E. A note on growth curves of rabbit lines selected on growth rate or litter size. *Anim. Prod.* **1993**, *57*, 332–334. [CrossRef]
39. Cifre, J.; Baselga, M.; Gómez, E.A.; de la Luz, G.M. Effect of embryo cryopreservation techniques on reproductive and growth traits in rabbits. *Ann. De Zootech.* **1999**, *48*, 15–24. [CrossRef]
40. Vicente, J.S.; Saenz-de-Juano, M.D.; Jiménez-Trigos, E.; Viudes-de-Castro, M.P.; Peñaranda, D.S.; Marco-Jiménez, F. Rabbit morula vitrification reduces early foetal growth and increases losses throughout gestation. *Cryobiology* **2013**, *67*, 321–326. [CrossRef]
41. Saenz-De-Juano, M.D.; Marco-Jimenez, F.; Schmaltz-Panneau, B.; Jimenez-Trigos, E.; Viudes-De-Castro, M.P.; Penaranda, D.S.; Jouneau, L.; Lecardonnel, J.; Lavara, R.; Naturil-Alfonso, C.; et al. Vitrification alters rabbit foetal placenta at transcriptomic and proteomic level. *Reproduction* **2014**, *147*, 789–801. [CrossRef]
42. Saenz-de-Juano, M.D.; Vicente, J.S.; Hollung, K.; Marco-Jiménez, F. Effect of embryo vitrification on. *PLoS ONE* **2015**, *10*, e0125157.
43. McLaren, A. Analysis of maternal effects on development in mammals. *J. Reprod. Fertil.* **1981**, *62*, 591–596. [CrossRef]
44. Cowley, D.E.; Pomp, D.; Atchley, W.R.; Eisen, E.J.; Hawkins-Brown, D. The impact of maternal uterine genotype on postnatal growth and adult body size in mice. *Genetics* **1989**, *122*, 193–203.
45. Atchley, W.R.; Logsdon, T.; Cowley, D.E.; Eisen, E.J. Uterine effects, epigenetics, and postnatal skeletal development in the mouse. *Evolution* **1991**, *45*, 891–909. [CrossRef]
46. Naturil-Alfonso, C.; Marco-Jiménez, F.; Jiménez-Trigos, E.; Saenz-de-Juano, M.; Viudes-de-Castro, M.; Lavara, R.; Vicente, J. Role of Embryonic and Maternal Genotype on Prenatal Survival and Foetal Growth in Rabbit. *Reprod. Domest. Anim.* **2015**, *50*, 312–320. [CrossRef] [PubMed]
47. Smith, C. Use of stored frozen semen and embryos to measure genetic trends in farm livestock. *Z. Für Tierzüchtung Und Züchtungsbiologie* **1977**, *94*, 119–130. [CrossRef]
48. Ralls, K. Mammals in which females are larger than males. *Q. Rev. Biol.* **1976**, *51*, 245–276. [CrossRef] [PubMed]
49. Pascual, M.; Pla, M.; Blasco, A. Effect of selection for growth rate on relative growth in rabbits. *J. Anim. Sci.* **2008**, *86*, 3409–3417. [CrossRef] [PubMed]
50. de la Fuente, L.F.; Rosell, J.M. Body weight and body condition of breeding rabbits in commercial units. *J. Anim. Sci.* **2012**, *90*, 3252–3258. [CrossRef]
51. Garcia-Dominguez, X.; Marco-Jiménez, F.; Peñaranda, D.S.; Diretto, G.; Garcia-Carpintero, V.; Cañizares, J.; Vicente, J. Long-term and transgenerational phenotypic, transcriptionaland metabolic effects in rabbit males born following vitrified embryo transfer. *Sci. Rep.* **2020**, *10*, 11313. [CrossRef]
52. Riggs, R.; Mayer, J.; Dowling-Lacey, D.; Chi, T.F.; Jones, E.; Oehninger, S. Does storage time influence postthaw survival and pregnancy outcome? An analysis of 11,768 cryopreserved human embryos. *Fertil. Steril.* **2010**, *93*, 109–115. [CrossRef]

53. Sanchez-Osorio, J.; Cuello, C.; Gil, M.A.; Parrilla, I.; Almiñana, C.; Caballero, I.; Roca, J.; Vazquez, J.M.; Rodriguez-Martinez, H.; Martinez, E.A. In vitro postwarming viability of vitrified porcine embryos: Effect of cryostorage length. *Theriogenology* **2010**, *74*, 486–490. [CrossRef]
54. Fogarty, N.M.; Maxwell, W.M.C.; Eppleston, J.; Evans, G. The viability of transferred sheep embryos after long-term cryopreservation. *Reprod. Fertil. Dev.* **2000**, *12*, 31–37. [CrossRef]
55. Mozdarani, H.; Moradi, S. Effect of vitrification on viability and chromosome abnormalities in 8-cell mouse embryos at various storage durations—PubMed. *Biol Res.* **2007**, *40*, 299–306. [CrossRef]
56. Lavara, R.; Baselga, M.; Vicente, J.S. Does storage time in LN2 influence survival and pregnancy outcome of vitrified rabbit embryos? *Theriogenology* **2011**, *76*, 652–657. [CrossRef] [PubMed]
57. Fang, Y.; Zeng, S.; Fu, X.; Jia, B.; Li, S.; An, X.; Chen, Y.; Zhu, S. Developmental competence in vitro and in vivo of bovine IVF blastocyst after 15 years of vitrification—PubMed. *Cryo Lett.* **2014**, *35*, 232–238.
58. Gómez, E.A.; Baselga, M.; Rafel, O.; García, M.L.; Ramón, J. Selection, diffusion and performances of six Spanish lines of meat rabbit. In *Proceedings of the 2. Internacional Conference on Rabbit Production in Hot Climates*; Testik, A., Baselga, M., Eds.; Cahiers Options Méditerranéennes: Zaragoza, Spain, 1999; Volume 41, pp. 147–152.
59. Pascual, J.J.; Cervera, C.; Blas, E.; Fernandez-Carmona, J. Effect of high fat diets on the performance and food intake of primiparous and multiparous rabbit does. *Anim. Sci.* **1998**, *66*, 491–499. [CrossRef]
60. Fortun-Lamothe, L.; Prunier, A.; Bolet, G.; Lebas, F. Physiological mechanisms involved in the effects of concurrent pregnancy and lactation on foetal growth and mortality in the rabbit. *Livest. Prod. Sci.* **1999**, *60*, 229–241. [CrossRef]
61. Rebollar, P.G.; Pérez-Cabal, M.A.; Pereda, N.; Lorenzo, P.L.; Arias-Álvarez, M.; García-Rebollar, P. Effects of parity order and reproductive management on the efficiency of rabbit productive systems. *Livest. Sci.* **2009**, *121*, 227–233. [CrossRef]
62. Rommers, J.M.; Kemp, B.; Meijerhof, R.; Noordhuizen, J.P.T.M. Rearing management of rabbit does: A review. *World Rabbit Sci.* **1999**, *7*, 125–138. [CrossRef]
63. Fortun-Lamothe, L.; Prunier, A. Effects of lactation, energetic deficit and remating interval on reproductive performance of primiparous rabbit does. *Anim. Reprod. Sci.* **1999**, *55*, 289–298. [CrossRef]
64. Fortun-Lamothe, L. Energy balance and reproductive performance in rabbit does. *Anim. Reprod. Sci.* **2006**, *93*, 1–15. [CrossRef]
65. Maertens, L.; Lebas, F.; Szendro, Z.S. Rabbit milk: A review of quantity, quality and non-dietary affecting factors. *World Rabbit Sci.* **2006**, *14*, 205–230. [CrossRef]

Article

Seminal Plasma Triggers the Differential Expression of the Glucocorticoid Receptor (*NR3C1/GR*) in the Rabbit Reproductive Tract

Mateo Ruiz-Conca [1,2,†], Jaume Gardela [1,2,†], Amaia Jauregi-Miguel [3], Cristina A. Martinez [1], Heriberto Rodríguez-Martinez [1], Manel López-Béjar [2,4,‡] and Manuel Alvarez-Rodriguez [1,2,*,‡]

1 Department of Biomedical and Clinical Sciences (BKV), Division of Children's and Women Health (BKH), Obstetrics and Gynecology, Linköping University, 58185 Linköping, Sweden; mateo.ruiz@uab.cat (M.R.-C.); jaume.gardela@uab.cat (J.G.); cristina.martinez-serrano@liu.se (C.A.M.); heriberto.rodriguez-martinez@liu.se (H.R.-M.)

2 Department of Animal Health and Anatomy, Veterinary Faculty, Universitat Autònoma de Barcelona, 08193 Bellaterra, Spain; manel.lopez.bejar@uab.cat or manel.lopezbejar@westernu.edu

3 Department of Biomedical and Clinical Sciences (BKV), Division of Molecular Medicine and Virology (MMV), Linköping University, 58185 Linköping, Sweden; amaya.jauregi.miguel@liu.se

4 College of Veterinary Medicine, Western University of Health Sciences, Pomona, CA 91766, USA

* Correspondence: manuel.alvarez-rodriguez@liu.se

† These authors contributed equally to this work.

‡ Both authors have contributed equally to the direction of this study.

Received: 2 October 2020; Accepted: 16 November 2020; Published: 19 November 2020

Simple Summary: Glucocorticoids are steroid hormones modulating different functions in mammals, including reproduction, that act through the glucocorticoid receptor, encoded by the gene called *NR3C1*. Here, we describe how the expression levels of the glucocorticoid receptor change along the different compartments of the female rabbit internal reproductive tract 20 h after insemination with sperm-free seminal plasma or natural mating (whole semen) (*Experiment 1*) and how these levels change at 10, 24, 36, 68, and 72 h post-mating, during specific stages over time, i.e., ovulation, fertilization and the interval of early embryo development to the morula stage occurs (*Experiment 2*). *NR3C1*-upregulation was found in the infundibulum at 20 h after all treatments, especially after sperm-free seminal plasma infusion compared to mating (*Experiment 1*). In *Experiment 2*, the receptor gene expression levels increased in a spatio-temporal sequence, corresponding to the assumed location of the rabbit embryos (particularly morulae) in the oviductal various segments and timepoints (particularly 72 h), compared to down-expression at uterine regions. We conclude that *NR3C1* may play a relevant role in the rabbit female reproductive tract.

Abstract: Rabbits are interesting as research animal models for reproduction, due to their condition of species of induced ovulation, with the release of endogenous gonadotropin-releasing hormone (GnRH) due to coitus. Glucocorticoid (GC) signaling, crucial for physiological homeostasis, is mediated through a yet unclear mechanism, by the GC receptor (NR3C1/GR). After mating, the female reproductive tract undergoes dynamic modifications, triggered by gene transcription, a pre-amble for fertilization and pregnancy. This study tested the hypothesis that when ovulation is induced, the expression of *NR3C1* is influenced by sperm-free seminal plasma (SP), similarly to after mating (whole semen), along the different segments of the internal reproductive tract of female rabbits. Semen (mating) was compared to vaginal infusion of sperm-free SP (*Experiment 1*), and changes over time were also evaluated, i.e., 10, 24, 36, 68, and 72 h post-mating, corresponding to specific stages, i.e., ovulation, fertilization, and the interval of early embryo development up to the morula stage (*Experiment 2*). All does were treated with GnRH to induce ovulation. Samples were retrieved from seven segments of the reproductive tract (from the cervix to infundibulum), at 20 h post-mating or sperm-free SP infusion (*Experiment 1*) or at 10, 24, 36, 68, and 72 h post-mating (*Experiment 2*).

Gene expression of *NR3C1* was analyzed by qPCR. Results showed an increase in *NR3C1* expression in the infundibulum compared to the other anatomical regions in the absence of spermatozoa when sperm-free SP infusion was performed (*Experiment 1*). Moreover, during the embryo transport through the oviduct, the distal isthmus was time-course upregulated, especially at 72 h, when morulae are retained in this anatomical region, while it was downregulated in the distal uterus at 68 h (*Experiment 2*). The overall results suggest that *NR3C1*, the GC receptor gene, assessed in the reproductive tract of does for the first time, shows differential expression changes during the interval of oviductal and uterine embryo transport that may imply a relevant role of the GC action, not only close to the site of ovulation and fertilization, but also in the endometrium.

Keywords: glucocorticoid receptor; gene expression; RT-qPCR; seminal plasma; female genital tract; rabbit

---

1. Introduction

Rabbits (*Oryctolagus cuniculus*) are wild animals originally from the south of Europe, which were domesticated and widely introduced around the world, nowadays present in every continent except Antarctica [1], and even considered as a pest in some areas [2]. These animals have been historically consumed in Mediterranean countries [3], while recently becoming important as production animals in some other regions [4]. They are also particularly well suited as model organisms for basic and applied reproductive experimental research, for their similarity to the chronology of human early embryonic development [5,6] and, especially, as a species characterized for copulation-induced ovulation as some felids [7] or camelids [8], best suited for experimental studies due to its early sexual maturation, short gestation, prolificacy and small size [9]. The use of assisted reproduction techniques (ARTs) in induced-ovulators is constrained by endocrine imbalances affecting the crucial steps of fertilization and early embryo development [10]. Therefore, the rabbit is an interesting animal model to improve intracytoplasmic sperm injection, embryo culture, embryo transfer, or cryopreservation [11], but also for commercial artificial insemination (AI) in rabbit intensive meat production [12].

Being an induced-ovulator, the rabbit requires the generation of genital-somatosensory signals during coitus to activate midbrain and brainstem noradrenergic neurons and generate the preovulatory peak of gonadotropin-releasing hormone (GnRH) [13,14]. The efficiency of natural mating, and also ARTs, relies on factors encompassing male and female parameters [15] and the interaction of both [16,17], hence improving the understanding of the effect of these factors in rabbit early development may improve fertility outcomes.

Although the role of ovarian steroid hormones progesterone and estrogens in the signaling pathways of reproductive stages is well-known, the importance of glucocorticoids (GCs) as regulators of reproduction is starting to be more recognized [18,19]. GCs are steroid hormones under the control of the hypothalamic–pituitary–adrenal axis, which are crucial for stress responses, behavior, and reproduction in mammals [20]. Even though GC production is essential for adequate physiology, they have commonly been assumed to be detrimental to reproductive performance and fertility, regarding the link between high GC levels and chronic stress [21]. However, GC basal levels have an important role in reproduction, comprising important reproductive events such as male–female mating interaction, oocyte maturation, early embryo development, fetus–mother communication, parturition, and lactation [19,22]. Although the underlying action mechanism of these hormones is complex and still not sufficiently understood, the GC receptor (NR3C1/GR) is assumed to play a key function in the mediation of GC action and gene transcription [23–25], including reproduction and embryo development [19,26–28]. The GCs bind to their receptor constituting a complex that can be transported to the nucleus, where they bind to GC response elements (GRE) of the DNA sequence, inducing the activation or repression of gene transcription [29,30], which can thereby modulate the changing female environment during the early embryo development stage [28,31–33]. Thus, GCs have been shown to influence the female reproductive tract, including the prostaglandin-mediated smooth muscle contractility movements [34,35], the corpus luteum formation and function [36], the Janus kinase/signal transducers and the activators of transcription (JAK/STAT) pathway, the immune response or the estrogen signaling, among others [37]. The effects of GC exposure to oocytes and preimplantational embryos are still not completely known, as whether they have a protective, innocuous, or harmful effect seems to greatly differ among mammalian species [38–41]. In rabbits, *NR3C1* has been recently postulated as a candidate gene implicated in reproductive seasonal differences between wild and domestic animals [42].

Since understanding the role of *NR3C1* in the reproductive tract is relevant for reproductive biology, we attempted to describe the GC receptor expression (*NR3C1*) of organ samples collected along the different anatomical segments of female rabbits internal reproductive tract, in response to natural mating or sperm-free seminal plasma (SP) infusion for the purpose of determining whether sperm-free SP was able to specifically affect gene expression similarly to mating, and how mating-induced *NR3C1* expression changes over time during the interval of early embryo development (10 h to 72 h).

2. Materials and Methods

2.1. Ethics Statement

The handling of the animals was performed according to the standards of animal care according to the Spanish Law (RD1201/2005), the European Directive (2010/63/EU; (BOE, 2005: 252:34367-91)) and the Directive 2010/63/EU of the European Parliament and of the Council of 22th September 2010 on the protection of animals used for scientific purposes (2010; 276:33-79). The Committee of Ethics and Animal Welfare of the Universitat Autònoma de Barcelona (Spain) approved this study (Expedient #517).

2.2. Animals

Adult New Zealand White rabbit bucks (n = 6) and does (n = 24), from 7 to 13 months old (mo), coming from an experimental farm of the Institut de Recerca i Tecnologia Agroalimentaries (IRTA, Torre Marimon, Spain) were used in this study. The animals were housed in individual cages for each rabbit (85 × 40 × 30 cm) equipped with plastic footrests, a feeder (restricted to 180 g/day of an all-mash pellet) and a nipple drinker (fresh water was always available). The environmental conditions were controlled, with a 16 h/8 h light/darkness photoperiod, temperature ranging from 15 to 20 °C during winter and from 20 to 26 °C during summer, and relative humidity between 60% to 75% was maintained by a forced ventilation system.

For the obtention of the ejaculate, the males were trained using an artificial vagina when they were 4.5 mo. A homemade polyvinyl chloride artificial vagina, containing water at 50 °C, was used. For this study, only one ejaculate per male was collected, discarding the ejaculates containing urine and/or calcium carbonate deposits.

2.3. Experimental Design

The experimental design used in this study is based on our previously described experimental approach [43].

2.3.1. Experiment 1: Analysis of Gene Expression Differences in the Does' Reproductive Tract, at 20 h Post-Mating (Whole Semen) or Seminal Plasma (SP) Infusion (Sperm-Free)

Gene expression analyses for *NR3C1* were performed in sequential segments of the female reproductive tracts (n = 9): endocervix (Cvx), endometrium (distal uterus: DistUt, proximal uterus: ProxUt), utero-tubal junction: UTJ, distal isthmus: Isth, ampulla: Amp, and infundibulum: Inf.

Tissue samples were collected at 20 h post-treatment with 0.03 mg GnRH im (intramuscular; Fertagyl®, Esteve Veterinaria, Barcelona, Spain) in all experimental groups (n = 9): post-mating (n = 3), post-SP-infusion (n = 3) and control, no mating or infusion, (n = 3, control group).

2.3.2. Experiment 2: Analysis of Gene Expression Differences in the Reproductive Tract of Mated Rabbit Females from 10 h Post-mating to up to 72 h Post-mating

A group of 15 rabbit does were sequentially euthanized at 10, 24, 36, 68, and 72 h post-mating (n = 3, time of collection). Reproductive tract sections (Cvx, DistUt, ProxUt, UTJ, Isth, Amp, and Inf) were recovered for gene expression analysis for *NR3C1*. The 10 h post-mating group was established as the reference group as this is the presumed time of ovulation in rabbits.

2.4. Mating and Semen Collection

The rabbit does included in the mating groups of *Experiment 1* and *2* were sequentially mated with two randomly selected bucks to decrease male-variation effects. The does additionally received an injection of 0.03 mg GnRH im previously to being mated to reinforce ovulation. Ovulation is expected at about 10 h after GnRH stimulation in all groups.

After the semen collection from the same rabbit bucks, as described above, the sperm-free SP was isolated after centrifugation at 2000× *g* for 10 min and checked for the absence of spermatozoa. The harvested sperm-free SP was immediately pooled for vaginal infusions of *Experiment 1*.

2.5. Collection of Tissues and Embryos

The does were euthanized by the administration of 600 mg pentobarbital sodium (Dolethal, Vetoquinol, Madrid, Spain) intravenously (marginal ear vein). Then, the samples of the female reproductive tracts were randomly chosen from the same lateral side (right), segmented and collected. The tissues of the oviductal segments were collected in toto. In *Experiment 2*, before segment-sectioning the internal reproductive tract, the entire oviduct was isolated from the uterus. In mated does, embryos were collected by flushing the oviduct (phosphate buffer saline supplemented with 5% fetal calf serum and 1% antibiotic–antimycotic solution), which were examined by number and developmental stage. The number of ovarian follicles and the number of embryos on each developmental stage were annotated and have been published elsewhere [43]. Briefly, at 24 h, 53.0 ± 40.2% of the embryos (2-4 cell stage) were recovered, at 36 h the embryo recovery rate was 84.1 ± 31.5% (8-cell stage), at 68 h it was 103.7 ± 17.1% (morula), and at 72 h, 104.8 ± 6.7% of the embryos (compacted morula) were retrieved, with respect to the total of ovulated follicles counted at each stage (Mean ± SD). Intervals of embryo development in the mated group were extrapolated to the sperm-free SP-infused does. All reproductive segments were stored in RNAlater solution at −80 °C.

2.6. Real Time Quantitative PCR Analyses

The TRIzol-based protocol was used for the total RNA extraction, as described elsewhere [43]. Briefly, in 1 mL TRIzol was used to mechanically disrupt the tissues (TissueLyser II with 7 mm stainless steel beads, Qiagen, Germany). The homogenized tissues underwent different centrifugation steps and were incubated with isopropanol and RNA precipitation solution (1.2 M NaCl and 0.8 M $Na_2C_6H_6O_7$) for RNA pellet obtaining. The RNA concentration and quality were determined from the absorbance of 260 nm measured by Thermo Scientific NanoDrop™ 2000, and the Agilent 2100 Bioanalyzer (Agilent Technologies, Palo Alto, CA, USA), respectively. The first strand cDNA synthesis was performed using the High-Capacity RNA-to-cDNA™ Kit (Applied Biosystems™, Foster City, CA, USA) and the samples were stored at −20 °C until further analyses. Quantitative PCR (qPCR) was performed in a Real-Time PCR Detection System (CFX96™; Bio-Rad Laboratories, Inc; Hercules, CA, USA) following the steps previously described [43]. Two technical replicates were used for each sample. Figure S1 depicts the melting curves and the peak curves of *β-ACTIN* and *NR3C1*. Efficiencies of the primers were calculated using five different concentrations of the same cDNA sample (serial dilutions of 1/5), using three technical replicates for each concentration. The gene relative expression levels were quantified using the Pfaffl method [44] and *β-ACTIN* as a housekeeping gene for cDNA normalization. The primer sequences, product sizes, and efficiencies are shown in Table 1. For the *β-ACTIN* gene, commercial gene-specific PCR primers for rabbit were used (PrimePCR™ SYBR® Green Assay: ACTB, Rabbit; Bio-Rad Laboratories, Inc; Hercules, CA, USA). The amplicons of qPCR were loaded into an agarose gel after mixing with GelRed® Nucleic Acid Gel Stain (Biotium, Fremont, CA, USA) to confirm the product sizes (Figure S2). After running, the gel was imaged by a gel imaging system (ChemiDoc XRS+ System, BioRad Laboratories, Inc; Hercules, CA, USA).

Table 1. Primers used for the quantitative PCR analyses.

Gene	Primer Sequence (5′-3′)	Product Size (bp)	Efficiency (%)
NR3C1	F: CACAACTCACCCCAACACTG R: CAGGAGGGTCATTTGGTCAT	212	89.6
β-ACTIN	F: commercial, not available R: commercial, not available	120	88.6

NR3C1: glucocorticoid receptor; *β-ACTIN*: beta-actin. F: forward, R: reverse, A: adenine, C: cytosine, G: guanine, T: thymine, bp: base pair.

2.7. Statistical Analyses

All data were processed with CFX Maestro™ 1.1 software version 4.1.2433.1219 (Bio-Rad Laboratories, Inc; Hercules, CA, USA) and were analyzed for normal distribution and homoscedasticity using the Shapiro–Wilk Normality test and Levene's test. Log(x) transformation was used to restore a normal distribution prior to analysis. The statistical analysis was conducted in R version 3.6.1. [45] with *nlme* [46] to perform linear mixed-effects (LME) models and *multcomp* [47] to perform pairwise comparisons adjusted by Tukey's test. Data are presented as median (minimum, maximum), unless otherwise stated. The threshold for significance was set at $p < 0.05$.

Treatments of the *Experiment 1* (negative control, mating (positive control) and sperm-free SP) were used as fixed effects and each individual doe as the random part of the model. Pairwise comparisons were adjusted by Tukey's test. A second LME model was used including the different sample collection times of *Experiment 2* (10, 24, 36, 68, and 72 h post-mating) as fixed effects and each individual doe as the random part of the model. Post-hoc comparisons were performed using Tukey's multiple comparisons test.

Finally, the differential expression changes in qPCR results among anatomical segments (Cvx to Inf) in the *Experiment 1* and *2* were further re-analyzed, using the UTJ as an arbitrary reference anatomical medial compartment among all samples examined, which is located in the middle of the tract. This was performed in order to compare gene expression changes, per gene, issued both by mating or sperm-free SP vaginal infusion, respectively, to control (*Experiment 1*) or by different times post-mating: 10, 24, 36, 68, and 72 h (*Experiment 2*), among the different tissues of the female reproductive tract. Each tissue was included as fixed effects and each individual doe as the random part of the LME model. As stated above, the analysis of differences among each tissue of the female reproductive tract was performed by the Tukey's multiple comparison test.

3. Results

3.1. Experiment 1: Differential Gene Expression in Rabbit Female Reproductive Tract at 20 h after Natural Mating or Infusion of Sperm-Free Seminal Plasma

The results of *Experiment 1* are shown in Figure 1, where differential expression changes in *NR3C1* in the different anatomical segments of the rabbit reproductive tract were analyzed 20 h after mating or SP infusion. First, expression changes in each segment were compared between the negative control and the treatments of natural mating and SP-infusion. Significative differences in *NR3C1* expression between natural mating and SP infusion were shown in Inf ($p < 0.05$) (G), where the sperm-free SP upregulated its expression in this anatomical segment. None of the rest of the treatments displayed any significant difference ($p > 0.05$).

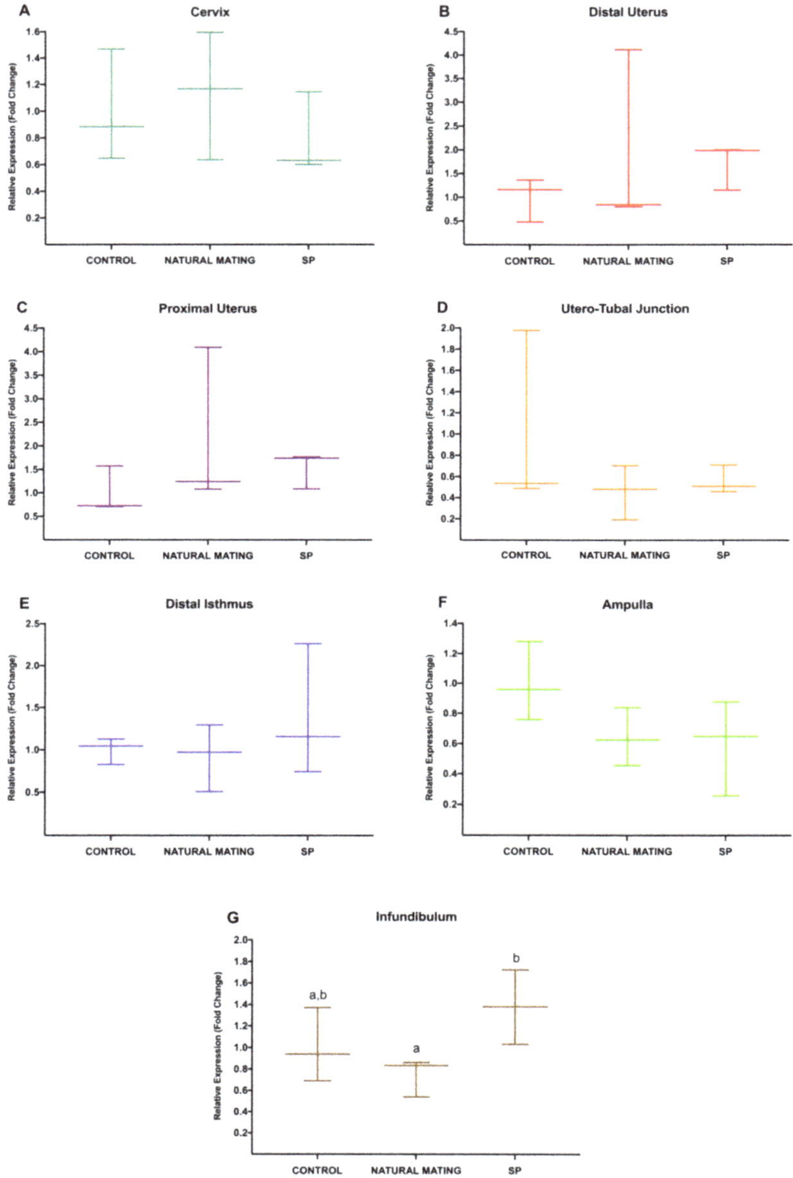

Figure 1. (**A–G**). Gene expression differences in *NR3C1* in the different anatomical segments (cervix to infundibulum; Experiment 1) (**A–G**), between negative control, natural mating, and seminal plasma (SP) treatments. Different letters indicate values that differed significantly between treatments in the same anatomical region ($p < 0.05$). The expression was relativized using the negative control as a reference. Median (minimum, maximum).

Second, the expression changes triggered by each treatment were compared between anatomical regions in the doe reproductive tract, represented in Figure 2. In the negative controls (A), significant *NR3C1* upregulation was reported in Amp and Inf, relative to the rest of the anatomical segments ($p < 0.05$; except Cvx, $p > 0.05$) and also a downregulation in the uterus (DistUt and ProxUt), relative to Cvx, Amp and Inf ($p < 0.05$). In the case of sperm-free SP infusion (C), significant upregulation of *NR3C1* expression was shown in Inf compared to the rest of the anatomical segments ($p < 0.05$). Moreover, upregulation in Inf compared to UTJ was also found ($p < 0.05$) in the natural mating group.

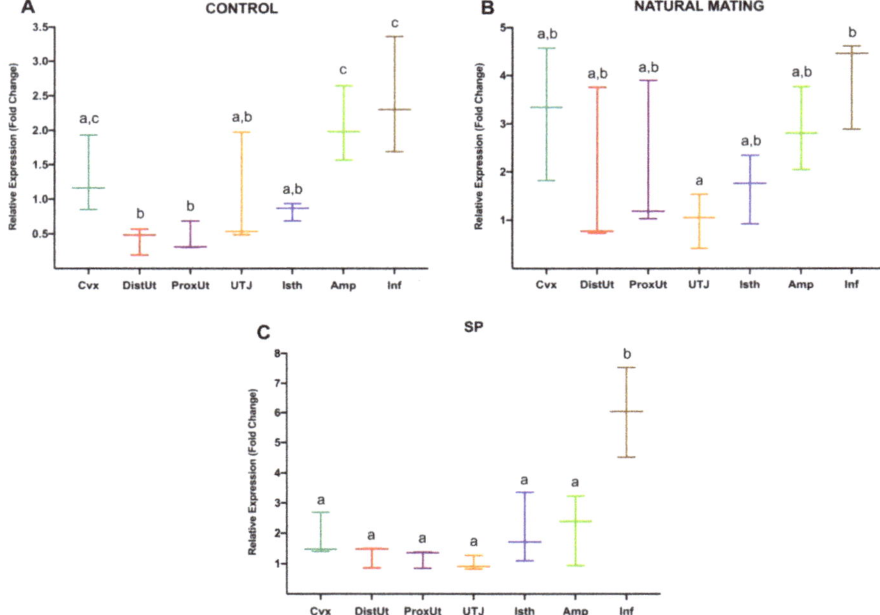

Figure 2. (**A–C**). Gene expression differences of *NR3C1* between anatomical segments (cervix, Cvx; distal uterus, DistUt; proximal uterus, ProxUt; utero-tubal junction, UTJ; distal isthmus, Isth; ampulla, Amp; infundibulum, Inf) in negative control group (**A**), natural mating (**B**), and sperm-free seminal plasma infusion (SP) (**C**). Different letters indicate values that differed significantly between anatomical regions in the same treatment ($p < 0.05$). The expression was relativized using the UTJ as a reference. Median (minimum, maximum).

3.2. Experiment 2: Differential Gene Expression in Rabbit Female Reproductive Tract from 10 h to up to 72 h in Response to Natural Mating

Figure 3 shows the results of *Experiment 2*, where 10 h post-mating was used to relativize the expression of the rest of the groups (24 h, 36 h, 68 h, 72 h post-mating). These timepoints are representative of embryo developmental stages. Thus, preovulatory stage, 2-4 cell embryo, 8-cell embryo, morula, and compacted morula correspond to 10, 24, 36, 68, and 72 h, respectively. In this sense, the number of ovarian follicles and embryos, and the present developmental stages, were evaluated and could be found in the aforementioned tissue collection description.

Significative differences in *NR3C1* expression were found in the DistUt (B), Isth (E), and Amp (F) at different times after mating. In the DistUt, a downregulation was found at 68 h when compared to *NR3C1* expression at 10 h post-mating ($p < 0.05$). Differently, in the Isth, *NR3C1* expression was upregulated, showing significant differences at 72 h ($p < 0.05$), when compared to 10 and 24 h post-mating. Moreover, in the Amp, the *NR3C1* gene was found to be significatively upregulated at

68h, when compared to the rest of the timepoints analyzed 10, 24, 36, and 72 h post-mating ($p < 0.05$). Upregulation at 36 h was also found when compared to 24 h in this anatomical segment ($p < 0.05$).

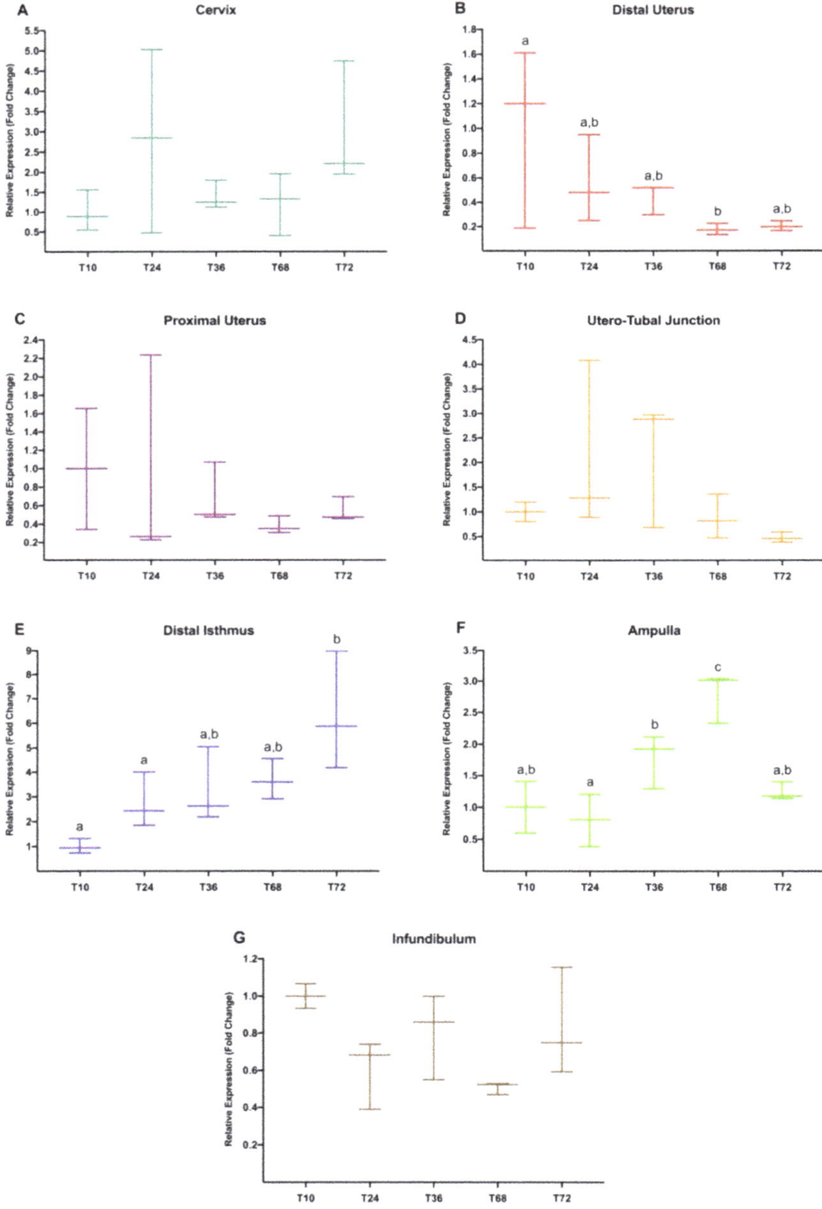

Figure 3. (**A–G**). *NR3C1* gene expression differences between each timepoint (T10, 10 h; T24, 24 h; T36, 36 h; T68, 68 h; and T72, 72 h; Experiment 2) post-natural mating in the different anatomical segments of the doe reproductive tract (cervix to infundibulum) (**A–G**). Different letters in the same anatomical region indicate values that differed significantly between timepoints ($p < 0.05$). The expression was relativized using the T10 as a reference. Median (minimum, maximum).

The expression changes triggered by mating in each timepoint were also compared among the different anatomical regions (Figure 4). At 10 h post-mating (A), *NR3C1* was upregulated in the Inf compared to the Isth ($p < 0.05$). At 36 h (C), the expression of this gene was upregulated in the Amp when compared to the DistUt, ProxUt, and UTJ ($p < 0.05$) and also upregulated in the Inf, compared to the DistUt ($p < 0.05$). Similarly, after 68 h post-mating (D), significative upregulation was found in the Amp compared to the rest of the anatomical regions ($p < 0.05$). Additionally, expression in the Cvx was upregulated compared to the DistUt and UTJ ($p < 0.05$), and *NR3C1* expression in the Isth was also upregulated when compared to the DistUt ($p < 0.05$). Finally, at 72 h post-mating (E), expression was higher in the oviduct (Isth, Amp, and Inf) and also the Cvx compared to the DistUt, ProxUt, and UTJ ($p < 0.05$). Additionally, the ProxUt was upregulated compared to the UTJ ($p < 0.05$).

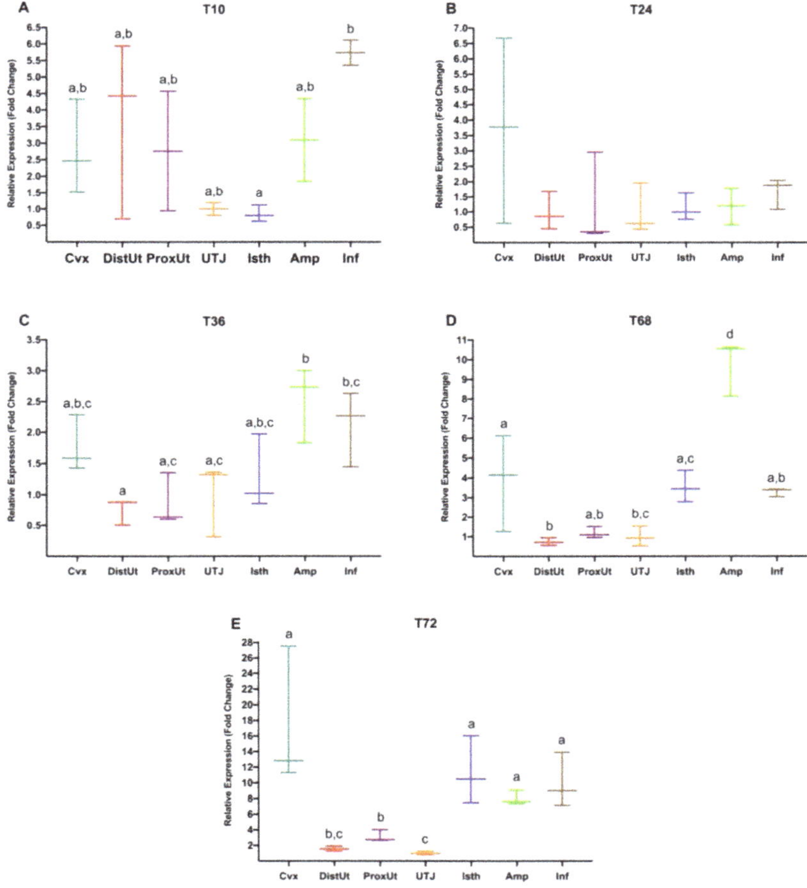

Figure 4. (**A–E**). *NR3C1* gene expression differences between anatomical segments (cervix, Cvx; distal uterus, DistUt; proximal uterus, ProxUt; utero-tubal junction, UTJ; distal isthmus, Isth; ampulla, Amp; infundibulum, Inf) in each of the timepoints 10 h (**A**, T10), 24 h (**B**, T24), 36 h (**C**, T36), 68 h (**D**, T68), and 72 h (**E**, T72) post-mating. Different letters in the same timepoint indicate values that differed significantly between anatomical regions ($p < 0.05$). The expression was relativized using the UTJ as a reference. Median [minimum, maximum].

4. Discussion

In the present study, we evaluated changes in the post-ovulatory expression of the GC receptor gene (*NR3C1*) in the internal reproductive tract (cervix to infundibulum) of the female rabbit. *NR3C1* gene expression differences relative to insemination treatments and anatomical regions were analyzed 20 h after sperm-free SP infusion or natural mating, with the purpose of comparing the effects exerted by the whole semen or the sperm-free SP portion of the ejaculate (*Experiment 1*), and additionally, after mating at different timepoints (10 h, 24 h, 36 h, 68 h, and 72 h) (*Experiment 2*), during specific stages, i.e., ovulation, fertilization, and the interval of early embryo development to the morula stage achievement.

In our study, at 20 h after treatment, when ovulation may have already taken place 10 h ago [48], we found an upregulation of the *NR3C1* expression in the infundibulum, the oviductal segment where the follicular fluid and the mature oocytes are released shortly after ovulation [49]. The promotion of the receptor expression in the infundibulum was observed in both natural mated and control ovulated females, but the *NR3C1* expression was significantly higher after sperm-free SP infusion, e.g., in the absence of spermatozoa. Moreover, the upregulation in this region was significantly higher than in any other of the reproductive tract segments sampled in the study. Hence, GC receptor expression in the infundibulum seems to be generally increased after ovulation, but SP, either provided by infusion or directly by mating, may promote *NR3C1* action at this location. In this sense, SP, which is not only related to the spermatozoa transport, but also to the modulation of the immune response exerted upon sperm- free SP contact in the female reproductive tract [16], has been proven to have a positive effect on fertility in a variety of species [50], including rodents [51], pigs [52], and humans [53], and may also play a role in the GC pathway in the reproductive tract [37]. SP is produced by sexual accessory glands together with secretions from the epididymis [54] and, in rabbits, it seems to play a role in spermatozoa protection, fertilization [55] and also immune modulation during development [56]. After natural mating and sperm-free SP infusion [17], the female reproductive tract undergoes a variety of modifications that influence the initiation of controlled inflammatory response, ovulation stimulation, changes in the transcription of genes related to reproductive stages [50] and early embryonic development [57,58]. Thus, even when the SP, deposited in the cervix, does not directly reach the upper regions of the oviduct such as the infundibulum, proteins of this fluid may be absorbed by the endometrium and achieve the ovary via lymphatic route or, alternatively, by activating signaling cascades [59] that may modify the ovulation in rabbit [60]. In this way, SP molecules may modulate the overall genetic expression on different anatomical segments of the reproductive tract and, in light of our results, SP may also promote the action of GCs in the infundibulum, mediated by *NR3C1*, 20 h after GnRH injection. The specific localization of this expression change produced at this particular timepoint may be linked to the effects of SP components in the inflammatory response produced in the ovulation process. In this sense, previous studies have identified specific SP factors directly involved in the induced-ovulation triggering, such as the nerve growth factor (β-NGF). In rabbits, this mechanism of ovulation induction seems to be more complex compared to what is found in other species [8,14,60,61]. In that sense, previous studies found local effects of SP in the ovary that increased the number of hemorrhagic anovulatory follicles [62,63], suggesting an indirect action of SP on the luteinizing hormone (LH) receptors [14,64]. The differences that we found between the SP and natural mating (also containing SP) treatments, where the expression in the infundibulum seemed higher than in the rest of the tissues, although not significant, are also intriguing. In this sense, those differences between both treatments could rely on the different physical stimulation that has been reported to play an important complementary role in the rabbit ovulation, that is not required in the camelids [63]. Thus, whether the specific effect on *NR3C1* expression we observed in the infundibulum is modulated by particular SP molecules and which those molecules remains are to be elucidated, but it might be plausible that β-NGF plays a role in the GC receptor promotion that is produced at the time of the ovulation in the infundibulum.

At that moment, shortly after the LH surge, ovulation starts and the rupture of mature ovarian follicles occurs [65]. After oocytes' release, an inflammatory-like response, very similar to other inflammatory reactions, is produced at this location [66,67], stimulated by a variety of changes in the reproductive tract including angiogenesis, vascular permeability, and exhaustive cellular differentiation, together with the production of mediators associated with inflammatory processes, such as steroids, prostaglandins, and cytokines, which may, in turn, be released to the infundibulum after the follicle rupture [65]. The GC levels can be locally regulated within the reproductive tract to serve precise functions depending on the reproductive stage temporal context [22], or the specific anatomical region [68]. Therefore, the GCs, steroid hormones well-known for their anti-inflammatory actions, are very present in the oviduct and may contribute to the phase transition by promoting healing and repair after ovulation [69]. This may also be supported by the direct detection of ten times higher total cortisol levels in the follicular fluid after the LH surge [70], together with increased levels of 11β-hydroxysteroid dehydrogenases (11β-HSD1) during the luteal phase compared to the follicular phase [71,72], responsible for the conversion of cortisone into active cortisol, which provides direct and indirect evidence of the important presence of GCs in this region.

Moreover, we found that the expression of the GC receptor is also modulated over time, corresponding to the different preimplantational embryo developmental stages, e.g., preovulatory, 2-4 cells embryo, 8-cell embryo, morula, and compacted morula that we checked, as some of the oviductal and uterine anatomical segments show very disparate expression values during the hours that were tested here (10, 24, 36, 68, and 72 h), which were only evaluated after natural mating. Here, the uterus displayed low values compared to what was found in the oviduct at 36, 68, and 72 h post-coitum. At 68 h, these uterine values were also decreased compared to the values recorded in the periovulatory phase (around 10 h), which may imply a GC action in the uterus during the period prior to embryo implantation. In this context, the widely known action of steroid hormones modulates the uterus to prepare a suitable environment for successful embryo implantation [73]. In that regard, the necessary proinflammatory effects of estradiol in the uterus, including edema, increased vascularization, and promotion of the immune antibacterial activity, have previously been shown to be antagonized by the action of GCs in different mammalian species including rat, baboon, and sheep [74,75], and may also apply to induced ovulation species, as in the case of the lagomorphs [14]. GCs have also been demonstrated to assist difficult estrogen action by blocking cell differentiation, development and growth in the uterus, which may inhibit embryo attachment [76], trophoblast invasion, and implantation [77–79]. However, even when high levels of GCs may be potentially detrimental for the action of estrogens, presumably predominant in the uterus at this stage, *NR3C1* uterus knockout mouse models resulted in impaired implantation, decidualization, and pregnancy [28], demonstrating that certain levels of GC regulation may be essential for an adequate uterine function.

Interestingly, an upregulation of GC action seems to be present along the rabbit oviduct at subsequent post-ovulatory timepoints according to our results. Thus, the *NR3C1* expression was found to be especially increased in the ampulla, and also in the infundibulum, at 36 h after mating and ovulation induction, when the early embryos are supposed to be transported through this oviductal region while performing the initial cell divisions (8-cell stage). At 68 h post-mating, when the embryo may be at the morula stage, the expression in the entire oviduct is also upregulated, with particularly high levels in the ampulla. This was also found at 72 h when the expression of the receptor was promoted in all the oviductal anatomical segments (distal isthmus, ampulla, and infundibulum) and also in the cervix. By this time, the morula stage is fully achieved and the embryos may have already reached the distal isthmus [80,81]. Thus, at 72 h, the levels of *NR3C1* in the distal isthmus were importantly high compared to the expression observed at the periovulatory stage (10 h after GnRH injection), and also to the expression at 24 h, shortly after fertilization is assumed to take place in upper oviductal regions [82]. This high value of *NR3C1* displayed in the distal isthmus may be explained by the previously described action of steroid hormones in smooth muscle [83]. In this tissue, where the receptor is very present [84,85], GC action may cause a decrease in the prostaglandin action [36].

Thereon, one of the functions of prostaglandins is the preimplantational embryo retention in the oviduct for approximately 3 days (72 h) by stimulating the oviductal smooth muscle contractility [34,86,87]. Moreover, RU486 (mifepristone), a glucocorticoid receptor antagonist [88], has been shown to increase the oviduct smooth muscle contractile frequency in rabbit [89]. Thus, the decrease in prostaglandins, favored by the action of steroid hormones, may be the cause of the isthmic sphincter relaxation that allows the morula embryos to enter into the uterus on its way down towards the implantation site [35]. In light of these results, GC action seems to be present along the oviduct in a distribution that may correspond to the assumed spatio-temporal location of the rabbit embryos regarding anatomical segments and timepoints, especially at the morula stage. After ovulation, the oocytes, together with follicular fluid, are released to the oviduct, where there is cross-talk between the gametes, the embryo, and the oviductal regions [90]. In this way, in vitro studies showed that cortisol production, mediated by 11β-HSD1, increased during maturation, and continued high during fertilization [91], and cortisol supplementation has improved blastocyst development rates in bovine [40], indicating that the GC activation may be part of the complex processes taken place during fertilization and early embryo development. In this sense, different results have been found regarding the effect of GCs in oocytes and early embryo development, as the influence of their presence seems to be species-specific [38]. To our knowledge, this is the first time that GC receptor levels have been described in the rabbit reproductive tract. In other species, high levels of cortisol did not affect oocyte metabolism in equine [39], while some harmful effects have been reported in pig [27,38], and different results have been shown in mice [38,92]. The levels of GC receptor are crucial in the GC regulation, however, the regulation of these hormones is complex and comprises steps that are still not completely described, involving a great number of molecules, such as 11β-HSD [38] and peptidyl-prolyl cis/trans isomerase FK506-binding proteins (i.e., FKBPs immunophilins) [93,94], among others. Thereby, the actions of GCs seem to variate among species and the importance of their regulation in reproduction is still far from being fully understood.

5. Conclusions

This is the first time that expression of *NR3C1*, the GC receptor gene, has been assessed in the internal reproductive tract of rabbits. Our results showed that, after sperm-free SP infusion, in the absence of spermatozoa, there is an increase in *NR3C1* expression in the infundibulum compared to natural mating. In the experiment over time, the differential expression of *NR3C1* was detected not only close to the site of ovulation and fertilization (ampulla and infundibulum), but also in the endometrium (distal uterus). The differential expressions are present over the interval during which early embryo development occurs, which may suggest a relevant role of the GC action, mediated by *NR3C1* on oviductal and uterine embryo transport. These results pave the way for further analysis that may elucidate the exact mechanism involved in the *NR3C1* action as well as its potential applications increasing the efficiency of the ARTs.

Supplementary Materials: The following are available online at http://www.mdpi.com/2076-2615/10/11/2158/s1, Figure S1. Melting temperatures of PCR products of *NR3C1* (blue) and *β-ACTIN* (green) in cervix tissue (36, 68, and 72 h post-mating groups) including negative template controls (red). For DNA-binding dyes (SYBR green), the fluorescence is brightest when the two strands of DNA anneal. Therefore, as the temperature rises towards the melting temperature (Tm), relative fluorescence units (RFU) decrease at a constant rate (constant slope). At the Tm, there is a dramatic reduction in the fluorescence with a noticeable change in slope, displayed in the Melt Curve graph. The rate of this change is determined by plotting the negative first regression of fluorescence versus temperature (-d(RFU)/dT), displayed in the Melt Peak graph. The greatest rate of change in fluorescence results in visible peaks and represents the Tm of the double-stranded DNA complexes: 83.5 °C for *NR3C1*, and 89.5 °C for *β-ACTIN*. Figure S2. Agarose gel displaying PCR product size (bp: base pair) of *β-ACTIN* (120 bp) and *NR3C1* (212 bp).

Author Contributions: Conceptualization, H.R.-M., M.L.-B. and M.A.-R.; methodology, M.L.-B. and M.A.-R.; software, M.R.-C., J.G., A.J.-M. and M.A.-R.; validation, A.J.-M.; formal analysis, J.G., A.J.-M. and M.A.-R.; investigation, M.R.-C., J.G. and C.A.M.; resources, H.R.-M., M.L.-B. and M.A.-R.; data curation, J.G. and M.R.-C.; writing—original draft preparation, J.G. and M.R.-C.; writing—review and editing, A.J.-M., C.A.M., H.R.-M., M.L.-B. and M.A.-R.; visualization, J.G. and M.R.-C.; supervision, M.L.-B. and M.A.-R.; project administration,

M.A.-R.; funding acquisition, H.R.-M., M.L.-B. and M.A.-R. All authors have read and agreed to the published version of the manuscript.

Funding: This research was funded by the Research Council FORMAS, Stockholm (Project 2017-00946 and Project 2019-00288), The Swedish Research Council (Vetenskaprådet, VR; project 2015-05919) and Juan de la Cierva Incorporación Postdoctoral Research Program (MICINN; IJDC-2015-24380). J.G. is supported by the Generalitat de Catalunya, Agency for Management of University and Research Grants co-financed with the European Social Found (grants for the recruitment of new research staff 2018 FI_B 00236) and M.R.-C. is supported by the Government of SpainMinistry of Education, Culture and Sports (Training programme for Academic Staff FPU15/06029). CA.M. is supported by the Seneca Foundation (20780/PD/18) Murcia (Spain).

Acknowledgments: We thank Annette Molbaek and Åsa Schippert, from the genomic Core Facility at LiU for expert assistance when running the bioanalyzer. We appreciate the kind support provided by Míriam Piles and Oscar Perucho and the technical staff from Torre Marimon—Institut de Recerca i Tecnologia Agroalimentàries (IRTA, Caldes de Montbui, Barcelona), and also Annaïs Carbajal and Sergi Olvera-Maneu for their kind assistance in sample handling.

Conflicts of Interest: The authors declare no conflict of interest. The funders had no role in the design of the study; in the collection, analyses, or interpretation of data; in the writing of the manuscript, or in the decision to publish the results.

References

1. Tablado, Z.; Revilla, E.; Palomares, F. Breeding like rabbits: Global patterns of variability and determinants of European wild rabbit reproduction. *Ecography* **2009**, *32*, 310–320. [CrossRef]
2. Roy-Dufresne, E.; Lurgi, M.; Brown, S.C.; Wells, K.; Cooke, B.; Mutze, G.; Peacock, D.; Cassey, P.; Berman, D.; Brook, B.W.; et al. The Australian national rabbit database: 50 year of population monitoring of an invasive species. *Ecology* **2019**, *100*, e02750. [CrossRef] [PubMed]
3. Petraci, M.; Cavani, C. Rabbit meat processing: Historical perspective to future directions. *World Rabbit Sci.* **2013**, *21*, 217–226.
4. Li, S.; Zeng, W.; Li, R.; Hoffman, L.C.; He, Z.; Sun, Q.; Li, H. Rabbit meat production and processing in China. *Meat Sci.* **2018**, *145*, 320–328. [CrossRef] [PubMed]
5. Püschel, B.; Daniel, N.; Bitzer, E.; Blum, M.; Renard, J.P.; Viebahn, C. The rabbit (*Oryctolagus cuniculus*): A model for mammalian reproduction and early embryology. *Cold Spring Harb. Protoc.* **2010**, *5*, 1–6.
6. Fischer, B.; Chavatte-Palmer, P.; Viebahn, C.; Santos, A.N.; Duranthon, V. Rabbit as a reproductive model for human health. *Reproduction* **2012**, *144*, 1–10. [CrossRef] [PubMed]
7. Jorge-Neto, P.N.; Luczinski, T.C.; Araújo, G.R.D.; Salomão Júnior, J.A.; Traldi, A.D.S.; Santos, J.A.M.D.; Requena, L.A.; Gianni, M.C.M.; Deco-Souza, T.D.; Pizzutto, C.S.; et al. Can jaguar (*Panthera onca*) ovulate without copulation? *Theriogenology* **2020**, *147*, 57–61. [CrossRef]
8. Silva, M.; Paiva, L.; Ratto, M.H. Ovulation mechanism in South American Camelids: The active role of β-NGF as the chemical signal eliciting ovulation in llamas and alpacas. *Theriogenology* **2020**, *150*, 280–287. [CrossRef]
9. Adams, G.; Ratto, M.; Silva, M.; Carrasco, R. Ovulation-inducing factor (OIF/NGF) in seminal plasma: A review and update. *Reprod. Domest. Anim.* **2016**, *51*, 4–17. [CrossRef]
10. Pelican, K.M.; Wildt, D.E.; Pukazhenthi, B.; Howard, J.G. Ovarian control for assisted reproduction in the domestic cat and wild felids. *Theriogenology* **2006**, *66*, 37–48. [CrossRef]
11. Garcia-Dominguez, X.; Marco-Jimenez, F.; Viudes-de-Castro, M.P.; Vicente, J.S. Minimally invasive embryo transfer and embryo vitrification at the optimal embryo stage in rabbit model. *J. Vis. Exp.* **2019**, e58055. [CrossRef] [PubMed]
12. Piles, M.; Tusell, L.; Lavara, R.; Baselga, M. Breeding programmes to improve male reproductive performance and efficiency of insemination dose production in paternal lines: Feasibility and limitations. *World Rabbit Sci.* **2013**, *21*, 61–75. [CrossRef]
13. Bakker, J.; Baum, M.J. Neuroendocrine regulation of GnRH release in induced ovulators. *Front. Neuroendocrinol.* **2000**, *21*, 220–262. [CrossRef] [PubMed]
14. Ratto, M.H.; Berland, M.; Silva, M.E.; Adams, G.P. New insights of the role of β-NGF in the ovulation mechanism of induced ovulating species. *Reproduction* **2019**, *157*, R199–R207. [CrossRef] [PubMed]
15. Brun, J.M.; Sanchez, A.; Ailloud, E.; Saleil, G.; Theau-Clément, M. Genetic parameters of rabbit semen traits and male fertilising ability. *Anim. Reprod. Sci.* **2016**, *166*, 15–21. [CrossRef] [PubMed]

16. Robertson, S.A. Seminal plasma and male factor signalling in the female reproductive tract. *Cell Tissue Res.* **2005**, *322*, 43–52. [CrossRef]
17. Parada-Bustamante, A.; Oróstica, M.L.; Reuquen, P.; Zuñiga, L.M.; Cardenas, H.; Orihuela, P.A. The role of mating in oviduct biology. *Mol. Reprod. Dev.* **2016**, *83*, 875–883. [CrossRef]
18. Whirledge, S.; Kisanga, E.P.; Taylor, R.N.; Cidlowski, J.A. Pioneer factors FOXA1 and FOXA2 assist selective glucocorticoid receptor signaling in human endometrial cells. *Endocrinology* **2017**, *158*, 4076–4092. [CrossRef]
19. Whirledge, S.; Cidlowski, J.A. Glucocorticoids and reproduction: Traffic control on the road to reproduction. *Trends Endocrinol. Metab.* **2017**, *28*, 399–415. [CrossRef]
20. Wang, J.-C.; Harris, C. *Glucocorticoid Signaling from Molecules to Mice to Man*; Springer: New York, NY, USA, 2015; ISBN 9781493928941.
21. Whirledge, S.; Cidlowski, J.A. A role for glucocorticoids in stress-impaired reproduction: Beyond the hypothalamus and pituitary. *Endocrinology* **2013**, *154*, 4450–4468. [CrossRef]
22. Whirledge, S.; Cidlowski, J.A. Glucocorticoids, stress, and fertility. *Minerva Endocrinol.* **2010**, *35*, 109–125. [PubMed]
23. Ratman, D.; Vanden Berghe, W.; Dejager, L.; Libert, C.; Tavernier, J.; Beck, I.M.; de Bosscher, K. How glucocorticoid receptors modulate the activity of other transcription factors: A scope beyond tethering. *Mol. Cell. Endocrinol.* **2013**, *380*, 41–54. [CrossRef]
24. Cain, D.W.; Cidlowski, J.A. Immune regulation by glucocorticoids. *Nat. Rev. Immunol.* **2017**, *17*, 233–247. [CrossRef] [PubMed]
25. Bekhbat, M.; Rowson, S.A.; Neigh, G.N. Checks and balances: The glucocorticoid receptor and NFkB in good times and bad. *Front. Neuroendocrinol.* **2017**, *46*, 15–31. [CrossRef] [PubMed]
26. Ruiz-Conca, M.; Gardela, J.; Alvarez-Rodríguez, M.; Mogas, T.; López-Béjar, M. Immunofluorescence analysis of NR3C1 receptor following cortisol exposure during bovine in vitro oocyte maturation. *Anim. Reprod. Sci.* **2019**, *16*, 753.
27. Yang, J.-G.; Chen, W.-Y.; Li, P.S. Effects of glucocorticoids on maturation of pig oocytes and their subsequent fertilizing capacity in vitro1. *Biol. Reprod.* **1999**, *60*, 929–936. [CrossRef]
28. Whirledge, S.D.; Oakley, R.H.; Myers, P.H.; Lydon, J.P.; DeMayo, F.; Cidlowski, J.A. Uterine glucocorticoid receptors are critical for fertility in mice through control of embryo implantation and decidualization. *Proc. Natl. Acad. Sci. USA* **2015**, *112*, 15166–15171. [CrossRef]
29. Wochnik, G.M.; Rüegg, J.; Abel, G.A.; Schmidt, U.; Holsboer, F.; Rein, T. FK506-binding proteins 51 and 52 differentially regulate dynein interaction and nuclear translocation of the glucocorticoid receptor in mammalian cells. *J. Biol. Chem.* **2005**, *280*, 4609–4616. [CrossRef]
30. Petta, I.; Dejager, L.; Ballegeer, M.; Lievens, S.; Tavernier, J.; de Bosscher, K.; Libert, C. The interactome of the glucocorticoid receptor and its influence on the actions of glucocorticoids in combatting inflammatory and infectious diseases. *Microbiol. Mol. Biol. Rev.* **2016**, *80*, 495–522. [CrossRef]
31. Simmons, R.M.; Satterfield, M.C.; Welsh, T.H.; Bazer, F.W.; Spencer, T.E. HSD11B1, HSD11B2, PTGS2, and NR3C1 expression in the peri-implantation ovine uterus: Effects of pregnancy, progesterone, and interferon Tau1. *Biol. Reprod.* **2010**, *82*, 35–43. [CrossRef]
32. Siemieniuch, M.J.; Majewska, M.; Takahashi, M.; Sakatani, M.; Łukasik, K.; Okuda, K.; Skarzynski, D.J. Are glucocorticoids auto- and/or paracrine factors in early bovine embryo development and implantation? *Reprod. Biol.* **2010**, *10*, 249–256. [CrossRef]
33. Majewska, M.; Lee, H.Y.; Tasaki, Y.; Acosta, T.J.; Szostek, A.Z.; Siemieniuch, M.; Okuda, K.; Skarzynski, D.J. Is cortisol a modulator of interferon tau action in the endometrium during early pregnancy in cattle? *J. Reprod. Immunol.* **2012**, *93*, 82–93. [CrossRef] [PubMed]
34. Lindblom, B.; Hamberger, L.; Ljung, B. Contractile patterns of isolated oviductal smooth muscle under different hormonal conditions. *Fertil. Steril.* **1980**, *33*, 283–287. [CrossRef]
35. Wånggren, K.; Stavreus-Evers, A.; Olsson, C.; Andersson, E.; Gemzell-Danielsson, K. Regulation of muscular contractions in the human Fallopian tube through prostaglandins and progestagens. *Hum. Reprod.* **2008**, *23*, 2359–2368. [CrossRef]
36. Andersen, Y.C. Possible new mechanism of cortisol action in female reproductive organs: Physiological implications of the free hormone hypothesis. *J. Endocrinol.* **2002**, *173*, 211–217. [CrossRef]

37. Ruiz-Conca, M.; Gardela, J.; Martínez, C.A.; Wright, D.; López-Bejar, M.; Rodríguez-Martínez, H.; Álvarez-Rodríguez, M. Natural mating differentially triggers expression of glucocorticoid receptor (NR3C1)-related genes in the preovulatory porcine female reproductive tract. *Int. J. Mol. Sci.* **2020**, *21*, 4437. [CrossRef]
38. Gong, S.; Sun, G.-Y.; Zhang, M.; Yuan, H.-J.; Zhu, S.; Jiao, G.-Z.; Luo, M.-J.; Tan, J.-H. Mechanisms for the species difference between mouse and pig oocytes in their sensitivity to glucorticoids. *Biol. Reprod.* **2017**, *96*, 1019–1030. [CrossRef]
39. Scarlet, D.; Ille, N.; Ertl, R.; Alves, B.G.; Gastal, G.D.A.; Paiva, S.O.; Gastal, M.O.; Gastal, E.L.; Aurich, C. Glucocorticoid metabolism in equine follicles and oocytes. *Domest. Anim. Endocrinol.* **2017**, *59*, 11–22. [CrossRef]
40. Da Costa, N.N.; Brito, K.N.L.; Santana, P.D.P.B.; Cordeiro, M.D.S.; Silva, T.V.G.; Santos, A.X.; Ramos, P.D.C.; Santos, S.D.S.D.; King, W.A.; Miranda, M.D.S.; et al. Effect of cortisol on bovine oocyte maturation and embryo development in vitro. *Theriogenology* **2016**, *85*, 323–329. [CrossRef]
41. Ruiz-Conca, M.; Alvarez-Rodriguez, M.; Mogas, T.; Gardela, J. NR3C1 expression in response to cortisol during bovine in vitro oocyte maturation. In Proceedings of the 15 th International Congress of the Spanish Society for Animal Reproduction (AERA), Toledo, Spain, 7–9 November 2019; p. 129.
42. Carneiro, M.; Piorno, V.; Rubin, C.-J.; Alves, J.M.; Ferrand, N.; Alves, P.C.; Andersson, L. Candidate genes underlying heritable differences in reproductive seasonality between wild and domestic rabbits. *Anim. Genet.* **2015**, *46*, 418–425. [CrossRef]
43. Gardela, J.; Jauregi-Miguel, A.; Martinez, C.A.; Rodriguez-Martinez, H.; Lopez-Bejar, M.; Alvarez-Rodriguez, M. Semen modulates the expression of NGF, ABHD2, VCAN, and CTEN in the reproductive tract of female rabbits. *Genes* **2020**, *11*, 758. [CrossRef] [PubMed]
44. Pfaffl, M.W. A new mathematical model for relative quantification in real-time RT-PCR. *Nucleic Acids Res.* **2001**, *29*, e45. [CrossRef] [PubMed]
45. Computing, R.F.S. *R: A Language and Environment for Statistical Computing*; R Core Team: Vienna, Austria, 2013.
46. Pinheiro, J.C.; DebRoy, S.S.; Sarkar, D.; Bates, D.M. Nlme: Linear and nonlinear mixed effects models. *R Package Version* **2020**, *3*, 111.
47. Hothorn, T.; Bretz, F.; Westfall, P. Simultaneous inference in general parametric models. *Biom. J.* **2008**, *50*, 346–363. [CrossRef] [PubMed]
48. Dukelow, W.R.; Williams, W.L. Survival of capacitated spermatozoa in the oviduct of the rabbit. *Reproduction* **1967**, *14*, 477–479. [CrossRef] [PubMed]
49. Talbot, P.; Geiske, C.; Knoll, M. Oocyte pickup by the mammalian oviduct. *Mol. Biol. Cell* **1999**, *10*, 5–8. [CrossRef] [PubMed]
50. Schjenken, J.E.; Robertson, S.A. The female response to seminal fluid. *Physiol. Rev.* **2020**, *100*, 1077–1117. [CrossRef]
51. Robertson, S.A. Seminal fluid signaling in the female reproductive tract: Lessons from rodents and pigs. *J. Anim. Sci.* **2007**, *85*, E36–E44. [CrossRef]
52. O'Leary, S.; Jasper, M.J.; Warnes, G.M.; Armstrong, D.T.; Robertson, S.A. Seminal plasma regulates endometrial cytokine expression, leukocyte recruitment and embryo development in the pig. *Reproduction* **2004**, *128*, 237–247. [CrossRef]
53. Sharkey, D.J.; Macpherson, A.M.; Tremellen, K.P.; Robertson, S.A. Seminal plasma differentially regulates inflammatory cytokine gene expression in human cervical and vaginal epithelial cells. *Mol. Hum. Reprod.* **2007**, *13*, 491–501. [CrossRef]
54. Castellini, C.; Mourvaki, E.; Cardinali, R.; Collodel, G.; Lasagna, E.; del Vecchio, M.T.; dal Bosco, A. Secretion patterns and effect of prostate-derived granules on the sperm acrosome reaction of rabbit buck. *Theriogenology* **2012**, *78*, 715–723. [CrossRef] [PubMed]
55. Bezerra, M.J.B.; Arruda-Alencar, J.M.; Martins, J.A.M.; Viana, A.G.A.; Viana Neto, A.M.; Rêgo, J.P.A.; Oliveira, R.V.; Lobo, M.; Moreira, A.C.O.; Moreira, R.A.; et al. Major seminal plasma proteome of rabbits and associations with sperm quality. *Theriogenology* **2019**, *128*, 156–166. [CrossRef] [PubMed]
56. Schjenken, J.E.; Robertson, S.A. Seminal fluid and immune adaptation for pregnancy - comparative biology in mammalian species. *Reprod. Domest. Anim.* **2014**, *49*, 27–36. [CrossRef] [PubMed]

57. Martinez, C.A.; Cambra, J.M.; Gil, M.A.; Parrilla, I.; Alvarez-Rodriguez, M.; Rodriguez-Martinez, H.; Cuello, C.; Martinez, E.A. Seminal plasma induces overexpression of genes associated with embryo development and implantation in day-6 porcine blastocysts. *Int. J. Mol. Sci.* **2020**, *21*, 3662. [CrossRef]
58. Martinez, C.A.; Cambra, J.M.; Parrilla, I.; Roca, J.; Ferreira-Dias, G.; Pallares, F.J.; Lucas, X.; Vazquez, J.M.; Martinez, E.A.; Gil, M.A.; et al. Seminal plasma modifies the transcriptional pattern of the endometrium and advances embryo development in pigs. *Front. Vet. Sci.* **2019**, *6*, 465. [CrossRef]
59. Waberski, D. Effects of semen components on ovulation and fertilization. *J. Reprod. Fertil. Suppl.* **1997**, *52*, 105.
60. Maranesi, M.; Petrucci, L.; Leonardi, L.; Piro, F.; Rebollar, P.G.; Millán, P.; Cocci, P.; Vullo, C.; Parillo, F.; Moura, A.; et al. New insights on a NGF-mediated pathway to induce ovulation in rabbits (*Oryctolagus cuniculus*). *Biol. Reprod.* **2018**, *98*, 634–643. [CrossRef]
61. Kershaw-Young, C.M.; Druart, X.; Vaughan, J.; Maxwell, W.M.C. β-Nerve growth factor is a major component of alpaca seminal plasma and induces ovulation in female alpacas. *Reprod. Fertil. Dev.* **2012**, *24*, 1093–1097. [CrossRef]
62. Silva, M.; Niño, A.; Guerra, M.; Letelier, C.; Valderrama, X.P.; Adams, G.P.; Ratto, M.H. Is an ovulation-inducing factor (OIF) present in the seminal plasma of rabbits? *Anim. Reprod. Sci.* **2011**, *127*, 213–221. [CrossRef]
63. Garcia-Garcia, R.M.; Masdeu, M.d.M.; Sanchez Rodriguez, A.; Millan, P.; Arias-Alvarez, M.; Sakr, O.G.; Bautista, J.M.; Castellini, C.; Lorenzo, P.L.; Rebollar, P.G. B-nerve growth factor identification in male rabbit genital tract and seminal plasma and its role in ovulation induction in rabbit does. *Ital. J. Anim. Sci.* **2018**, *17*, 442–453. [CrossRef]
64. Bomsel-Helmreich, O.; Huyen, L.V.N.; Durand-Gasselin, I. Effects of varying doses of HCG on the evolution of preovulatory rabbit follicles and oocytes. *Hum. Reprod.* **1989**, *4*, 636–642. [CrossRef] [PubMed]
65. Duffy, D.M.; Ko, C.; Jo, M.; Brannstrom, M.; Curry, T.E. Ovulation: Parallels With inflammatory processes. *Endocr. Rev.* **2019**, *40*, 369–416. [CrossRef] [PubMed]
66. Richards, J.A.S.; Liu, Z.; Shimada, M. Immune-like mechanisms in ovulation. *Trends Endocrinol. Metab.* **2008**, *19*, 191–196. [CrossRef]
67. Espey, L.L. Ovulation as an inflammatory reaction—A hypothesis. *Biol. Reprod.* **1980**, *22*, 73–106. [CrossRef] [PubMed]
68. Gross, K.L.; Cidlowski, J.A. Tissue-specific glucocorticoid action: A family affair. *Trends Endocrinol. Metab.* **2008**, *19*, 331–339. [CrossRef] [PubMed]
69. Myers, M.; Lamont, M.C.; van den Driesche, S.; Mary, N.; Thong, K.J.; Hillier, S.G.; Duncan, W.C. Role of luteal glucocorticoid metabolism during maternal recognition of pregnancy in women. *Endocrinology* **2007**, *148*, 5769–5779. [CrossRef]
70. Harlow, C.R.; Jenkins, J.M.; Winston, R.M.L. Increased follicular fluid total and free cortisol levels during the luteinizing hormone surge. *Fertil. Steril.* **1997**, *68*, 48–53. [CrossRef]
71. Thurston, L.M.; Abayasekara, D.R.E.; Michael, A.E. 11β-hydroxysteroid dehydrogenase expression and activities in bovine granulosa cells and corpora lutea implicate corticosteroids in bovine ovarian physiology. *J. Endocrinol.* **2007**, *193*, 299–310. [CrossRef]
72. Chin, E.; Jonas, K.C.; Abayasekara, D.R.E.; Michael, A.E. Expression of 11β-hydroxysteroid dehydrogenase (11βHSD) proteins in luteinizing human granulosa-lutein cells. *BioScientifica* **2003**, *178*, 127–135.
73. Gong, H.; Jarzynka, M.J.; Cole, T.J.; Jung, H.L.; Wada, T.; Zhang, B.; Gao, J.; Song, W.C.; DeFranco, D.B.; Cheng, S.Y.; et al. Glucocorticoids antagonize estrogens by glucocorticoid receptor-mediated activation of estrogen sulfotransferase. *Cancer Res.* **2008**, *68*, 7386–7393. [CrossRef]
74. Rhen, T.; Grissom, S.; Afshari, C.; Cidlowski, J.A. Dexamethasone blocks the rapid biological effects of 17β-estradiol in the rat uterus without antagonizing its global genomic actions. *FASEB J.* **2003**, *17*, 1849–1870. [CrossRef] [PubMed]
75. Witorsch, R.J. Effects of elevated glucocorticoids on reproduction and development: Relevance to endocrine disruptor screening. *Crit. Rev. Toxicol.* **2016**, *46*, 420–436. [CrossRef] [PubMed]
76. Ryu, J.S.; Majeska, R.J.; Ma, Y.; LaChapelle, L.; Guller, S. Steroid regulation of human placental integrins: Suppression of α2 integrin expression in cytotrophoblasts by glucocorticoids. *Endocrinology* **1999**, *140*, 3904–3908. [CrossRef] [PubMed]

77. Bitman, J.; Cecil, H.C. Differential inhibition by cortisol of estrogen-stimulated uterine responses. *Endocrinology* **1967**, *80*, 423–429. [CrossRef]
78. Johnson, D.C.; Dey, S.K. Role of histamine in implantation: Dexamethasone inhibits estradiol-induced implantation in the rat. *Biol. Reprod.* **1980**, *22*, 1136–1141. [CrossRef] [PubMed]
79. Rhen, T.; Cidlowski, J.A. Estrogens and glucocorticoids have opposing effects on the amount and latent activity of complement proteins in the rat uterus. *Biol. Reprod.* **2006**, *74*, 265–274. [CrossRef]
80. Higgins, B.D.; Kane, M.T. Inositol transport in mouse oocytes and preimplantation embryos: Effects of mouse strain, embryo stage, sodium and the hexose transport inhibitor, phloridzin. *Reproduction-Cambridge* **2003**, *125*, 111–118. [CrossRef]
81. Greenwald, G.S. A study of the transport of ova through the rabbit oviduct. *Fertil. Steril.* **1961**, *12*, 80–95. [CrossRef]
82. Cole, H.H.; Cupps, P.T. *Reproduction in Domestic Animals*, 2nd ed.; Academic Press: New York, NY, USA, 1969.
83. Barton, B.E.; Herrera, G.G.; Anamthathmakula, P.; Rock, J.K.; Willie, A.M.; Harris, E.A.; Takemaru, K.I.; Winuthayanon, W. Roles of steroid hormones in oviductal function. *Reproduction* **2020**, *159*, R125–R137. [CrossRef]
84. Goodwin, J.E.; Zhang, J.; Geller, D.S. A critical role for vascular smooth muscle in acute glucocorticoid-lnduced hypertension. *J. Am. Soc. Nephrol.* **2008**, *19*, 1291–1299. [CrossRef]
85. Rog-Zielinska, E.A.; Thomson, A.; Kenyon, C.J.; Brownstein, D.G.; Moran, C.M.; Szumska, D.; Michailidou, Z.; Richardson, J.; Owen, E.; Watt, A.; et al. Glucocorticoid receptor is required for foetal heart maturation. *Hum. Mol. Genet.* **2013**, *22*, 3269–3282. [CrossRef]
86. Spilman, C.H.; Harper, M.J.K. Effects of prostaglandins on oviductal motility and egg transport. *Gynecol. Obstet. Investig.* **1975**, *6*, 186–205. [CrossRef] [PubMed]
87. Blair, W.D.; Beck, L.R. In vivo effects of prostaglandin F2α and E2 on contractility and diameter of the rabbit oviduct using intraluminal transducers. *Biol. Reprod.* **1977**, *16*, 122–127. [CrossRef] [PubMed]
88. Peeters, B.W.M.M.; Ruigt, G.S.F.; Craighead, M.; Kitchener, P. Differential effects of the new glucocorticoid receptor antagonist ORG 34517 and RU486 (mifepristone) on glucocorticoid receptor nuclear translocation in the AtT20 cell line. *Ann. N. Y. Acad. Sci.* **2008**, *1148*, 536–541. [CrossRef] [PubMed]
89. Xi, M.; Yang, L.Z.; Hsieh, C.M.; Y Shen, L.L. The effect of RU486 on contractility of rabbit oviduct smooth muscle. *Sheng Li Xue Bao Acta Physiol. Sin.* **1996**, *48*, 277–283.
90. Fernandez-Fuertes, B.; Rodríguez-Alonso, B.; Sánchez, J.M.; Simintiras, C.A.; Lonergan, P.; Rizos, D. Looking at the big picture: Understanding how the oviduct's dialogue with gametes and the embryo shapes reproductive success. *Anim. Reprod.* **2018**, *15*, 751–764. [CrossRef]
91. Tetsuka, M.; Tanakadate, M. Activation of hsd11b1 in the bovine cumulus-oocyte complex during ivm and ivf. *Endocr. Connect.* **2019**, *8*, 1029–1039. [CrossRef]
92. Andersen, C.Y. Effect of glucocorticoids on spontaneous and follicle-stimulating hormone induced oocyte maturation in mouse oocytes during culture. *J. Steroid Biochem. Mol. Biol.* **2003**, *85*, 423–427. [CrossRef]
93. Scammell, J.G.; Denny, W.B.; Valentine, D.L.; Smiths, D.F. Overexpression of the FK506-binding immunophilin FKBP51 is the common cause of glucocorticoid resistance in three New World primates. *Gen. Comp. Endocrinol.* **2001**, *124*, 152–165. [CrossRef]
94. Ratajczak, T.; Cluning, C.; Ward, B.K. Steroid receptor-associated immunophilins: A gateway to steroid signalling. *Clin. Biochem. Rev.* **2015**, *36*, 31–52.

Publisher's Note: MDPI stays neutral with regard to jurisdictional claims in published maps and institutional affiliations.

Article

Efficiency of Tris-Based Extender Steridyl for Semen Cryopreservation in Stallions

Elena Nikitkina *, Artem Musidray, Anna Krutikova, Polina Anipchenko, Kirill Plemyashov and Gennadiy Shiryaev

Russian Research Institute for Farm Animal Genetics and Breeding—Branch of the L.K. Ernst Federal Science Center for Animal Husbandry, Moskovskoye sh. 55A, St. Petersburg, Pushkin 196625, Russia; 13linereg@mail.ru (A.M.); anntim2575@mail.ru (A.K.); aps.vet.93@yandex.ru (P.A.); kirill060674@mail.ru (K.P.); gs-2027@yandex.ru (G.S.)

* Correspondence: nikitkinae@yandex.ru; Tel.: +7-9213039976

Received: 2 September 2020; Accepted: 2 October 2020; Published: 4 October 2020

Simple Summary: The cryopreservation and long-term storage of semen is one of the methods for accelerated improvement of the genetic qualities of animals. However, horse breeders prefer to use fresh or chilled semen, as the fertilizing capacity of frozen equine semen is much lower. It is important to find extenders, or a combination of extenders, that will improve semen survival after freezing. It is also important that the extender can be easily and simply prepared for use. Steridyl is a concentrate to which you just need to add sterilized water. This extender was developed for ruminants. In this study we tested Steridyl for freezing stallion semen. The motility, morphology, energy metabolism, DNA damage, and fertility of sperm frozen in Steridyl were evaluated. As a result, Steridyl was shown to be a good extender for equine semen freezing.

Abstract: The fertilizing ability of stallion sperm after freezing is lower than in other species. The search for the optimal extender, combination of extenders, and the freezing protocol is relevant. The aim of this study was to compare lactose-chelate-citrate-yolk (LCCY) extender, usually used in Russia, and Steridyl® (Minitube) for freezing sperm of stallions. Steridyl is a concentrated extender medium for freezing ruminant semen. It already contains sterilized egg yolk. Semen was collected from nine stallions, aged from 7 to 12 years old. The total and progressive motility of sperm frozen in Steridyl was significantly higher than in semen frozen in LCCY. The number of spermatozoa with normal morphology in samples frozen in LCCY was 60.4 ± 1.72%, and with Steridyl, 72.4 ± 2.10% ($p < 0.01$). Semen frozen in Steridyl showed good stimulation of respiration by 2.4-DNP, which indicates that oxidative phosphorylation was retained after freezing–thawing. No differences among the extenders were seen with the DNA integrity of spermatozoa. Six out of ten (60%) mares were pregnant after artificial insemination (AI) by LCCY frozen semen, and 9/12 (75%) by Steridyl frozen semen. No differences among extenders were seen in pregnancy rate. In conclusion, Steridyl was proven to be a good diluent for freezing stallion semen, even though it was developed for ruminants.

Keywords: semen freezing; semen extender; stallions; semen quality; fertilizing ability; Tris

---

1. Introduction

Artificial insemination is one of the main methods of genetically improving the horse population, due to the accelerated use of the best stallions. Frozen semen has a number of advantages over fresh or chilled semen. Frozen semen preserves the horse's genetics for a long time, even after death, allowing the material to be transported over long distances, around the world. In modern assisted reproductive technology programs such as in vitro fertilization (IVF) and intracytoplasmic sperm injection (ICSI), cryopreserved sperm are mainly used. Artificial insemination of horses is widespread. However, breeders prefer to use fresh or chilled semen [1,2]. The fertilizing ability of stallion sperm after freezing is lower than in other species [3,4]. One of the reasons is the selection of stallions based on the exterior sports quality, and productivity, and not focused on reproductive qualities [4,5]. Sperm freezing in horses is very important for the preservation of valuable, rare, and endangered genotypes [6–8].

The search for the optimal extender, combination of extenders, and the freezing protocol is relevant. Semen extenders affect semen quality, especially after freezing [6,9]. Extenders usually are composed of sugars, electrolytes, egg yolk, and skim milk [10]. Skim milk-based extenders are effective in freezing stallion semen. Lipoproteins in skim milk protect spermatozoa from cold shock [10]. Chicken yolk is also one of the main components of the medium. The role of egg yolk in the cryopreservation of sperm is to be a resistance factor, which helps to protect against cold shock, and a storage factor, which helps to maintain viability [5]. The phospholipid, cholesterol, and the low-density lipoprotein content of chicken egg yolk have been identified as the protective components [11]. Most commercial extenders require the addition of fresh egg yolk. However, milk and egg yolk are biological fluids, and may contain components that are unfavorable for stallion semen. For example, α-lactoglobulin has been shown to be deleterious to the survival of equine sperm [12]. It has been reported that the source of egg yolk in Tris-based extenders affects semen motility in chilled dromedary camel ejaculates [13], and on post-thawing sperm motility and fertility in buffalo [14]. This could be attributed to the differences in fatty acid, phospholipids, cholesterol, and lipoprotein levels in egg yolk [15].

Steridyl® (Minitube, Germany) is concentrated extender medium for freezing of bull semen, and semen of other ruminants. Steridyl is a Tris based extender. The major benefit of Steridyl is that it already contains sterilized egg yolk in the concentrate. This eliminates the time-consuming preparation of fresh egg yolk. In this study we tried to freeze equine semen, diluted with Steridyl. We chose lactose-chelate-citrate-yolk (LCCY) extender for control, as it is used in Russia in most cases, and is recommended by the requirements of the Russian Federation.

The aim of the study was to compare LCCY extender, usually used in Russia, and Steridyl for freezing sperm in stallions.

2. Materials and Methods

2.1. Ethics Statement

The principles of laboratory animal care were followed, and all procedures were conducted according to the ethical guidelines of the L.K. Ernst Federal Science Center for Animal Husbandry and the Law of the Russia Federation on Veterinary Medicine No. 4979-1 (14 May 1993).

2.2. Animals

Nine stallions aged from 7 to 12 years old were used for this study. Semen was routinely collected 1–2 times a week during breeding season (March–June 2019).

2.3. Chemicals and Extenders Preparation

The chemicals used for the preparation of lactose-chelate-citrate-yolk (LCCY) extender and centrifugation solution were ordered from Sigma-Aldrich (Sigma, Saint Louis, MO, USA). Egg yolk was collected from chickens of the bioresource collection "Genetic Collection of Rare and Endangered Chicken Breeds" (RRIFAGB, Pushkin, St. Petersburg, Russia). Steridyl contains Tris, citric acid, sugar, buffers, glycerol, purified water, irradiated sterile egg yolk, and antibiotics (Tylosin, Gentamicin, Spectinomycin, Lincomycin). The chemicals proportion is the company's trade secret. Lactose-chelate-citrate-yolk (LCCY) extender consisted of 321 mM lactose, 3 mM ethylenediaminetetraacetic acid dinatrium salt (Trilon B), 3 mM sodium citrate, 0.95 mM sodium bicarbonate, 20% freshly prepared egg yolk, and 0.4 mg/mL gentamicin. Steridyl was prepared through a 1:1.5 dilution with sterilized water. The centrifugation solution consisted of 204 mM lactose, 25 mM glucose, 3 mM ethylenediaminetetraacetic acid dinatrium salt (Trilon B), 0.4 mM magnesium sulfate, 21 mM sodium chloride, and 14 mM potassium citrate.

2.4. Semen Collection and Preparation

The semen was collected with a Hannover artificial vagina (Minitüb GmbH, Tiefenbach, Germany). A total of 39 ejaculates (3–4 ejaculates per stallion) were collected. Immediately after collection, the semen was divided into two equal parts. One part (P1) was diluted by lactose-chelate-citrate-yolk (LCCY) extender, and a second part (P2) was diluted by centrifugation extender. Dilution ratio was 1:1. Sperm concentration, total (TM) and progressive motility (PM) was evaluated by a computer-assisted sperm analysis (CASA) in a Mackler chamber at 37 °C. The Argus CASA system (ArgusSoft LTD., St. Petersburg, Russia) and a Motic BA 410 microscope (Motic, Hong Kong, China) were used. Ejaculates with volume less than 30mL, concentration less than 100 × 106 sperm/mL, and total motility less than 60% were excluded from the study. A total of 8 ejaculates were excluded from the study. A total of 31 ejaculates were chosen for cryopreservation.

2.5. Cryopreservation

The diluted samples were centrifuged for 8 min at 600 g, the supernatant was eliminated and the P1 was resuspended in LCCY medium containing 3.5% glycerol, and the P2 was resuspended in Steridyl® medium (Minitüb GmbH, Tiefenbach, Germany). Final concentration was 200×10^6 cells/mL. Semen was loaded into 0.5 mL straws and equilibrated at +5 °C for 120 min. The straws were frozen in liquid nitrogen vapor at −110 °C for 12 min, and stored in a liquid nitrogen tank.

2.6. Semen Evaluation after Thawing

After at least 24 h, semen was thawed at 37 °C for 30 s, and the contents of the straw were emptied into a 1.5 mL microcentrifuge tube. Semen was evaluated for total and progressive motility as previously described.

Morphology was assessed using Diff-Quick kit (Abris+, St. Petersburg, Russia) stained smears and CASA (ArgusSoft–module Morphology). The morphology of at least 250 spermatozoa was examined on each slide at 1000× magnification with immersion oil.

We used a sperm chromatin dispersion (SCD) test [16] to assay DNA fragmentation in semen. GoldCyto DNA kit (Guangzhou, China) was used for the SCD test. A total of 100 spermatozoa per sample were measured by CASA (ArgusSoft–module DNA fragmentation).

Oxidative phosphorylation (OXPHOS) was assessed by the reaction of cellular respiration rate (CR) on adding 2,4-dinitrophenol (2.4-DNP) to the semen sample [17]. The addition of 2.4-DNP disrupts the proton gradient by carrying protons across a membrane, and uncouples proton pumping from ATP synthesis, because it carries protons across the inner mitochondrial membrane. As a result it increases respiration rate. If the respiration rate increases, there is good OXPHOS. We evaluated the CR using an ion meter "Expert-001MTX" and Clarke electrode (Research and Production Company "Econix-Expert", Moscow, Russia). Next, 100 µL of semen was added to the chamber with 1 mL of 11% lactose, and the rate of decrease in oxygen concentration was measured. Then, 10 µL 2.4-DNP was added. The ratio of the respiration rate with 2.4-DNP to the respiration rate before adding 2.4-DNP was found. Oxidative phosphorylation can be considered good when the rate of cellular respiration increases two or more times after the addition of 2.4-DNP.

2.7. Artificial Insemination (AI) and Pregnancy Control

Ten mares were inseminated with LCCY frozen sperm and twelve mares were inseminated with Steridyl frozen sperm. The mares were in good health. Artificial insemination was carried out in accordance with the breeding program. Doses frozen in Steridyl or in LCCY from the selected stallions were used randomly. AIs were performed by transcervical injection of the dose of thawed sperm, with preliminary control of follicle development 0–6 h after ovulation. The dose of frozen sperm was

250 million spermatozoa with PM. Pregnancy was determined by ultrasound examination on the 14th day after ovulation. The number of pregnant mares was counted in each AI cycle.

2.8. Statistical Analysis

For statistical analysis, the computer software IBM-SPSS Statistics 19 (IBM, Armonk, NY, USA) and Statistica 10 (TIBCO Software Inc., Palo Alto, United States) were used. All data were normally distributed (Kolmogorov–Smirnov test). Data were analyzed by ANOVA. The data were expressed as means ± standard error of the mean. All pairwise multiple comparisons between means were conducted by *t*-test. We compared pregnancy results by z-test for two proportions. Differences were considered statistically significant at $p < 0.05$.

3. Results

Semen used in this study was in the acceptable range for sperm motility and morphology after dilution. Total motility, progressive motility, and spermatozoa of normal morphology of semen diluted in Steridyl or LCCY extenders are shown in Table 1. No differences between the two extenders were found.

Table 1. Quality parameters (means ± SEM) in stallion semen samples (n = 31) diluted in Steridyl and lactose-chelate-citrate-yolk (LCCY) extenders.

	Steridyl	LCCY
Total Motility, %	83.2 ± 1.56	83.6 ± 1.97
Progressive Motility, %	75.3 ± 1.93	75.3 ± 1.38
Spermatozoa of Normal Morphology, %	75.8 ± 2.03	76.2 ± 1.82

n—number of samples for each extender.

Characteristics of semen quality frozen in Steridyl or LCCY extenders are shown in Table 2. The post thaw total and progressive motility of semen frozen with Steridyl were higher than when frozen with LCCY ($p < 0.05$). Six stallions classified as 'good freezers' had 41.2 ± 1.13% progressive motility in samples frozen with Steridyl, and 37.2 ± 1.02% progressive motility in samples frozen with LCCY. Three stallions classified as "poor freezers" had 29.7 ± 2.93% progressive motility in samples frozen with Steridyl, and 18.0 ± 0.73% progressive motility in samples frozen with LCCY. The number of spermatozoa with normal morphology in samples frozen in LCCY was lower than with Steridyl ($p < 0.01$). No differences among extenders were seen with the DNA integrity of spermatozoa.

Table 2. Quality parameters (means ± SEM) in stallion semen samples (n = 31) frozen/thawed in Steridyl and LCCY extenders.

	Steridyl	LCCY
Total Motility, %	43.1 ± 1.86 ^{a}	39.6 ± 0.93 ^{b}
Progressive Motility, %	36.3 ± 2.14 ^{a}	31.7 ± 1.13 ^{b}
Spermatozoa of Normal Morphology, %	72.4 ± 2.10 ^{c}	60.4 ± 1.72 ^{d}
Spermatozoa with Damaged DNA, %	27.2 ± 6.16	29.3 ± 4.32
Pregnancy Rates (Pregnancy Confirmation) at Day 14, %	75 ± 12.5	60 ± 15.5

n—number of samples for each extender. ^{a,b} Differences are significant for $p < 0.05$. ^{c,d} Differences are significant for $p < 0.01$.

Pregnancy rates (pregnancy confirmation) at day 14 are shown in Table 2. A total of 6/10 (60%) mares were pregnant after artificial insemination (AI) by LCCY frozen semen, and 9/12 (75%) by Steridyl frozen semen. No statistical difference was found between groups.

The increase of cellular respiration rate after the addition of 2.4-DNP was 1.92 ± 0.04 times in LCCY frozen sperm, and 2.21 ± 0.07 times in Steridyl frozen sperm (Figure 1). Oxidative phosphorylation was better in Steridyl than in LCCY ($p = 0.0005$).

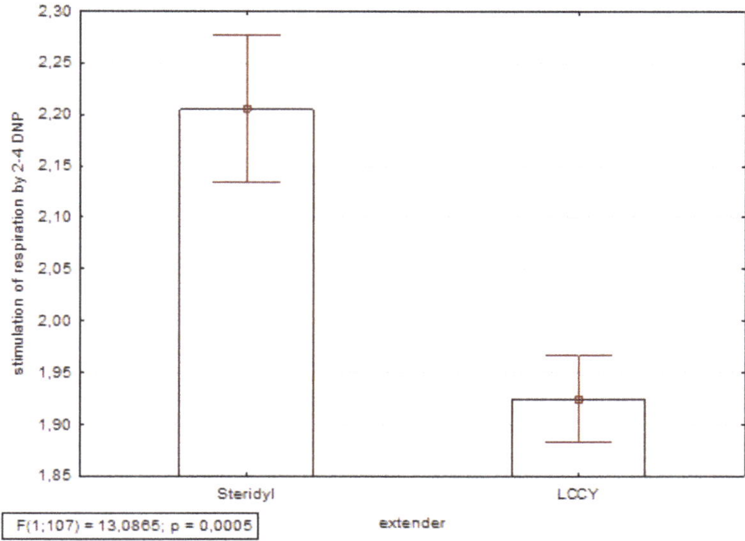

Figure 1. Stimulation of respiration by 2.4-dinitrophenol (2.4 DNP) in Steridyl frozen sperm and LCCY frozen sperm. The means ± SEM are presented.

4. Discussion

There are many factors that affect the quality of frozen semen in stallions. These are freezing protocols, extender formulation, and type of cryoprotectant. The composition of extender is one of the main factors in determining the viability of frozen semen [18]. Although most extenders for freezing equine semen provide sperm viability and pregnancy rates, new freezing extenders or modifications of freezing protocols are still required [4,18–20]. Stallion semen freezing protocols vary from laboratory to laboratory, using different extenders, both commercial and homemade, and different cryoprotectants. Often the protocols and extenders are selected individually for the stallion. It is reported that only 30–40% of stallions produce sperm with good cryoresistance [5,21]. Finding the extender that is easy to prepare and suitable for most stallions is important.

Egg yolk is a common component of most semen extenders for domestic animals. It has been shown to have a positive effect on sperm, and to protect acrosomes and the plasma membranes against cold shock [22]. When using commercial diluents, you have to add the egg yolk prior to use. The chicken yolk must be added before using most extenders. It is not possible to use yolks of the same quality, so the quality of the extender may vary. In this study, we used the commercial extender Steridyl to freeze semen from stallions. Steridyl was developed to freeze ruminant semen. Its main advantage is that it already contains sterile egg yolk, and you just need to add sterile water to the concentrate [23].

The results of our study showed that total and progressive motility of frozen semen, on average, met the requirements for sperm use for AI [2,24]. It is generally accepted that, on average, even under ideal conditions, 40–50% of spermatozoa do not survive freezing [25]. Vidament suggested that 35% post thawing motility is sufficient for insemination of mares [20]. Loomis and Graham suggested that 30% is enough. However, many stallions do not have even 30% motility after thawing [2]. Stallions with less than 20% motility after freezing may be qualified as "poor freezers'". The total and progressive motility of sperm frozen in Steridyl was significantly higher than in semen frozen in LCCY. Stallions classified as "good freezers" had higher motility in both extenders than stallions classified as "poor freezers". One "poor freezer" had less than 20% progressive motility in both extenders. Two "poor freezers" had less than 20% progressive motility in LCCY frozen samples, and 35% or more in Steridyl frozen samples. So Steridyl can be used in "poor freezers".

Usually, the assessment of motility is the first and main parameter by which to judge the survival of sperm after freezing [26]. However, deeper tests are also needed. Sometimes sperm with high abnormal morphology has good motility [27]. In our studies, sperm frozen in Steridyl had better morphology than in LCCY. Although some authors have noted that the extenders do not affect the morphology [26,28]. The main cell injuries that were encountered in our study were in the tail and neck of the spermatozoa. There were thickened necks, twisted and broken tails, swollen and wrinkled acrosomes, and missing acrosomes. Stallions differed in sperm morphology after cryopreservation. Some stallions had 55–58% spermatozoa of normal morphology in LCCY semen, and 60–65% in Steridyl frozen semen. Some stallions had more than 70% spermatozoa of normal morphology in both extenders.

The intensity of energy metabolism is a very important criterion to assess sperm quality [29,30]. Oxidative phosphorylation (OXPHOS) is the metabolic pathway in which cells use enzymes to oxidize nutrients, thereby releasing energy which is used to produce adenosine triphosphate (ATP). Glycolysis predominates in bull and ram semen. But stallion spermatozoa is mainly dependent on the mitochondrial oxidative phosphorylation pathway to produce ATP [31,32]. Mitochondria are one of the cell structures most sensitive to the damaging effects of low temperatures [29,33]. Respiratory response to 2.4-DNP supplementation is a good OXPHOS test. The positive correlation between conception rate in cows and stimulation of the respiration by 2.4-DNP of bovine sperm after freezing was found ($r = 0.62$, $p < 0.05$) [34]. In our studies, semen frozen in Steridyl showed good stimulation of the respiration by 2.4-DNP, which indicates that oxidative phosphorylation was retained after freezing–thaw. We also observed individual variation between stallions and ejaculates in the response of cellular respiration to 2.4-DNP additions. In five samples, frozen both in Steridyl and in LCCY, respiration stimulation with 2,4 dinitrophenol was one, which means that there was no reaction to dinitrophenol, and cellular respiration was without oxidative phosphorylation.

Many authors have suggested that DNA fragmentation affects fertility [8,35–38]. The sperm DNA was thought to be "dysfunctional" until fertilization took place. During the process of spermatogenesis, a high degree of compaction of sperm chromatin and associated nucleoproteins occurs. This degree of compaction is necessary to protect the DNA of the sperm during transport through the male and female reproductive tract, and for proper fertilization and embryo development [35]. Sperm DNA quality is associated with early embryonic death [39]. When evaluating sperm frozen in Steridyl and LCCY, DNA fragmentation was quite high. Our data are consistent with data of other authors. Sperm frozen in Steridyl and LCCY were not significantly different.

Fertility is the main indicator of sperm quality. Fertility is influenced by many factors, one of which is the extender [4,6,18,26]. In our study, semen frozen in both diluents had good fertility. No statistical difference was found between groups, but pregnancy confirmation on day 14 was a little higher in Steridyl frozen sperm. A very small number of inseminations were performed, and this low number might be the reason why there were not differences in the pregnancy rate between extenders.

This study showed a high tolerance of equine semen for cryopreservation in Steridyl. The quality of the semen frozen in Steridyl was superior to the quality of the semen frozen in the LCCY in almost all studied parameters. Perhaps this is due to the quality of the yolk, or perhaps Steridyl has a better effect on the sperm ability to survive during freezing procedure. At the same time, individual variability in sperm quality after freezing was observed in different stallions.

Steridyl is Tris-based extender. Tris-based extenders have been successfully used to freeze the semen of other animals [22,26]. However, previous work indicates poor equine sperm survival after freezing with Tris [26]. The authors point to the toxic effect of Tris-buffer due to its ability to penetrate the sperm membrane and alter intracellular metabolism. However, in our study, sperm metabolism was not severely impaired, and according to the reaction of the cellular respiration rate to the addition of 2.4-dinitrophenol, metabolism was better than in LCCY frozen semen. Sugar-lactose, glucose, fructose are also added to the diluent. Sugars are energy substrates, and they also protect sperm from osmotic shock and the formation of ice crystals inside and outside cells. Steridyl contains fructose and LCCY contains lactose. Pojprasath et al. noted in their study that fructose was less effective than glucose and sorbitol in protecting stallion sperm during cryopreservation [18]. In our study, the fructose containing diluent was better than lactose. The exact composition of Steridyl is not known, Steridyl's composition and the ratio of its components may have better protective properties than other studied Tris-based equine extenders.

5. Conclusions

Steridyl has proven to be a good diluent for freezing stallion semen, although it was developed for ruminants. Steridyl can be recommended for semen freezing of both "good freezer" and "poor freezer" stallions.

Author Contributions: Conceptualization, E.N. and A.M.; formal analysis, G.S.; investigation, E.N., A.M., A.K. and P.A.; writing—original draft preparation, E.N. and A.M; writing—review and editing, K.P.; project administration, K.P. All authors have read and agreed to the published version of the manuscript.

Funding: This research was funded by Ministry of Science and Higher Education of the Russian Federation, project No. AAAA-A18-118021990006-9.

Acknowledgments: Authors acknowledge Is. Shapiev for consultation on the assessment of energy metabolism in semen.

Conflicts of Interest: The authors declare no conflict of interest.

References

1. Aurich, J.E. Artificial Insemination in Horses-More than a Century of Practice and Research. *J. Equine Vet. Sci.* **2012**, *32*, 458–463. [CrossRef]
2. Loomis, P.R.; Graham, J.K. Commercial semen freezing: Individual male variation in cryosurvival and the response of stallion sperm to customized freezing protocols. *Anim. Reprod. Sci.* **2008**, *105*, 119–128. [CrossRef] [PubMed]
3. Blottner, S.; Warnke, C.; Tuchscherer, A.; Heinen, V.; Torner, H. Morphological and functional changes of stallion spermatozoa after cryopreservation during breeding and non-breeding season. *Anim. Reprod. Sci.* **2001**, *65*, 75–88. [CrossRef]
4. Contreras, M.J.; Treulen, F.; Arias, M.E.; Silva, M.; Fuentes, F.; Cabrera, P.; Felmer, R. Cryopreservation of stallion semen: Effect of adding antioxidants to the freezing medium on sperm physiology. *Reprod. Domest. Anim.* **2020**, *55*, 229–239. [CrossRef] [PubMed]
5. Clulow, J.R.; Maxwell, W.M.C.; Evans, G.; Morris, L.H.A. A comparison of duck and chicken egg yolk for cryopreservation of stallion sperm. *Aust. Vet. J.* **2007**, *85*, 232–235. [CrossRef]
6. Scherzer, J.; Fayrer-Hosken, R.A.; Aceves, M.; Hurley, D.J.; Ray, L.E.; Jones, L.; Heusner, G.L. Freezing equine semen: The effect of combinations of semen extenders and glycerol on post-thaw motility. *Aust. Vet. J.* **2009**, *87*, 275–279. [CrossRef]
7. Atroshchenko, M.M.; Bragina, E.E.; Zaitsev, A.M.; Kalashnikov, V.V.; Naumenkova, V.A.; Kudlaeva, A.M.; Nikitkina, E.V. Conservation of genetic resources in horse breeding and major structural damages of sperm during semen cryopreservation in stallions. *Nat. Conserv. Res.* **2019**, *4*, 78–82. [CrossRef]
8. Atroshchenko, M.M.; Arkhangelskaya, E.; Isaev, D.A.; Stavitsky, S.B.; Zaitsev, A.M.; Kalaschnikov, V.V.; Leonov, S.; Osipov, A.N. Reproductive characteristics of thawed stallion sperm. *Animals* **2019**, *9*, 1099. [CrossRef]
9. Alvarenga, M.A.; Papa, F.O.; Landim-Alvarenga, F.C.; Medeiros, A.S.L. Amides as cryoprotectants for freezing stallion semen: A review. *Anim. Reprod. Sci.* **2005**, *89*, 105–113. [CrossRef]
10. Alghamdi, A.S.; Troedsson, M.H.T.; Xue, J.L.; Crabo, B.G. Effect of seminal plasma concentration and various extenders on postthaw motility and glass wool-Sephadex filtration of cryopreserved stallion semen. *Am. J. Vet. Res.* **2002**, *63*, 880–885. [CrossRef]
11. Pace, M.M.; Graham, E.F. Components in egg yolk which protect bovine spermatozoa during freezing. *J. Anim. Sci.* **1974**, *39*, 1144–1149. [CrossRef] [PubMed]

12. Batellier, F.; Vidament, M.; Fauquant, J.; Duchamp, G.; Arnaud, G.; Yvon, J.M.; Magistrini, M. Advances in cooled semen technology. *Anim. Reprod. Sci.* **2001**, *68*, 181–190. [CrossRef]
13. Panahi, F.; Niasari-Naslaji, A.; Seyedasgari, F.; Ararooti, T.; Razavi, K.; Moosavi-Movaheddi, A.A. Supplementation of tris-based extender with plasma egg yolk of six avian species and camel skim milk for chilled preservation of dromedary camel semen. *Anim. Reprod. Sci.* **2017**, *184*, 11–19. [CrossRef] [PubMed]
14. Naz, S.; Umair, M.; Iqbal, S. Ostrich egg yolk improves post thaw quality and in vivo fertility of Nili Ravi buffalo (Bubalus bubalis) bull spermatozoa. *Theriogenology* **2019**, *126*, 140–144. [CrossRef] [PubMed]
15. Swelum, A.A.A.; Saadeldin, I.M.; Ba-Awadh, H.; Al-Mutary, M.G.; Moumen, A.F.; Alowaimer, A.N.; Abdalla, H. Efficiency of commercial egg yolk-free and egg yolk-supplemented tris-based extenders for dromedary camel semen cryopreservation. *Animals* **2019**, *9*, 999. [CrossRef]
16. Fernández, J.L.; Muriel, L.; Rivero, M.T.; Goyanes, V.; Vazquez, R.; Alvarez, J.G. The sperm chromatin dispersion test: A simple method for the determination of sperm DNA fragmentation. *J. Androl.* **2003**, *24*, 59–66. [CrossRef]
17. Moroz, L.G.; Shapiev, I. Polarographic method for assessing energy metabolism in sperm by stimulating respiration with 2.4-dinitrophenol. *Bull. State Sci. Inst. All-Russian Res. Inst. Farm Anim. Genet. Breed.* **1978**, *33*, 28–30.
18. Pojprasath, T.; Lohachit, C.; Techakumphu, M.; Stout, T.; Tharasanit, T. Improved cryopreservability of stallion sperm using a sorbitol-based freezing extender. *Theriogenology* **2011**, *75*, 1742–1749. [CrossRef]
19. Vidament, M.; Dupere, A.M.; Julienne, P.; Evain, A.; Noue, P.; Palmer, E. Equine frozen semen: Freezability and fertility field results. *Theriogenology* **1997**, *48*, 907–917. [CrossRef]
20. Vidament, M. French field results (1985–2005) on factors affecting fertility of frozen stallion semen. *Anim. Reprod. Sci.* **2005**, *89*, 115–136. [CrossRef]
21. Šichtař, J.; Bubeníčková, F.; Sirohi, J.; Šimoník, O. How to increase post-thaw semen quality in poor freezing stallions: Preliminary results of the promising role of seminal plasma added after thawing. *Animals* **2019**, *9*, 414. [CrossRef]
22. El-Shamaa, I.; El-Seify, E.S.; Hussein, A.; El-Sherbieny, M.; El-Sharawy, M. A comparison of duck and chicken egg yolk for cryopreservation of Egyptian buffalo bull spermatozoa Scientific Papers. *Sci. Pap. Ser. D Anim. Sci.* **2012**, *55*, 109–113.
23. Minitube Insemination Results When Applying Triladyl® or Steridyl as Preservation Media for Bull Semen. *Technical Report*; Minitube: Tiefenbach, Germany, 2012; Volume 2.
24. Barrier Battut, I.; Kempfer, A.; Becker, J.; Lebailly, L.; Camugli, S.; Chevrier, L. Development of a new fertility prediction model for stallion semen, including flow cytometry. *Theriogenology* **2016**, *86*, 1111–1131. [CrossRef]
25. Prien, S. Cryoprotectants & Cryopreservation of Equine Semen: A Review of Industry Cryoprotectants and the Effects of Cryopreservation on Equine Semen Membranes. *J. Dairy Vet. Anim. Res.* **2016**, *3*. [CrossRef]
26. Alamaary, M.S.; Haron, A.W.; Ali, M.; Hiew, M.W.H.; Adamu, L.; Peter, I.D. Effects of four extenders on the quality of frozen semen in Arabian stallions. *Vet. World* **2019**, *12*, 34–40. [CrossRef]
27. Varner, D.D. Approaches to Breeding Soundness Examination and Interpretation of Results. *J. Equine Vet. Sci.* **2016**, *43*, S37–S44. [CrossRef]
28. Teodora, V.; Groza, M.I. The Effect of Different Freezing Procedures on Sperm Head Morphometry in Stallions. *Bull. Univ. Agric. Sci. Vet. Med. Cluj-Napoca Vet. Med.* **2008**, *65*, 146–151. [CrossRef]
29. Nikitkina, E.S.I. Assessment of the respiratory activity in equine sperm. Reproduction in Domestic Animals. *Reprod. Domest. Anim.* **2014**, *49*, 49–50.
30. Moraes, C.R.; Meyers, S. The sperm mitochondrion: Organelle of many functions. *Anim. Reprod. Sci.* **2018**, *194*, 71–80. [CrossRef]
31. Davila, M.P.; Muñoz, P.M.; Bolaños, J.M.G.; Stout, T.A.E.; Gadella, B.M.; Tapia, J.A.; Balao Da Silva, C.; Ortega Ferrusola, C.; Peña, F.J. Mitochondrial ATP is required for the maintenance of membrane integrity in stallion spermatozoa, whereas motility requires both glycolysis and oxidative phosphorylation. *Reproduction* **2016**, *152*, 683–694. [CrossRef]
32. Gibb, Z.; Lambourne, S.R.; Aitken, R.J. The paradoxical relationship between stallion fertility and oxidative stress. *Biol. Reprod.* **2014**, *91*. [CrossRef]

33. García, B.M.; Moran, A.M.; Fernández, L.G.; Ferrusola, C.O.; Rodriguez, A.M.; Bolaños, J.M.G.; Da Silva, C.M.B.; Martínez, H.R.; Tapia, J.A.; Peña, F.J. The mitochondria of stallion spermatozoa are more sensitive than the plasmalemma to osmotic-induced stress: Role of c-Jun N-terminal kinase (JNK) pathway. *J. Androl.* **2012**, *33*, 105–113. [CrossRef] [PubMed]
34. Nikitkina, E.; Shapiev, I.; Moroz, L.K.V.P. Correlation between the level of respiratory stimulation by 2,4-dnp and fertilizing ability of frozen sperm. *Reprod. Domest. Anim.* **2016**, *51*, 121–122.
35. Evenson, D.; Jost, L. Sperm chromatin structure assay is useful for fertility assessment. *Methods Cell Sci.* **2000**, *22*, 169–189. [CrossRef] [PubMed]
36. Hernández-Avilés, C.; Zambrano-Varón, J.; Jiménez-Escobar, C. Current trends on stallion semen evaluation: What other methods can be used to improve our capacity for semen assessement? *J. Vet. Androl.* **2019**, *4*, 1–19.
37. Benchaib, M.; Braun, V.; Lornage, J.; Hadj, S.; Salle, B.; Lejeune, H.; Guérin, J.F. Sperm DNA fragentation decreases the pregnancy rate in an assisted reproductive technique. *Hum. Reprod.* **2003**, *18*, 1023–1028. [CrossRef]
38. Morrell, J.M.; Johannisson, A.; Dalin, A.M.; Hammar, L.; Sandebert, T.; Rodriguez-Martinez, H. Sperm morphology and chromatin integrity in Swedish warmblood stallions and their relationship to pregnancy rates. *Acta Vet. Scand.* **2008**, *50*. [CrossRef]
39. Burruel, V.; Klooster, K.L.; Chitwood, J.; Ross, P.J.; Meyers, S.A. Oxidative damage to rhesus macaque spermatozoa results in mitotic arrest and transcript abundance changes in early embryos. *Biol. Reprod.* **2013**, *89*. [CrossRef]

Article

Perfluorooctane Sulfonate (PFOS) and Perfluorohexane Sulfonate (PFHxS) Alters Protein Phosphorylation, Increase ROS Levels and DNA Fragmentation during In Vitro Capacitation of Boar Spermatozoa

Iván Oseguera-López [1], Serafín Pérez-Cerezales [2,†], Paola Berenice Ortiz-Sánchez [1], Oscar Mondragon-Payne [3], Raúl Sánchez-Sánchez [2], Irma Jiménez-Morales [4], Reyna Fierro [4] and Humberto González-Márquez [4,*]

1. Doctorado en Ciencias Biológicas y de la Salud, Universidad Autónoma Metropolitana, Mexico City 09340, Mexico; ivanoslo@yahoo.com.mx (I.O.-L.); biobere.fis@gmail.com (P.B.O.-S.)
2. Departamento de Reproducción Animal, Instituto Nacional de Investigación y Tecnología Agraria y Alimentaria, 28040 Madrid, Spain; perez.serafin@inia.es (S.P.-C.); raulss@inia.es (R.S.-S.)
3. Maestría en Biología Experimental, Universidad Autónoma Metropolitana, Mexico City 09340, Mexico; oscar_payne@hotmail.com
4. Departamento de Ciencias de la Salud, Universidad Autónoma Metropolitana-Iztapalapa, Mexico City 09340, Mexico; jimi@xanum.uam.mx (I.J.-M.); reyna@xanum.uam.mx (R.F.)

* Correspondence: hgm@xanum.uam.mx; Tel.: +52-55-5804-6557

† Deceased.

Received: 5 October 2020; Accepted: 15 October 2020; Published: 21 October 2020

Simple Summary: Perfluorinated compounds are synthetic chemicals, with a wide variety of applications like firefighting foams, food packaging, additives in paper and fabrics to avoid dyes. Perfluorooctane sulfonate and perfluorohexane sulfonate are globally distributed, and contaminates air, water, food, and dust, have toxic effects and bioaccumulate. Significant levels of these compounds have found in blood serum, breast milk, and semen of occupationally exposed and unexposed people, as well as in blood serum and organs of the domestic, farm, and wild animals. The present study seeks to analyze the toxic effects and possible alterations caused by the presence of these compounds in boar sperm during the in vitro capacitation, due to their toxicity, worldwide distribution, and lack of information in spermatozoa physiology during pre-fertilization processes.

Abstract: Perfluorooctane sulfonate (PFOS) and perfluorohexane sulfonate (PFHxS) are toxic and bioaccumulative, included in the Stockholm Convention's list as persistent organic pollutants. Due to their toxicity, worldwide distribution, and lack of information in spermatozoa physiology during pre-fertilization processes, the present study seeks to analyze the toxic effects and possible alterations caused by the presence of these compounds in boar sperm during the in vitro capacitation. The spermatozoa capacitation was performed in supplemented TALP-Hepes media and mean lethal concentration values of 460.55 μM for PFOS, and 1930.60 μM for PFHxS were obtained. Results by chlortetracycline staining showed that intracellular Ca^{2+} patterns bound to membrane proteins were scarcely affected by PFOS. The spontaneous acrosome reaction determined by FITC-PNA was significantly reduced by PFOS and slightly increased by PFHxS. Both toxic compounds significantly alter the normal capacitation process from 30 min of exposure. An increase in ROS production was observed by flow cytometry and considerable DNA fragmentation by the comet assay. The immunocytochemistry showed a decrease of tyrosine phosphorylation in proteins of the equatorial and acrosomal zone of the spermatozoa head. In conclusion, PFOS and PFHxS have toxic effects on the sperm, causing mortality and altering vital parameters for proper sperm capacitation.

Keywords: boar spermatozoa; perfluorinated compounds; PFHxS; PFOS; spermatozoa toxicology

1. Introduction

Perfluorinated compounds (PFCs) are characterized by a fully fluorinated hydrophobic linear carbon chain attached to one or more hydrophilic head groups [1]. Due to their properties as hydro-oil repellents and surfactants that are resistant to chemical and biological degradation, as well as their thermal stability [2–4], they are used widely in applications and in such products as paints, lubricants, stain repellents, additives for paper products, and aqueous film-forming foams used to fight electrical fires [5,6]. Due to their toxicity, bioaccumulation, and environmental persistence, they appear on the Stockholm Convention's list of Persistent Organic Pollutants [5,7].

Perfluorooctane sulfonate (PFOS) and perfluorohexane sulfonate (PFHxS) have been detected in human populations worldwide. The highest PFOS and PFHxS levels in blood were reported in retired fluorochemical production workers at 799 ng/mL (range, 14–3,490), and 290 ng/mL (range, 16–1,295), respectively, with calculated half-lives for elimination in serum (arithmetic and geometric mean) of 5.4–4.8, and 8.5–7.3 years, respectively [8]. However, PFCs residues are not present only in humans, as studies have detected them in farm animals like boars, cows, and chickens, among others [9,10]. PFOS in the amounts of ≤1780 and ≤28.6 (μg/kg) were found in liver and muscle samples of domestic and wild boars [11]. Other authors found 0.37 ng/mL in blood samples, and 54 ng/g in liver samples, though the PFHxS residues detected were not significant [9]. The domestic boar is a useful farm animal because of its economic importance. It is also considered an appropriate model for several areas of medical research, such as metabolic and infectious studies [12]. Boars are closely related to humans in terms of anatomy, genetics and physiology, so they are an excellent animal model for study, as genetically modified models can be created for use in modeling human reproduction and pathologies [13,14].

PFCs affect reproductive physiology in several species. In humans, PFOS and PFHxS have been shown to reduce morphologically normal spermatozoa, while PFOS increased spermatozoa tail abnormalities and decreased testosterone levels and motility [15–17]. In mice [18], PFOS diminished serum testosterone concentrations and epididymal spermatozoa counts, while in rats it damaged the blood-testis barrier function by disrupting the tight junction-permeability barrier of Sertoli cells [19,20]. In zebra fish, it decreased spermatozoa quality and had an estrogenic effect that increased estradiol levels but decreased testosterone in the juvenile phase [21].

Studies have shown the effects of PFCs on spermatozoa quality and morphology, but their impact on pre-fertilization processes such as capacitation and the acrosome reaction have not been described. Capacitation is essential for the successful binding of the spermatozoa to the oocyte and, therefore, fertilization. This process is characterized by biochemical changes that occur inside the female tract. Upon completing their capacitation, spermatozoa can perform exocytosis from the acrosome, called the acrosome reaction. This can be induced in vitro by chemical and biological agents like zona pellucida proteins, calcium ionophores, glycosaminoglycans, and progesterone [22,23].

Due to the global distribution of PFOS and PFHxS, and the lack of information on their effects on spermatozoa physiology during pre-fertilization processes, the aim of this study was to analyze their possible toxicity and physiological alterations on boar spermatozoa during the in vitro capacitation.

2. Materials and Methods

2.1. Boar Spermatozoa Samples and Incubation Media

All chemicals were purchased from the Sigma Chemical Company (St. Louis, MO, USA), unless otherwise indicated. All experimental procedures were performed according to institutional and European regulations. Semen samples were obtained from the sperm-enriched fraction of the ejaculates of 8 healthy, fertile and multi-breed boars (Duroc, Pietrain and hybrids Large White × Landrace) between 1 and 4 years of age, using the gloved-hand method. The gelatinous fraction was filtered and the post-sperm fraction was not used. All the samples used were obtained in the months of October to May between 8:30 and 10 a.m. and each pig was given 7 days rest between each sample. All assays were carried out with technical duplicates between 3 and 4 h after sample collection. For the standardization of the techniques and to ensure that there were no individual effects, at least 4 samples were obtained per boar. The samples employed were classified as normozoospermic according to established criteria: viability and motility > 80%, concentration > 200 × 10^6 spermatozoa/mL, and morphological abnormalities < 15% [24]. To determine the concentration, a 1:1000 dilution of the washed sample was diluted in water, a 10 µL aliquot was placed in a Neubauer chamber. The Eosin-Nigrosin stain, described below was used to determine mortality and morphological abnormalities. Sperm motility was determined subjectively. All samples were observed with an optical microscope Nikon Eclipse E400 (Nikon, Tokyo, Japan) at a magnification of 400×. The data from the initial evaluations of the samples used in this protocol are described in Table 1.

Table 1. Initial sperm evaluation data.

Samples at T0	Assay (%)				
	Eosin-Nigrosin		CTC Stain		FITC-PNA
Variable	Mortality	Abnormalities	Capacitated	sAR	sAR
$n = 6$	5.69 ± 1.64	1.86 ± 0.38	6 ± 1.33	1.86 ± 0.56	2.31 ± 0.33

sAR = Spontaneous Acrosome Reaction. All samples used were classified as normozoospermic according previously mentioned criteria.

After initial evaluation, the samples were centrifuged at 600× g for 5 min. The resulting pellet was rinsed with temperate phosphate-buffered saline (PBS) twice and re-suspended in 1 mL of TALP-HEPES capacitation medium (3.1 mM KCl, 100 mM NaCl, 0.29 mM $NaH_2PO_4 \cdot H_2O$, 10 mM Hepes, 2.5 mM $NaHCO_3$, 21.6 mM sodium lactate, 2.1 mM $CaCl_2 \cdot 2H_2O$, 1.5 mM $MgCl_2 \cdot 6H_2O$, and 10 µg/mL phenol red as the pH indicator). On the day of analyses, the medium was supplemented with 1 mM of sodium pyruvate and 6 mg/mL of bovine serum albumin fraction V, and then adjusted to pH 7.4. To induce capacitation, an aliquot of 5× 10^6 washed spermatozoa was transferred to 1 mL of TALP-HEPES and incubated for 4 h at 39 °C in a semi-humid atmosphere, with or without PFCs (concentrations indicated below) [25–27].

The spermatozoa experimental groups were:

(1) For the LC_{50} determination.

  a. T0, spermatozoa washed and not capacitated.
  b. Controls, incubated only in capacitation medium, and to ensure that nontoxic effect of DMSO, used as a diluent of PFCs, 10 µL of DMSO was added in another aliquot.
  c. Capacitated in the presence of PFOS (1000, 2000 and 3000 µM) and PFHxS (1000, 2500 and 5000 µM).

(2) For CTC stain and acrosome status by FITC-PNA.

  a. T0, spermatozoa washed and not capacitated.
  b. Controls, incubated only in capacitation medium, and to ensure that nontoxic effect of DMSO, used as a diluent of PFCs, 10 µL of DMSO was added in another aliquot.
  c. Capacitated in the presence of sub-lethal fractions (1/5 LC_{50}, 1/2 LC_{50}, LC_{50}) of PFOS and PFHxS.

(3) For tyrosine phosphorylation, reactive oxygen species (ROS), and comet assays only LC_{50} concentration of each PFCs were tested. The controls in these last tests were spermatozoa capacitated without the toxics.

(4) Incubation times for experiments above were of 4 h, except in the ROS-determination experiments, where the samples were incubated for 30 and 120 min (Figure 1).
(5) For microscopic studies, at least 200 cells were analyzed per slide.

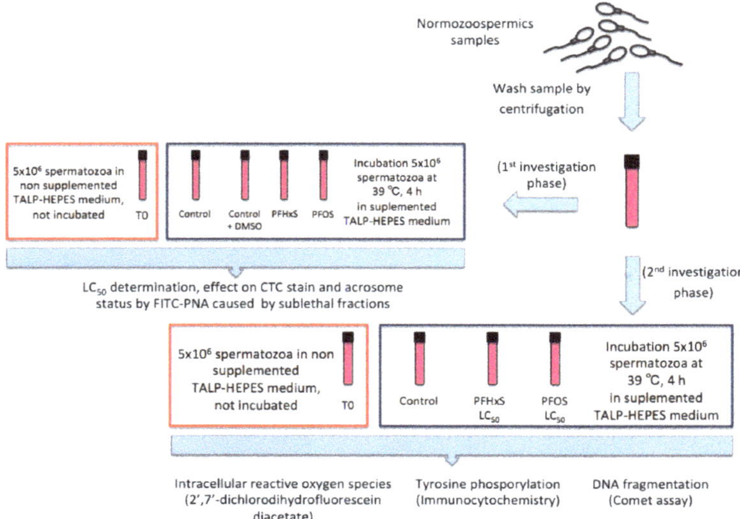

Figure 1. General experimental diagram was designed in two phases. (i) In the first, the LC_{50} values of both PFCs were calculated and the effect of sublethal fractions (1/5 LC_{50}, 1/2 LC_{50}, and LC_{50}) on sperm capacitation by CTC staining and sAR by FITC-PNA were analyzed. (ii) In the second phase, the effect of the LC_{50} of both PFCs on ROS levels were determined by 2′, 7′-dichlorodihydrofluorescein diacetate, tyrosine phosphorylation by immunocytochemistry and DNA fragmentation with the comet assay.

Live and dead sperm populations were considered for data collection in all the techniques mentioned.

2.2. Determination of Mean Lethal Concentration (LC_{50})

The spermatozoa were incubated in capacitation medium supplemented with 1000, 2000 and 3000 μM of PFOS, and 1000, 2500 and 5000 μM of PFHxS under the conditions described previously. The PFCs concentrations used were selected to get a range including the maximal mortality, and thus being able to establish the mean lethal dose. Six ejaculated samples from different boars were used, and were all handled independently. To demonstrate the non-toxicity of the DMSO, the control samples were incubated in the capacitation medium, and in the same medium supplemented with DMSO.

2.3. Spermatozoa Viability

Viability was evaluated by eosin-nigrosin staining. A 5 μL drop containing approximately 25,000 spermatozoa was placed on a glass slide temperated at 37 °C, mixed with 5 μL of stain solution (0.67% eosin Y and 10% nigrosin), incubated by 30 s and smeared. Finally, the slide was allowed to air dry at 37 °C. A parallel Hoechst-Propidium Iodide (PI) technique was used. In this case, a 5 μL drop containing approximately 25,000 spermatozoa was placed on a glass slide, mixed with 10 μM of Hoechst, and incubated for 2 min at room temperature (RT). After that, 10 μM of PI were added and the sample was observed immediately under fluorescence microscopy.

2.4. Capacitation and the Spontaneous Acrosome Reaction (sAR)

The capacitation and sAR processes were measured with 1/5 LC_{50}, 1/2 LC_{50} and LC_{50} of the PFCs. The amounts of non-capacitated spermatozoa (fluorescence throughout the head), capacitated spermatozoa (fluorescence in the acrosome zone), and spermatozoa with sAR (fluorescence in the post-equatorial zone) (Figure 2) were determined by the chlortetracycline (CTC) staining method. A 5 μL drop, containing approximately 25,000 spermatozoa were dropped onto a pre-heated glass slide and mixed with 5 μL of 750 μM of CTC prepared in a buffer containing 20 mM of Tris, 130 mM of NaCl and 5 mM of L-Cysteine. After 30 s, this was mixed with 5 μL of 0.2% glutaraldehyde solubilized in Tris buffer (0.5 mM pH 7.4), mounted with FluoroMount™ (F4680), and gently pressed with a coverslip [28]. Six independent samples were analyzed.

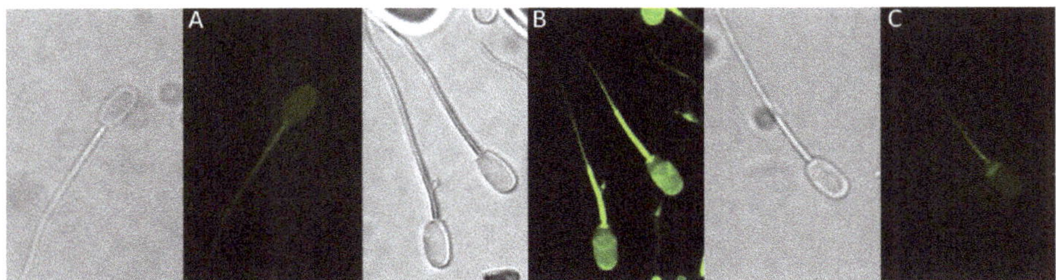

Figure 2. CTC stain patterns. 3 different patters were observed with the use of CTC stain. (**A**) Non-Capacitated, fluorescence throughout the spermatozoa head. (**B**) Capacitated, intense fluorescence in the equatorial and acrosomal zone. (**C**) Acrosomal reacted, fluorescence in equatorial and sometimes in the post-equatorial zone.

Hoechst-FITC-PNA was used as an alternative method for measuring sAR. For this procedure, a 20-μL sample droplet was placed on a glass slide and air dried at 37 °C. It was then immersed for 30 s in 100% methanol at −20 °C, air dried inside an extraction chamber and stored in a dry atmosphere at RT for later processing. The slide was washed twice for 5 min in PBS, the excess PBS was removed, and 20 μL of 15 μg/mL FITC-PNA prediluted in 5 μg/mL Hoechst were added. This was incubated for 30 min in a wet chamber under darkness, washed for 10 min in double distilled water, dried at 37 °C, and mounted with FluoroMount™ (F4680). Six independent samples were done.

2.5. Evaluation of Tyrosine Phosphorylation

To determine tyrosine phosphorylation, immunocytochemistry was performed as described previously with some modifications [29]. Samples of 5×10^6 spermatozoa/mL were washed twice with PBS by centrifugation, then fixed with 2% formaldehyde (Merck, Calbiochem) for 10 min at 37 °C, and permeabilized with 90% methanol (PanReac Quimica, S.A.U., Spain) for 30 min on ice. The samples were rinsed twice with 500 μL of incubation buffer (IB) (PBS 1% BSA), re-suspended in 90 μL of IB, and incubated for 10 min at RT to block them. Next, 5 μL (1:20) of PTyr antibody (anti human-mouse phosphotyrosine, Invitrogen 14-5001-82) were added to each sample and incubated overnight at 4 °C. Samples were rinsed and re-suspended in 95 μL of IB after adding 5 μL (1:20) of secondary antibody (Alexa 488 goat anti-mouse Invitrogen A11029), and incubated for 1 h at RT. Afterwards, they were rinsed with IB by centrifugation and re-suspended in 50 μL of PBS. A 5 μL droplet was placed on a glass slide, mixed with 10 μM of Hoechst to contrast nuclei, and observed under fluorescence microscopy. Three independent samples were done. By fluorescence microscopy a total of 6 phosphorylation patterns were found; no fluorescence and fluorescence in the flagellum, the equatorial segment, the equatorial segment and flagellum, the acrosome and flagellum, the acrosome alone, and the equatorial segment and flagellum (Figure 3). Fluorescence was classified in 4 patterns: no fluorescence in the head (NFH); and fluorescence in the acrosome area (AF), the equatorial head zone (EHZ), and the equatorial + acrosome zone (AE). Images of the patterns observed were captured with a confocal microscope (Leica TCS-SP2, Leica, Germany). To determine fluorescence intensity by flow cytometry, the remaining sample was re-suspended in IB to 500 μL.

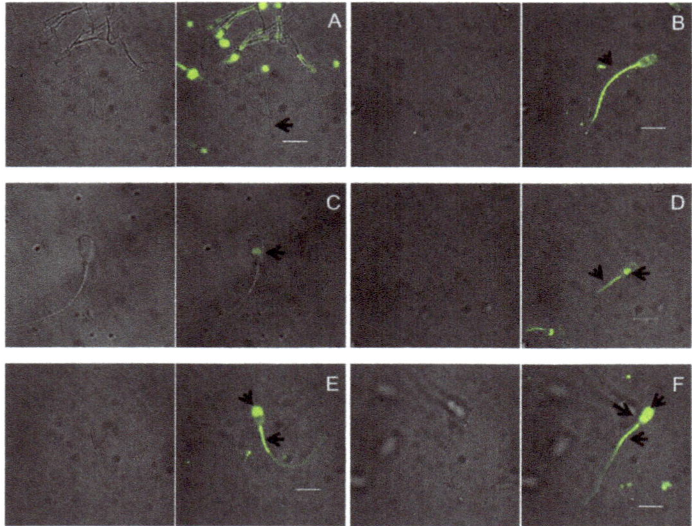

Figure 3. Fluorescence micrographs of various phosphorylation patterns in tyrosine residues of boar sperm capacitated in vitro, the fluorescence was granted by the use of fluorophore Alexa 488. Six different patterns were found, (**A**) No fluorescence. (**B**) Fluorescence in flagellum. (**C**) Fluorescence in equatorial segment. (**D**) Fluorescence in equatorial segment and flagellum. (**E**) Fluorescence in acrosome and flagellum. (**F**) Fluorescence in acrosome, equatorial segment and flagellum. Patterns were classified as: no fluorescence on head (NFH), fluorescence in acrosome (AF), equatorial head segment (EF), and acrosome plus equatorial head segment (AE). —, 1 μm.

2.6. *Intracellular Reactive Oxygen Species (ROS) Levels*

To determine ROS levels, 2′,7′-dichlorodihydrofluorescein diacetate (H2DCFDA) was used, as reported previously, with some modifications [30]. On the day of the experiment, non-capacitated and capacitated spermatozoa were mixed with Hoechst and H2DCFDA to a final concentration of 1×10^6 spermatozoa/mL with 10 μM of Hoechst and 200 μM of H2DCFDA. The spermatozoa were incubated for 30 min at 25 °C. Fluorescence was assessed by flow cytometry. Three independent samples were done.

2.7. *Single-Cell Gel Electrophoresis (SCGE), or the Comet Assay*

The alkaline version of the SCGE was used to measure DNA fragmentation, as described previously [31]. A suspension of 85 μL of 0.5% low-melting point (LMP) agarose in PBS containing 20,000 spermatozoa was placed on a slide previously covered with 1% agarose and covered with a 22 × 22 mm coverslip. For gelation of the LMP agarose, the slides were left overnight in a wet box at 4 °C. After removing the coverslips, the slides were incubated at 37 °C for 1 h in lysis solution (2 M NaCl, 55 mM EDTA-Na2, 8 mM Tris, 4% Triton X-100, 0.1% SDS, 1 mM DTT, and 0.5 mg/mL of proteinase K, pH 8). Next, they were washed twice in alkaline (0.3 M NaOH, 1 mM EDTA-Na2, pH 12) electrophoresis solution, placed in the electrophoresis cell, and filled with the same solution to cover them with about 1 cm of solution. Electrophoresis was run for 10 min at 25 V. The slides were then neutralized in 0.4 M of Tris-HCl pH 7.5 for 10 min. After that, they were fixed in methanol for

3 min and left to air dry prior to staining with ethidium bromide for observation under fluorescence microscopy at 400×. The comets were digitalized with a Nikon 5100 digital camera (Nikon, Japan) coupled to the microscope in manual mode, with a constant configuration. At least 150 comet images were analyzed using the free software, Casplab 1.2.3 beta2 (CaspLab.com) [32]. Four independent samples were done.

2.8. *Flow Cytometry*

For the immunofluorescence assay of tyrosine phosphorylation and the fluorescent probe for intracellular ROS, mean fluorescence intensity (FI) was determined by flow cytometry in a FACSCanto II flow cytometer (BD Biosciences, San Jose, CA, USA). The spermatozoa were gated in the FSC/SSC dot plot to exclude debris, and then confirmed by analyzing nuclear staining with Hoechst using a 405 nm laser and blue filter (450/50 nm). A sample of 5×10^6 spermatozoa was transferred to a BD Falcon 5 mL round-bottom tube and placed in the flow cytometer at a flow rate of 10 µL/min to count a total of 1×10^4 spermatozoa per determination. For flow cytometer compensation, control samples with only one fluorochrome were prepared with Hoechst for nuclei staining, H2DCFDA for ROS determination, and Alexa 488 for the secondary antibody, in order to determine tyrosine phosphorylation. For both FI analyses, a 488 nm laser and green filter (530/30 nm) were used. The data acquired were analyzed by logarithmic representation using FlowJo software (Becton–Dickinson, USA). Three independent samples were analyzed for each study. Trained personnel performed equipment calibration using the Long Clean mode as per the manufacturer's instructions. Six independent samples were done.

2.9. *Fluorescence Microscopy*

All samples analyzed by fluorescence microscopy were observed in a Nikon Optiphot-2 microscope (Nikon, Tokyo, Japan). The B-2A filter was used for the CTC, FITC-PNA, and Alexa 488 procedures with the secondary antibody in immunocytochemistry, while the UV-2A filter was utilized for fluorescence emitted by Hoechst. The G-2A filter was employed for the viability test with propidium iodide and the comet assay with ethidium bromide.

2.10. *Statistical Analyses*

The results are expressed as mean ± SD. The Probit test [33] was applied to determine the LC_{50}. To establish differences on the viability, capacitation process, AR, SCGE, as well as tyrosine phosphorylation adhesion, the data were subjected to one-way analysis of variance (ANOVA) Two-way ANOVA was used for the statistical analysis of the effect on ROS production. Mean pairwise comparisons were computed with a Tukey's HSD test with a p value ≤ 0.05. Statistical analysis was performed with the IBM SPSS software (IBM SPSS Statistics, version 20 for Mac Os).

3. Results

3.1. *Determination of the LC_{50} of PFOS and PFHxS*

No significant difference was observed between controls. PFOS showed more spermatozoa toxicity than PFHxS, as 80% mortality was obtained with 1000 µM of PFOS, while 2500 µM were required for PFHxS (Figure 4). The LC_{50} of both compounds was calculated by Probit at 460.55 µM for PFOS and 1930.60 µM for PFHxS.

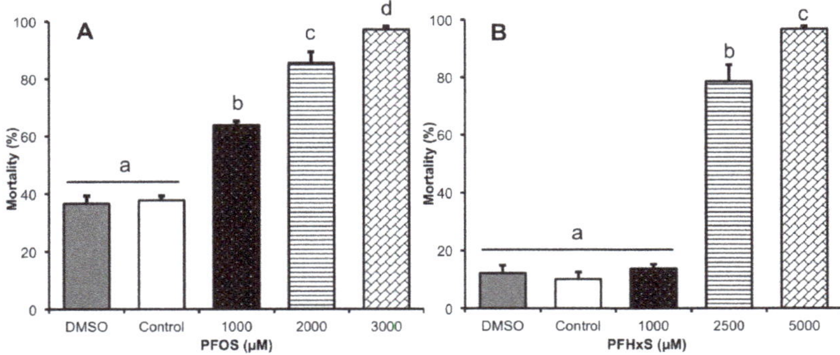

Figure 4. Mortality rates during boar spermatozoa in vitro capacitation process. (**A**) PFOS mortality increases significantly at all experimental concentrations evaluated. (**B**) PFHxS mortality is not significant at a concentration of 1000 µM, but it increases significantly at concentrations of 2500 and 5000 µM. Controls tested do not showed significant differences between them. Mortality data plotted correspond to eosin-nigrosin stain. All spermatozoa stained in pink by eosin (eosin positive) was considered dead, $n = 6$. $^{a, b, c, d}$ Different lowercase letters indicate the existence of significant differences between treatments ($p < 0.05$).

3.2. Effect of Sub-Lethal Fractions on Boar Spermatozoa Capacitation

The sub-lethal concentrations used were 1/5 LC_{50} (92 µM for PFOS and 386 µM for PFHxS), 1/2 LC_{50} (230 µM for PFOS and 965 µM for PFHxS) and LC_{50} (460 µM for PFOS and 1930 µM for PFHxS). We considered spermatozoa capacitated when they showed fluorescence in the acrosome. PFOS significantly decreased spermatozoa capacitation, from 70.28% ± 4.07 of capacitated spermatozoa in the control to 48.32 ± 3.43% in 1/5 LC_{50}, (−22%, $p < 0.001$), 40.03 ± 2.82% in 1/2 LC_{50} (−30%, $p < 0.001$), and 31.97 ± 1.83% in LC_{50} (−38%, $p < 0.001$). There was no effect upon adding DMSO to the TALP-HEPES medium, as this produced 70.61 ± 4.37% of spermatozoa capacitation (Figure 5A). PFHxS showed no effect on spermatozoa capacitation, determined by the fluorescence pattern of chlortetracycline (Figure 5B), since 69.7–71.6% of the spermatozoa were capacitated.

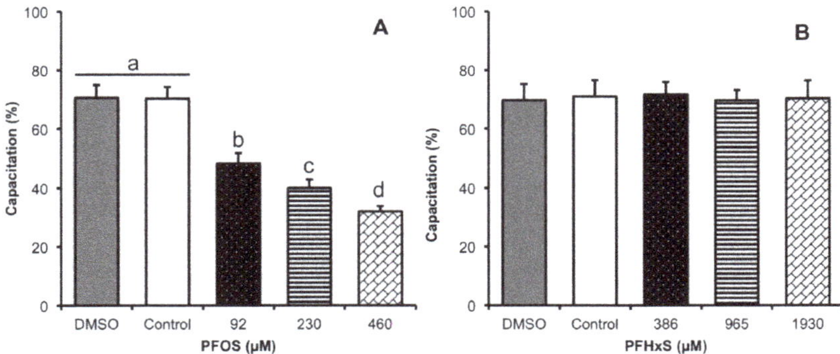

Figure 5. Effect of PFCs on sperm capacitation process. (**A**) The presence of PFOS significantly decreases the percentage of capacitated sperm at all sublethal concentrations tested. No alteration was observed in the sperm capacitation process in the presence of PFHxS as determined by the CTC technique, $n = 6$ (**B**). $^{a, b, c, d}$ Different lowercase letters indicate the existence of significant differences between treatments ($p < 0.05$). Sperm considered capacitated were CTC positive with pattern B (Figure 2). The sub-lethal concentrations used were 1/5 LC_{50} (92 µM for PFOS and 386 µM for PFHxS), 1/2 LC_{50} (230 µM for PFOS and 965 µM for PFHxS) and LC_{50} (460 µM for PFOS and 1930 µM for PFHxS).

3.3. Subsection Effect of Sub-Lethal Concentrations of PFCs on the sAR Process

The sAR process, evaluated by CTC, showed no significant differences at any of the sub-lethal concentrations of PFOS and PFHxS (Figure 6A,B), so we used a PNA lectin-binding assay to further test the effect of PFCs on acrosome status and integrity. The FITC-PNA technique found that PFOS significantly reduced sAR from 11.47 ± 1.74% in the control to 7.45 ± 1.73% (−4.02%, $p = 0.005$) in 1/2 LC_{50}, and 7.09 ± 1.59% (−4.38% $p = 0.001$) in LC_{50}. No significant differences were seen in 1/5 LC_{50} compared to controls (Figure 6C). PFHxS, in contrast, significantly increased sAR in 1/5 LC_{50}, from 10.46 ± 1.14% in the control to 14.77 ± 1.52% (4.31% $p = 0.001$), and to 15.12 ± 1.75% (4.66% $p < 0.001$) in 1/2 LC_{50}, and 15.39 ± 1.13% (4.93% $p < 0.001$) in LC_{50} (Figure 6D).

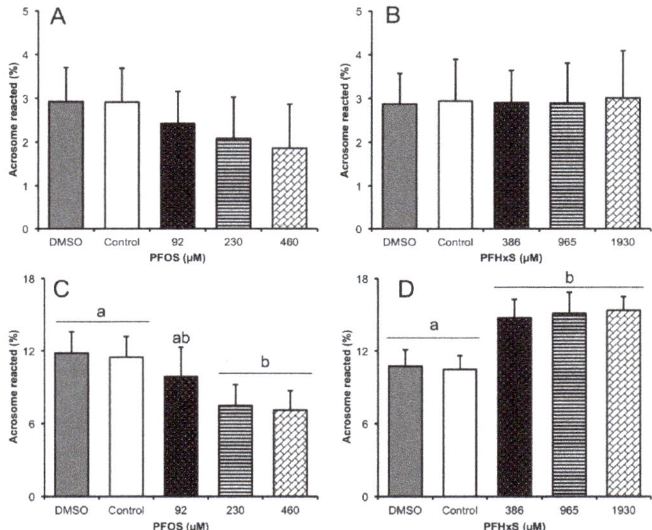

Figure 6. Effect of PFCs on spontaneous acrosome reaction (sAR) $n = 6$. (**A**,**B**), determined by CTC; (**C**,**D**), by FITC-PNA. (**A**) PFOS shows a tendency to decrease sAR, but it is not significant in any concentration. (**B**) PFHxS does not affect sAR in any of the concentrations analyzed. (**C**) PFOS shows a significant decrease from 1/2 LC_{50}, the 1/5 LC_{50} does not differ significantly from the control. (**D**) PFHxS, significantly increases the acrosome reacted in all conditions tested, there was no significant difference between controls. [a, b] Different lowercase letters indicate significant differences between treatments ($p < 0.05$).

3.4. Tyrosine Phosphorylation Determination

Tyrosine phosphorylation was evaluated by two mechanisms; first, FI was assessed by flow cytometry, then the fluorescence pattern was measured by fluorescence microscopy.

3.4.1. Flow Cytometry

The phosphorylation of tyrosine residues in spermatozoa proteins after capacitation in the absence of PFCs was significantly higher (control = 711.67 ± 120.02 FI) than before capacitation (T0 = 398 ± 23.30 FI) ($p = 0.003$). No significant differences were observed in the samples capacitated in the presence of PFCs, as results were 726 ± 67.45 FI and 610 ± 38.57 FI in LC_{50} of PFHxS and PFOS, respectively (Figure 7).

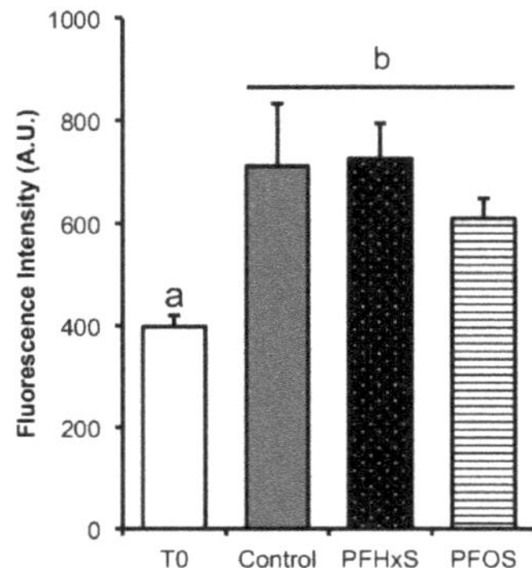

Figure 7. Flow cytometry of tyrosine phosphorylation in boar sperm. The fluorescence intensity obtained by fluorophore Alexa 488 were determined as arbitrary units (AU) did not show significant differences in any of the capacitation conditions tested. A significant difference was observed with the control at time T0, $n = 3$. [a, b] Different lowercase letters indicate the existence of significant differences between treatments ($p < 0.05$).

3.4.2. Fluorescence Microscopy

The NFH pattern was present at T0 (before capacitation) in 9.51 ± 1.03% of the spermatozoa. After capacitation, there were no significant differences in the presence of this pattern in the TALP-HEPES control (12.15 ± 0.80%). After incubation with PFOS LC_{50}, a significant increase was observed (12.69 ± 1.50%, $p = 0.025$) compared to T0, though no differences were observed for the other PFOS concentrations or PFHxS.

The AF pattern in T0 was 14.42 ± 0.75%, but after capacitation the control decreased significantly to 5.24 ± 1.8% ($p < 0.001$). The spermatozoa capacitated with PFOS at LC_{50} diminished 4.5 times to 3 ± 1.46% ($p < 0.001$), while with PFHxS at LC_{50} the decrease was 14-fold, to 1.28 ± 1.29% ($p < 0.001$).

EHZ was the most abundant pattern in all conditions, obtaining 75.62 ± 0.98% in T0. This pattern decreased significantly in the spermatozoa capacitated under the control conditions, to 65.38 ± 1.98% ($p < 0.001$). Compared to the control, exposure to LC_{50} of PFOS or PFHxS was significantly higher with results of 71.06 ± 2.50% ($p = 0.027$) and 81.6 ± 1.87% ($p < 0.001$), respectively.

The AE pattern significantly increased after capacitation, from 0.45 ± 0.79 in T0 to 17.23 ± 0.74 in the control ($p < 0.001$). Exposure to LC_{50} of PFOS or PFHxS, however, significantly decreased this AE pattern compared to the control to 13.26 ± 0.73 ($p = 0.003$), and 7.32 ± 1.25 ($p < 0.001$), respectively (Table 1). These results seem to indicate that PFCs slow the capacitation process by modifying the phosphorylation of tyrosine residues (Table 2).

Table 2. Effect of PCFs in tyrosine residues phosphorylation patterns in boar sperm capacitation.

Treatment	**Tyrosine Phosphorylation Patterns**			
	NFH	AF	EHZ	AE
T0	9.51 ± 1.03 [a]	14.42 ± 0.75 [a]	75.62 ± 0.98 [a]	0.45 ± 0.79 [a]
Control	12.15 ± 0.80 [ab]	5.24 ± 1.48 [bc]	65.38 ± 1.98 [b]	17.23 ± 0.74 [b]
PFOS	12.69 ± 1.50 [b]	3.00 ± 1.46 [bc]	71.06 ± 2.50 [a]	13.26 ± 0.73 [c]
PFHxS	9.81 ± 0.69 [a]	1.28 ± 1.29 [c]	81.6 ± 1.87 [c]	7.32 ± 1.25 [d]

Sperm capacitation was performed in Talp-Hepes medium at 39 °C for 4 h. Both compounds used in LC_{50}, for PFOS 460.55 µM and PFHxS 1930.60 µM. Treatments: T0, non-capacitated sperm sample; Control, capacitated; PFOS, Perfluorooctane Sulfonate y PFHxS, Perfluorohexane Sulfonate; PFC, perfluorinated compounds, $n = 3$. Patterns: NFH, No fluorescence on sperm head; AF, acrosome fluorescence; EHZ; equatorial head zone fluorescence; AE, acrosomal and equatorial fluorescence. [a, b, c, d] Different letters in same column represent significant statistical differences.

3.5. Effect of PFCs on ROS Production

The LC_{50} of PFOS and PFHxS caused an over-production of ROS at 30 min of treatment, generating them at 1.8 and 2.3 times, respectively, to 355.67 ± 70.55 ($p = 0.028$) and 459 ± 20.00 arbitrary fluorescence intensity units (A.U.) ($p = 0.003$), compared to the control at 181 ± 74.51. After 2 h of incubation in the capacitation medium, the LC_{50} of PFHxS showed significant differences compared to the control ($p = 0.006$). Upon comparing the same condition at different times, we found that the LC_{50} of PFOS and PFHxS showed significant differences, with values for PFOS of 196.67 ± 47.65 at T0, 355.67 ± 70.55 at 30 min ($p < 0.001$) and 279 ± 94.73 at 2 h ($p = 0.002$). In the case of PFHxS, 203.33 ± 38.19 at T0, 459 ± 20 at 30 min ($p < 0.001$), and 417 ± 37.72 at 2 h ($p = 0.001$) (Figure 8).

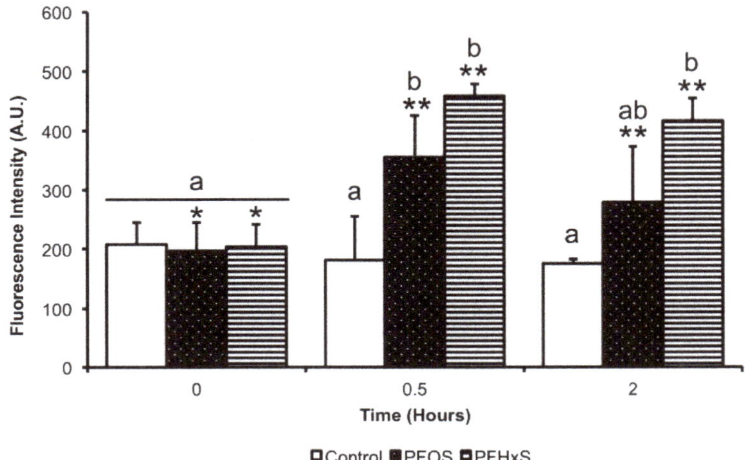

Figure 8. Determination of intracellular ROS by the use of 2′,7′-dichlorodihydrofluorescein diacetate, as a result of exposure to PFCs, $n = 3$. An increase in the fluorescence intensity (AU) of ROS was observed during the first 0.5 h of exposure to PFCs. At 2 h exposure, a decrease in ROS fluorescence intensity was recorded, but still significantly higher on PFHxS treatment, compared to the control. Both, PFOS and PFHxS shows significant differences between same treatment at different times. [a, b] Different lowercase letters indicate the existence of significant differences between treatments at same time evaluated ($p < 0.05$). Different asterisk numbers indicate significant differences between the sametreatment at different times evaluated ($p < 0.05$).

3.6. Effect of PFCs on DNA Fragmentation

Because the PFCs studied increased ROS production after 30 min, we decided to determine whether this increase generated sperm DNA fragmentation (Figure 9). However, only 5.07 ± 1.42% of the spermatozoa in T0 showed fragmentation, and no significant differences were found in relation to the capacitation control, though there were significant increases, to 13.40 ± 1.12 ($p = 0.001$) with PFOS and 24.2 ± 3.69 ($p < 0.001$) with PFHxS.

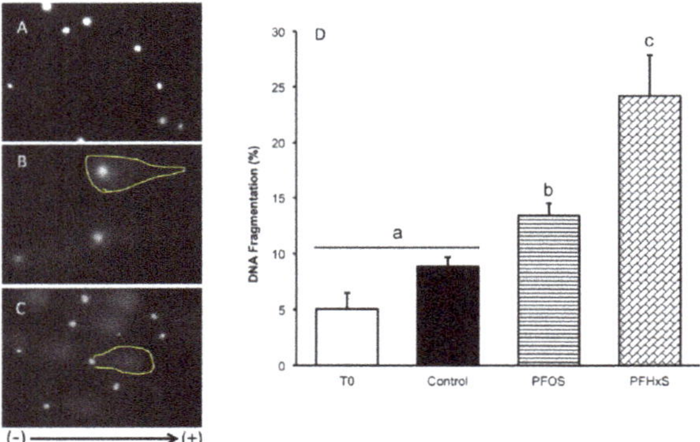

Figure 9. Determination of DNA fragmentation by comet assay, $n = 4$. (**A**) Sample in T0 with low level of fragmentation. (**B**) Sample with medium level of fragmentation after exposure to PFOS, and (**C**) Sample with high level of fragmentation due to exposure to PFHxS. The arrow indicates the direction of migration of DNA fragments to the positive pole, allowing the formation of the comet, the ethidium bromide allows the observation by fluorescence microscopy, yellow line indicates the tail comet area. The graph in (**D**) shows a significant increase in DNA fragmentation, about five and three times, after exposure to PFHxS and PFOS respectively compared to T0. No significant differences between T0 and the Control were detected. [a, b, c] Different lowercase letters indicate the existence of significant differences between treatments ($p < 0.05$).

4. Discussion

Although, many studies have evaluated the effect of PFCs on reproduction, most are of the in vivo type, or involved exposed persons. Very few have evaluated the toxic mechanisms of PFCs. Louis et al. (2015) [34] evaluated the relation between seven PFCs and the parameters of human semen quality, finding poor quality with six of the PFCs evaluated, in the form of abnormal heads, coiled and double tails, and more immature spermatozoa. Their results agree with those from Joensen et al. (2009) [15], who concluded that the combination of PFOA and PFOS decreased the amount of morphologically normal spermatozoa. The toxic effect of PFCs has not been extensively studied, especially not in relation to the physiology of spermatozoa capacitation. For this reason, and to discuss the mechanisms that are altered by the compounds analyzed, we compared the effects of PFCs studied in other cellular and animal models. The present experiment evaluated the effect of PFOS and PFHxS on the capacitation and sAR of boar spermatozoa in vitro. It seems that these two perfluorinated compounds analyzed are not as toxic as expected, as their LC_{50} is in the range of 0.5–2 mM, though this result may be attributable to the fact that the spermatozoa were incubated in the PFCs for only 30–240 min.

Slotkin et al. (2008) [35] evaluated the in vitro neurotoxicity of PFOS, PFOA (perfluorooctanoic acid), PFOSA (perfluorooctane sulfonamide) and PFBS (perfluorobutane sulfonate) in undifferentiated and differentiated PC12 cells. They concluded that each PFC exerted a distinct effect on those cells. PFOSA had the strongest effect, followed by PFOS, PFBS, and finally, PFOA. PFOSA was found to decrease DNA synthesis and cell viability, and to enhance oxidative stress. The authors proposed that the toxic effect of PFOSA could be associated with its greater hydrophobicity, which allowed it to easily access the cell membrane. In the boar spermatozoa analyzed in our study, we observed that PFOS is more toxic than PFHxS, a finding that correlates with its greater hydrophobicity.

The in vitro boar sperm capacitation methodology gave optimal results, obtaining levels of capacitation, similar to others previously reported [27,36]. The CTC technique is useful for differentiating among non-capacitated, capacitated, and AR spermatozoa in normal in vitro sperm, based on the CTC fluorescence patterns when they are in contact with intracellular Ca^{2+} bound to membrane proteins [37]. The results of this assay were also unexpected in light of the literature, for the Ca^{2+} patterns were scarcely affected by PFOS and not at all by PFHxS. Moreover, no changes were observed in sAR with either PFC. A study conducted with hBMSCs cells showed that PFHxS increased calcium transport [38], and that PFOS had a negative effect on suppressed synaptogenesis and inhibited neurite growth caused by abnormal regulation of calcium in the hippocampus [39]. As that work demonstrates, both compounds modify calcium regulation, and likely damage Ca^{2+} channels or Ca^{2+} internal redistribution, as observed in our study model, where exposure to PFHxS had a more evident effect. As all these results show, PFCs alter the transport and distribution of calcium in different cell models. We suspected that this could also affect the ion in spermatozoa, so we decided to use FITC-PNA to determine the status of the acrosome of our samples. As is well-known, the acrosome is a large secretory vesicle biochemically similar to a lysosome, and forms as a product of the Golgi apparatus [40]. It contains two membranes, one internal the other external, and one principle component is the sugar galactosyl β-1,3 N-acetyl galactosamine, which binds specifically to the peanut agglutinin conjugated with fluorescein isothiocyanate (FITC-PNA) that allows observations of acrosome integrity [41,42]. Using this method, we observed opposite effects of PFOS and PFHxS, as the spermatozoa with sAR were seen to decrease significantly when exposed to 1/2 CL_{50} of PFOS, but exposure to 1/5 of the LC_{50} of PFHxS increased this significantly. As mentioned above, PFOS is more toxic than PFHxS and may kill spermatozoa before they can perform the acrosome reaction. PFHxS in contrast, might take longer to kill cells, thus allowing them to complete sAR.

Boar spermatozoa are divided into 2 parts, the head and the flagellum. The latter is further divided into middle, principle and terminal sections. Signals received in the plasma membrane trigger activation of the signaling cascades for the hyperactivation of mobility. The tyrosine phosphorylation is essential for fertilizing the oocyte, since it is involved in spermatozoa hypermotility, the acrosome reaction, and gamete fusion [43,44].

Our analysis of tyrosine phosphorylation by flow cytometry did not produce significant differences under any conditions tested. However, analyses by fluorescence microscopy did reveal significant differences in proteins in the acrosome and equatorial zone of the spermatozoa head when incubated with the LC_{50} of PFHxS. In addition, incubation with the LC_{50} of PFOS and PFHxS caused significant inhibition in the pattern of full tyrosine phosphorylation (acrosome zone + head equatorial zone). These results seem to suggest that PFCs slow down the capacitation process by modifying tyrosine phosphorylation and inhibiting the change of EHZ and AZ to the AE pattern. This effect of PFCs on spermatozoa tyrosine phosphorylation has not been reported previously. Gao et al. (2017) [20] determined that PFOS damages the Sertoli cells by disturbing actin cytoskeleton by down-regulating p-Akt1-S473 and p-Akt2-S474. They added that SC79 (an Akt1/2 activator) blocked PFOS-induced Sertoli cell injuries by rescuing PFOS-induced F-actin disorganization. Qiu et al. (2016) [45] proposed that the target of PFOS is p38/ATF2, and that this is associated with an increase in phosphorylation in a dose- and time-dependent manner related to perturbations of the blood-testis barrier.

The effect of PFCs on ROS production and its cellular effect have been studied in, for example, SH-SY5Y neuroblast cells, where 25 μM of PFOS significantly generated ROS and caused neurotoxic effects [46]. In liver cells, PFOS caused cytotoxicity associated with ROS and lipid peroxidation with depletion of hepatocyte glutathione. Antioxidants and ROS scavengers inhibited this cytotoxicity [47]. PFHxS had apoptotic effects on cerebellar granule cells and increased the activation of ERK1/2, JNK, and p38 MAPK, but antioxidant treatment blocked these effects [48].

ROS are involved in regulating spermatozoa processes, as observations have shown that an increase in the concentration of these compounds can enhance spermatozoa capacitation, while adding antioxidant substances reduces capacitation [49]. In our model, both compounds significantly increased ROS after 30 min of exposure compared to controls. At 2 h of exposure, ROS decreased, but when analyzed with the LC_{50} of PFHxS the level was still significantly higher than in the control. Despite this increase, the boar spermatozoa exposed to both PFCs did not significantly increase in vitro spermatozoa capacitation. Although, ROS are necessary for sperm functions, some authors have shown that boar sperm is extremely sensitive to these compounds, and have related this to alterations in motility, acrosome integrity, and lipid peroxidation [50]. It may be that damage to the mitochondrial membrane by PFCs increases intracellular ROS levels and affects the membrane, proteins and sperm DNA [51].

As we observed, ROS production was considerable with both compounds. Studies have demonstrated that high ROS concentrations can cause DNA damage [52]. In the present case, we observed significant damage caused by exposure to both compounds, as PFHxS showed approximately 20%, and PFOS 10%, compared to controls, which presented only around 5%. Genomic damage may, therefore, be related to the increased production of ROS caused by the PFCs analyzed herein.

5. Conclusions

Perfluorinated compounds appear to damage sperm through different metabolic pathways. It is well known that increased ROS seems to be related to male infertility, and as mentioned above, boar sperm is susceptible to these compounds. Besides, the necessary processes for sperm capacitation, such as tyrosine phosphorylation, are strongly related to an optimal ROS concentration, so an alteration in this balance leads to damage or changes. Therefore, the alterations caused by perfluorinated in this process could cause a lack of sperm-oocyte recognition and prevent fertilization. Furthermore, the increase in ROS significantly damages DNA. The mobilization of intracellular calcium is another possible route of damage. PFHxS exposure affected the normal calcium patterns distribution, so it was impossible to determine capacitation and sAR using the CTC technique.

Author Contributions: Conceptualization, S.P.-C., R.F. and H.G.-M.; Data curation, I.O.-L., P.B.O.-S. and H.G.-M.; Formal analysis, I.O.-L., S.P.-C., R.S.-S., I.J.-M. and R.F.; Investigation, S.P.-C., P.B.O.-S., O.M.-P. and R.F.; Methodology, I.O.-L., P.B.O.-S. and O.M.-P.; Project administration, I.J.-M. and R.F.; Resources, S.P.-C., R.F. and H.G.-M.; Software, H.G.-M.; Supervision, S.P.-C. and H.G.-M.; Writing—original draft, I.O.-L.; Writing—review & editing, S.P.-C., P.B.O.-S., O.M.-P., R.S.-S., I.J.-M., R.F. and H.G.-M. All authors have read and agreed to the published version of the manuscript.

Funding: This research was funded by CONACyT, Mexico fellowships 283833 to I.O.-L., 593020 to P.B.O.-S., 940985 to O.M.-P.; Ramón y Cajal contract from the Spanish Ministry of Science and Innovation (RYC-2016-20147) to S.P.-C. This study was partially supported by CONACyT, Mexico, Grants 52877-Z/66953 and 0105961, and Spain, Grant RTI2018-096736-A-I00 from the Spanish Ministry of Science and Innovation.

Acknowledgments: We thank Alfonso Gutiérrez-Adán for his invaluable support during the mobility stay of I.O-L., Ernesto Gómez-Fidalgo for his assistance with the biological samples, and Sonia Heras and Leslye Sámano-Hernández for their critical review of this work. The present is dedicated to the memory of Serafín Pérez-Cerezales, an invaluable person, mentor and friend.

Conflicts of Interest: The authors declare that there is no interest conflict.

References

1. Petrovic, M.; Farré, M.; Eljarrat, E.; Diaz, S.; Barceló, D. Chapter 14—Environmental Analysis: Emerging Pollutants. In *Liquid Chromatography*; Fanali, S., Haddad, P.R., Poole, C.F., Schoenmakers, P., Lloyd, D., Eds.; Elsevier: Amsterdam, The Netherlands, 2013; pp. 389–410. [CrossRef]
2. Lau, C.; Anitole, K.; Hodes, C.; Lai, D.; Pfahles-Hutchens, A.; Seed, J. Perfluoroalkyl Acids: A Review of Monitoring and Toxicological Findings. *Toxicol. Sci.* **2007**, *99*, 366–394. [CrossRef] [PubMed]
3. Guerranti, C.; Ancora, S.; Bianchi, N.; Perra, G.; Fanello, E.L.; Corsolini, S.; Fossi, M.C.; Focardi, S.E. Perfluorinated compounds in blood of *Caretta caretta* from the Mediterranean Sea. *Mar. Pollut. Bull.* **2013**, *73*, 98–101. [CrossRef]
4. Kato, K.; Wong, L.-Y.; Basden, A.; Calafat, A.M. Effect of temperature and duration of storage on the stability of polyfluoroalkyl chemicals in human serum. *Chemosphere* **2012**, *91*, 115–117. [CrossRef]
5. Dorman, F.L.; Reiner, E.J. Chapter 28—Emerging and Persistent Environmental Compound Analysis. In *Gas Chromatography*; Poole, C.F., Ed.; Elsevier: Amsterdam, The Netherlands, 2012; pp. 647–677. [CrossRef]
6. Lindstrom, A.B.; Strynar, M.J.; Libelo, E.L. Polyfluorinated Compounds: Past, Present, and Future. *Environ. Sci. Technol.* **2011**, *45*, 7954–7961. [CrossRef]

7. Reiner, E.J.; Jobst, K.J.; Megson, D.; Dorman, F.L.; Focant, J.-F. Chapter 3—Analytical Methodology of POPs. In *Environmental Forensics for Persistent Organic Pollutants*; O'Sullivan, G., Sandau, C., Eds.; Elsevier: Amsterdam, The Netherlands, 2014; pp. 59–139. [CrossRef]
8. Olsen, G.W.; Burris, J.M.; Ehresman, D.J.; Froehlich, J.W.; Seacat, A.M.; Butenhoff, J.L.; Zobel, L.R. Half-Life of Serum Elimination of Perfluorooctanesulfonate, Perfluorohexanesulfonate, and Perfluorooctanoate in Retired Fluorochemical Production Workers. *Environ. Health Perspect.* **2007**, *115*, 1298–1305. [CrossRef] [PubMed]
9. Guruge, K.S.; Manage, P.M.; Yamanaka, N.; Miyazaki, S.; Taniyasu, S.; Yamashita, N. Species-specific concentrations of perfluoroalkyl contaminants in farm and pet animals in Japan. *Chemosphere* **2008**, *73*, S210–S215. [CrossRef] [PubMed]
10. Vestergren, R.; Orata, F.; Berger, U.; Cousins, I.T. Bioaccumulation of perfluoroalkyl acids in dairy cows in a naturally contaminated environment. *Environ. Sci. Pollut. Res.* **2013**, *20*, 7959–7969. [CrossRef]
11. Stahl, T.; Falk, S.; Failing, K.; Berger, J.; Georgii, S.; Brunn, H. Perfluorooctanoic Acid and Perfluorooctane Sulfonate in Liver and Muscle Tissue from Wild Boar in Hesse, Germany. *Arch. Environ. Contam. Toxicol.* **2011**, *62*, 696–703. [CrossRef] [PubMed]
12. Bassols, A.; Costa, C.; Eckersall, P.D.; Osada, J.; Sabrià, J.; Tibau, J. The pig as an animal model for human pathologies: A proteomics perspective. *Proteom. Clin. Appl.* **2014**, *8*, 715–731. [CrossRef]
13. Meurens, F.; Summerfield, A.; Nauwynck, H.; Saif, L.; Gerdts, V. The pig: A model for human infectious diseases. *Trends Microbiol.* **2012**, *20*, 50–57. [CrossRef]
14. Mordhorst, B.R.; Prather, R.S. Pig Models of Reproduction. In *Animal Models and Human Reproduction*; Constantinescu, G., Schatten, H., Eds.; John Wiley & Sons, Inc.: Hoboken, NJ, USA, 2017; pp. 213–234. [CrossRef]
15. Joensen, U.N.; Bossi, R.; Leffers, H.; Jensen, A.A.; Skakkebæk, N.E.; Jørgensen, N. Do Perfluoroalkyl Compounds Impair Human Semen Quality? *Environ. Health Perspect.* **2009**, *117*, 923–927. [CrossRef] [PubMed]
16. Joensen, U.N.; Veyrand, B.; Antignac, J.-P.; Jensen, M.B.; Petersen, J.H.; Marchand, P.; Skakkebæk, N.E.; Andersson, A.-M.; Le Bizec, B.; Jørgensen, N. PFOS (perfluorooctanesulfonate) in serum is negatively associated with testosterone levels, but not with semen quality, in healthy men. *Human Reprod.* **2012**, *28*, 599–608. [CrossRef]
17. Toft, G.; Jönsson, B.; Lindh, C.; Giwercman, A.; Spano, M.; Heederik, D.; Lenters, V.; Vermeulen, R.; Rylander, L.; Pedersen, H.; et al. Exposure to perfluorinated compounds and human semen quality in arctic and European populations. *Human Reprod.* **2012**, *27*, 2532–2540. [CrossRef] [PubMed]
18. Wan, H.T.; Zhao, Y.G.; Wong, M.H.; Lee, K.-F.; Yeung, W.S.; Giesy, J.P.; Wong, C.K.C. Testicular Signaling is the Potential Target of Perfluorooctanesulfonate-Mediated Subfertility in Male Mice1. *Biol. Reprod.* **2011**, *84*, 1016–1023. [CrossRef] [PubMed]
19. Wan, H.-T.; Mruk, D.D.; Wong, C.K.C.; Cheng, C.Y. Perfluorooctanesulfonate (PFOS) Perturbs Male Rat Sertoli Cell Blood-Testis Barrier Function by Affecting F-Actin Organization via p-FAK-Tyr407: An in Vitro Study. *Endocrinology* **2014**, *155*, 249–262. [CrossRef] [PubMed]
20. Gao, Y.; Chen, H.; Xiao, X.; Lui, W.Y.; Lee, W.M.; Mruk, D.D.; Cheng, C.Y. Perfluorooctanesulfonate (PFOS)-induced Sertoli cell injury through a disruption of F-actin and microtubule organization is mediated by Akt1/2. *Sci. Rep.* **2017**, *7*, 1110. [CrossRef] [PubMed]
21. Chen, J.; Wang, X.; Ge, X.; Wang, D.; Wang, T.; Zhang, L.; Tanguay, R.L.; Simonich, M.; Huang, C.; Dong, Q. Chronic perfluorooctanesulphonic acid (PFOS) exposure produces estrogenic effects in zebrafish. *Environ. Pollut.* **2016**, *218*, 702–708. [CrossRef]
22. Ickowicz, D.; Finkelstein, M.; Breitbart, H. Mechanism of sperm capacitation and the acrosome reaction: Role of protein kinases. *Asian J. Androl.* **2012**, *14*, 816–821. [CrossRef]
23. Meyers, S.A. Chapter 5—Sperm Physiology. In *Equine Breeding Management and Artificial Insemination*, 2nd ed.; Samper, J.C., Ed.; W.B. Saunders: Saint Louis, MO, USA, 2009; pp. 47–55. [CrossRef]
24. Alkmin, D.V.; Martinez-Alborcia, M.J.; Parrilla, I.; Vazquez, J.M.; Martinez, E.A.; Roca, J. The nuclear DNA longevity in cryopreserved boar spermatozoa assessed using the Sperm-Sus-Halomax. *Theriogenology* **2013**, *79*, 1294–1300. [CrossRef]
25. Yoshida, M.; Ishizaki, Y.; Kawagishi, H. Blastocyst formation by pig embryos resulting from in-vitro fertilization of oocytes matured in vitro. *J. Reprod. Fertil.* **1990**, *88*, 1–8. [CrossRef]

26. Maravilla-Galván, R.; Fierro, R.; González-Márquez, H.; Gomez-Arroyo, S.; Jiménez, I.; Betancourt, M. Effects of Atrazine and Fenoxaprop-Ethyl on Capacitation and the Acrosomal Reaction in Boar Sperm. *Int. J. Toxicol.* **2009**, *28*, 24–32. [CrossRef]
27. Ded, L.; Dostalova, P.; Zatecka, E.; Dorosh, A.; Dvořáková-Hortová, K.; Peknicova, J. Fluorescent analysis of boar sperm capacitation process in vitro. *Reprod. Biol. Endocrinol.* **2019**, *17*, 1–11. [CrossRef]
28. Fraser, L.R.; Herod, J.E. Expression of capacitation-dependent changes in chlortetracycline fluorescence patterns in mouse spermatozoa requires a suitable glycolysable substrate. *J. Reprod. Fertil.* **1990**, *88*, 611–621. [CrossRef]
29. Fàbrega, A.; Puigmulé, M.; Yeste, M.; Casas, I.; Bonet, S.; Pinart, E. Impact of epididymal maturation, ejaculation and in vitro capacitation on tyrosine phosphorylation patterns exhibited of boar (Sus domesticus) spermatozoa. *Theriogenology* **2011**, *76*, 1356–1366. [CrossRef] [PubMed]
30. Guthrie, H.D.; Welch, G.R. Determination of intracellular reactive oxygen species and high mitochondrial membrane potential in Percoll-treated viable boar sperm using fluorescence-activated flow cytometry. *J. Anim. Sci.* **2006**, *84*, 2089–2100. [CrossRef]
31. Pérez-Cerezales, S.; Laguna-Barraza, R.; De Castro, A.C.; Sánchez-Calabuig, M.J.; Cano-Oliva, E.; De Castro-Pita, F.J.; Montoro-Buils, L.; Pericuesta, E.; Fernández-González, R.; Gutiérrez-Adán, A. Sperm selection by thermotaxis improves ICSI outcome in mice. *Sci. Rep.* **2018**, *8*, 1–14. [CrossRef] [PubMed]
32. Końca, K.; Lankoff, A.; Banasik, A.; Lisowska, H.; Kuszewski, T.; Góźdź, S.; Koza, Z.; Wojcik, A. A cross-platform public domain PC image-analysis program for the comet assay. *Mutat. Res. Mol. Mech. Mutagen.* **2003**, *534*, 15–20. [CrossRef]
33. Brownlee, K.A.; Finney, D.J.; Tattersfield, F. Probit Analysis: A Statistical Treatment of the Sigmoid Response Curve. *J. Am. Stat. Assoc.* **1952**, *47*, 687. [CrossRef]
34. Louis, G.M.B.; Chen, Z.; Schisterman, E.F.; Kim, S.; Sweeney, A.M.; Sundaram, R.; Lynch, C.D.; Gore-Langton, R.E.; Barr, D.B. Perfluorochemicals and Human Semen Quality: The LIFE Study. *Environ. Health Perspect.* **2015**, *123*, 57–63. [CrossRef]
35. Slotkin, T.A.; MacKillop, E.A.; Melnick, R.L.; Thayer, K.A.; Seidler, F.J. Developmental Neurotoxicity of Perfluorinated Chemicals Modeled in Vitro. *Environ. Health Perspect.* **2008**, *116*, 716–722. [CrossRef]
36. Jiménez, I.; Fierro, R.; González-Márquez, H.; Mendoza-Hernández, G.; Romo, S.; Betancourt, M. Carbohydrate Affinity Chromatography Indicates that Arylsulfatase-A from Capacitated Boar Sperm has Mannose and N-Acetylglucosamine/Sialic Acid Residues. *Arch. Androl.* **2006**, *52*, 455–462. [CrossRef] [PubMed]
37. Ward, C.R.; Storey, B.T. Determination of the time course of capacitation in mouse spermatozoa using a chlortetracycline fluorescence assay. *Dev. Biol.* **1984**, *104*, 287–296. [CrossRef]
38. Pan, Y.; Qin, H.; Liu, W.; Zhang, Q.; Zheng, L.; Zhou, C.; Quan, X. Effects of chlorinated polyfluoroalkyl ether sulfonate in comparison with perfluoroalkyl acids on gene profiles and stemness in human mesenchymal stem cells. *Chemosphere* **2019**, *237*, 124402. [CrossRef]
39. Liao, C.-Y.; Li, X.-Y.; Wu, B.; Duan, S.; Jiang, G.-B. Acute Enhancement of Synaptic Transmission and Chronic Inhibition of Synaptogenesis Induced by Perfluorooctane Sulfonate through Mediation of Voltage-Dependent Calcium Channel. *Environ. Sci. Technol.* **2008**, *42*, 5335–5341. [CrossRef] [PubMed]
40. Wassarman, P.M. Fertilization, Mammalian. In *Encyclopedia of Genetics*; Brenner, S., Miller, J.H., Eds.; Academic Press: New York, NY, USA, 2001; pp. 690–695. [CrossRef]
41. Kopf, G.S. Acrosome. In *Encyclopedia of Genetics*; Brenner, S., Miller, J.H., Eds.; Academic Press: New York, NY, USA, 2001; pp. 3–4. [CrossRef]
42. Rajabi-Toustani, R.; Akter, Q.S.; Almadaly, E.A.; Hoshino, Y.; Adachi, H.; Mukoujima, K.; Murase, T. Methodological improvement of fluorescein isothiocyanate peanut agglutinin (FITC-PNA) acrosomal integrity staining for frozen-thawed Japanese Black bull spermatozoa. *J. Veter. Med Sci.* **2019**, *81*, 694–702. [CrossRef] [PubMed]
43. Naz, R.K.; Rajesh, P.B. Role of tyrosine phosphorylation in sperm capacitation/acrosome reaction. *Reprod. Biol. Endocrinol.* **2004**, *2*, 75. [CrossRef]
44. Jones, R.; James, P.S.; Oxley, D.; Coadwell, J.; Suzuki-Toyota, F.; Howes, E.A. The Equatorial Subsegment in Mammalian Spermatozoa Is Enriched in Tyrosine Phosphorylated Proteins. *Biol. Reprod.* **2008**, *79*, 421–431. [CrossRef]

45. Qiu, L.; Qian, Y.; Liu, Z.; Wang, C.; Qu, J.; Wang, X.; Wang, S. Perfluorooctane sulfonate (PFOS) disrupts blood-testis barrier by down-regulating junction proteins via p38 MAPK/ATF2/MMP9 signaling pathway. *Toxicology* **2016**, *373*, 1–12. [CrossRef]
46. Chen, N.; Li, J.; Li, D.; Yang, Y.; He, D. Chronic Exposure to Perfluorooctane Sulfonate Induces Behavior Defects and Neurotoxicity through Oxidative Damages, In Vivo and In Vitro. *PLoS ONE* **2014**, *9*, e113453. [CrossRef]
47. Khansari, M.R.; Yousefsani, B.S.; Kobarfard, F.; Faizi, M.; Pourahmad, J. In vitro toxicity of perfluorooctane sulfonate on rat liver hepatocytes: Probability of distructive binding to CYP 2E1 and involvement of cellular proteolysis. *Environ. Sci. Pollut. Res.* **2017**, *24*, 23382–23388. [CrossRef]
48. Lee, Y.J.; Choi, S.-Y.; Yang, J.-H. PFHxS induces apoptosis of neuronal cells via ERK1/2-mediated pathway. *Chemosphere* **2014**, *94*, 121–127. [CrossRef] [PubMed]
49. Ford, W. Regulation of sperm function by reactive oxygen species. *Human Reprod. Updat.* **2004**, *10*, 387–399. [CrossRef]
50. Awda, B.J.; Mackenzie-Bell, M.; Buhr, M. Reactive Oxygen Species and Boar Sperm Function1. *Biol. Reprod.* **2009**, *81*, 553–561. [CrossRef] [PubMed]
51. Cocuzza, M.; Sikka, S.C.; Athayde, K.S.; Agarwal, A. Clinical relevance of oxidative stress and sperm chromatin damage in male infertility: An evidence based analysis. *Int. Braz. J. Urol.* **2007**, *33*, 603–621. [CrossRef] [PubMed]
52. Kothari, S.; Thompson, A.; Agarwal, A.; Du Plessis, S.S. Free radicals: Their beneficial and detrimental effects on sperm function. *Indian J. Exp. Boil.* **2010**, *48*, 425–435.

Article

The Presence of D-Penicillamine during the In Vitro Capacitation of Stallion Spermatozoa Prolongs Hyperactive-Like Motility and Allows for Sperm Selection by Thermotaxis

Sara Ruiz-Díaz [1,2,†], Ivan Oseguera-López [3,†], David De La Cuesta-Díaz [1], Belén García-López [1], Consuelo Serres [4], Maria José Sanchez-Calabuig [4], Alfonso Gutiérrez-Adán [1,*] and Serafin Perez-Cerezales [1]

1 Department of Animal Reproduction, National Institute for Agriculture and Food Research and Technology (INIA), 28040 Madrid, Spain; sruizd@clinicatambre.com (S.R.-D.); davedlcd@gmail.com (D.D.L.C.-D.); belen.garcia@inia.es (B.G.-L.); serafin.perez@inia.es (S.P.-C.)
2 Mistral Fertility Clinics S.L., Clínica Tambre, 28002 Madrid, Spain
3 Unidad Iztapalapa, Universidad Autónoma Metropolitana, Ciudad de México 09340, Mexico; ivanoslo@yahoo.com.mx
4 Departamento de Medicina y Cirugía Animal, Facultad de Veterinaria, Universidad Complutense de Madrid (UCM), 28040 Madrid, Spain; cserres@ucm.es (C.S.); msanch26@ucm.es (M.J.S.-C.)
* Correspondence: agutierr@inia.es
† These authors contributed equally.

Received: 9 July 2020; Accepted: 18 August 2020; Published: 21 August 2020

Simple Summary: Capacitation of stallion semen in vitro is still a suboptimal procedure. The main objective of this study was the use of thermotaxis as a novel method for sperm selection and determining the most adequate media for maintaining frozen/thawed horse sperm longevity in vitro. Our results show that the most common media (Whitten's) used in this species is not the best for capacitating the semen in terms of hyperactive-like motility, and tyrosine phosphorylation being synthetic human tubal fluid supplemented with D-penicillamine is the most adequate in preserving these parameters during 180 min of incubation. Therefore, this media (with and without D-penicillamine) was chosen for performing thermotaxis. The selection conditions were a gradient of 3 °C of difference (35–38 °C) for 1 h. The results revealed that the selected fraction showed higher levels of tyrosine phosphorylation in the whole flagellum and lower levels of DNA fragmentation when compared to the unselected fraction (kept at 37 °C) when human tubal fluid with D-penicillamine was used. These results are promising for improving the in vitro embryo production rates in these species by improving the sperm selection methodology.

Abstract: Assisted reproductive technologies (ARTs) in the horse still yield suboptimal results in terms of pregnancy rates. One of the reasons for this is the lack of optimal conditions for the sperm capacitation in vitro. This study assesses the use of synthetic human tubal fluid (HTF) supplemented with D-penicillamine (HTF + PEN) for the in vitro capacitation of frozen/thawed stallion spermatozoa by examining capacitation-related events over 180 min of incubation. Besides these events, we explored the in vitro capacity of the spermatozoa to migrate by thermotaxis and give rise to a population of high-quality spermatozoa. We found that HTF induced higher levels of hyperactive-like motility and protein tyrosine phosphorylation (PTP) compared to the use of a medium commonly used in this species (Whitten's). Also, HTF + PEN was able to maintain this hyperactive-like motility, otherwise lost in the absence of PEN, for 180 min, and also allowed for sperm selection by thermotaxis in vitro. Remarkably, the selected fraction was enriched in spermatozoa showing PTP along the whole flagellum and lower levels of DNA fragmentation when compared to the unselected fraction (38% ± 11% vs 4.4% ± 1.1% and 4.2% ± 0.4% vs 11% ± 2% respectively, *t*-test $p < 0.003$, $n = 6$). This procedure of in vitro capacitation of frozen/thawed stallion spermatozoa in HTF + PEN followed by in vitro sperm selection by thermotaxis represents a promising sperm preparation strategy for in vitro fertilization and intracytoplasmic sperm injection in this species.

Keywords: stallion sperm; capacitation; penicillamine; thermotaxis; selection

1. Introduction

Despite nearly two decades of efforts, in vitro fertilization (IVF) in the horse remains unavailable, and intracytoplasmic sperm injection (ICSI) in this species still yields suboptimal results. One of the main causes of these limitations are suboptimal in vitro conditions for sperm capacitation, preventing successful IVF [1] and low ICSI outcomes [2].

Synthetic media successfully used for the in vitro capacitation of sperm in other mammalian species are also being tested in the stallion. Thus, reports exist of capacitation-related events that occur in equine spermatozoa under different incubation conditions employing the media Biggers–Whitten–Whittingham (BWW), Tyrode's, Whitten's, or human tubal fluid [3–8]. However, the available data indicate that using these media, spermatozoa are not fully capacitated, with the consequence that in vitro fertilization remains elusive in this species [1]. One of the main events occurring during sperm capacitation in mammals is the phosphorylation of multiple proteins at their tyrosine residues [9]. Specifically, protein tyrosine phosphorylation (PTP) in the sperm flagellum has been related to the hyperactive motility of sperm, and both these factors are considered hallmarks of mammalian sperm capacitation [10,11]. However, while PTP in the equine spermatozoon flagellum is elevated under different capacitating conditions in vitro, so far all attempts to establish its relationship with hyperactive motility have had limited success [12]. Romero-Aguirregomezcorta [13] showed that the type of hyperactivated motion induced in vitro by species-specific hyperactivation agonists significantly differed in stallion in relation to human or ram spermatozoa, and led to the absence of a rheotactic response in stallion sperm. As an explanation, the authors suggested a different role of sperm hyperactivation in the horse. However, we propose here that it could be the suboptimal in vitro conditions for stallion spermatozoa that prevent a true or complete hyperactivation response.

The proportions of mammalian spermatozoa that acquire a capacitated status at a given time point can be low (roughly around 10%) [14–16]. Further, in humans, this capacitated state of spermatozoa is transient (>50 min to <4 h) and occurs only once in a sperm's lifespan [14]. Harrison [17] described capacitation as a series of positive destabilizing events that eventually lead to sperm death. Later on, Aitken et al. [18] related the physiological production of reactive oxygen species (ROS) to capacitation, followed by apoptosis and sperm senescence. The concept of sperm death as a consequence of capacitation implies that too-high or too-rapid induction of capacitation can shorten the lifespan to the extent that fertilization is prevented [19]. Aitken et al. [20] suggested that the reduced lifespan of spermatozoa incubated in vitro is the outcome of an excessive production of free radicals that eventually provokes lipid peroxidation and the generation of electrophilic cytotoxic aldehydes. To counteract this effect, these authors showed that the addition of penicillamine (PEN), a molecule that neutralizes these aldehydes and slows down their production, improved the maintenance of motility of fresh equine spermatozoa using a non-capacitating medium. In earlier work, Pavlok [21] established that the addition of PEN to the capacitating medium significantly prolonged the lifespan of frozen/thawed bovine spermatozoa, maintaining their fertilization ability for at least 8 h. Accordingly, we hypothesized that capacitated stallion spermatozoa incubated under in vitro conditions for capacitation acquire PTP in their flagella, rapidly losing their viability. This means that physiological hyperactivation is prevented or defective, at least for a short time, and as a consequence, fertilizing ability is lost. This effect could be even more pronounced in some horses because of the initial low quality of their semen [2]. To test this hypothesis, herein we examined the effect of supplementing the capacitating medium with PEN to prolong the lifespan of capacitated equine spermatozoa.

Recently, we proposed that capacitated spermatozoa could be selected from the whole pool of spermatozoa by an in vitro thermotaxis assay [22]. Sperm thermotaxis, defined as the ability of spermatozoa to navigate within a temperature gradient towards the warmer temperature, seems to be exclusive to capacitated spermatozoa [23,24]. In recent work, we found that the DNA of human and mouse spermatozoa selected by thermotaxis in vitro is of very high integrity when compared to the DNA of unselected spermatozoa [22]. Further, the use of these selected spermatozoa significantly increased successful ICSI outcomes in mice. For selection by thermotaxis, capacitated spermatozoa need to preserve their motility during migration within the temperature gradient. This study aimed to evaluate if PEN could prolong the lifespan of frozen/thawed sperm in different capacitating media, thus allowing spermatozoa migration by thermotaxis. Also, this selection could serve to obtain a sperm fraction of high genetic quality for its use in both IVF and ICSI.

The objectives of this study were: (i) to assess the effect of penicillamine supplementation on capacitation-related events during incubation under capacitating conditions in frozen/thawed stallion spermatozoa, (ii) to examine the capacity of these spermatozoa to migrate by thermotaxis, and (iii) to determine DNA integrity and tyrosine phosphorylation in the spermatozoa selected by thermotaxis.

2. Materials and Methods

2.1. Reagents

All reagents were purchased from Sigma–Aldrich (Saint Louis, MO, USA) unless specified otherwise.

2.2. Experimental Design

In an initial experiment, we examined the effect of incubating frozen/thawed stallion spermatozoa processed by density gradient centrifugation (DGC) in Whitten's medium (WHI) and synthetic human tubal fluid (HTF), two media commonly used for the capacitation of mammalian spermatozoa. Because in our preliminary experiments HTF induced more signs of capacitation (confirmed in the experiments shown here), we supplemented it with 750 μM of penicillamine (HTF + PEN), as this concentration has been shown to prolong sperm motility in the horse [20]. Over an incubation period of 180 min, we evaluated the sperm integrity and capacitation, analyzing the plasma membrane integrity, acrosomal exocytosis, protein tyrosine phosphorylation, and motility (total motility and motion kinetics). The percentage of motile spermatozoa was determined after DGC (time 0) and at 30 and 180 min of incubation. Plasma membrane integrity, acrosomal exocytosis, and protein tyrosine phosphorylation were analyzed at time 0 and after 180 min of incubation.

After 30 min of incubation, HTF and HTF + PEN induced higher capacitation levels than WHI. Therefore, in the second experiment, spermatozoa were selected by thermotaxis after 30 min of incubation employing an in vitro system previously used in mouse and human sperm [22]. This selection was conducted for 60 min using a gradient from 35 to 38 °C (see the section below for details). Next, we determined the percentage of migration by thermotaxis. As migration by thermotaxis was only achieved using HTF + PEN, we analyzed in these samples the PTP of the migrating spermatozoa, non-migrating spermatozoa (those that did not migrate in the in vitro system), and unselected spermatozoa (aliquot incubated in parallel for 90 min at 37 °C in 5% CO_2). In addition, DNA fragmentation was examined in the migrating and unselected spermatozoa.

2.3. Semen Collection and Cryopreservation

Semen was collected from six fertile purebred Lusitano stallions aged 3 to 13 years housed at the Centro de Selección y Reproducción Animal (CENSYRA) using an artificial vagina (Hannover model, Minitüb, Landshut, Germany). All experimental procedures were performed according to institutional and European regulations. A nylon in-line filter (Animal Reproduction Systems, Chino, CA, USA) was used to eliminate the gel fraction. The sperm-rich fraction was diluted 1:2 (*v:v*) in INRA96 medium (IMV, L'Aigle, France) and subsequently processed for cryopreservation. Diluted ejaculates were centrifuged for 10 min at 900× *g* and the supernatant discarded. The sperm pellet was re-suspended in an egg yolk-based freezing extender (Gent, Minitube Ibérica, Tarragona, Spain) to obtain a final concentration of 200 × 10^6 sperm/mL, loaded into straws (0.5 mL) and sealed using sealing balls. Subsequently, the straws were equilibrated for 20 min at 4 °C and frozen by exposure to liquid nitrogen vapor at 4 cm above the liquid nitrogen level for 20 min. At the end of the cryopreservation process, the straws were submerged into liquid nitrogen at −196 °C where they were stored until analysis.

2.4. Sperm Sample Preparation and Incubation

Cryopreserved samples were thawed at 37 °C for 45 s in a water bath and processed by DGC. The contents of two straws were recovered into a microtube and transferred to a 15 mL centrifuge tube on top of 500 µL of equipure ™ (Nidacon, Mölndal, Sweden) and centrifuged for 20 min at 400× *g*. Next, the supernatant was discarded and each pellet was resuspended in one of the following media: (i) WHI (100 mM NaCl, 4.7 mM KCl, 4.8 mM L-lactic acid hemicalcium salt, 1.2 mM $MgCl_2 \times 6H_2O$, 5.5 mM glucose, 22 mM HEPES, and 1.0 mM pyruvic acid), (ii) HTF (2.04 mM $CaCl_2 \times 2H_2O$, 101.6 mM NaCl, 4.69 mM KCl, 0.37 mM KH_2PO_4, 0.2 mM $MgSO_4 \times 7H_2O$, 21.4 mM sodium lactate, 0.33 mM sodium pyruvate, and 2.78 mM glucose), or (iii) HTF supplemented with 750 µM of penicillamine (HTF + PEN). All media were supplemented with 25 mM $NaHCO_3$, 4 mg/mL of bovine serum albumin (BSA), 100 U/mL penicillin, 50 µg/mL streptomycin SO_4, and 0.001% (*w/v*) phenol red (pH = 7.4 and 280–300 mOsm/kg). Before their use, the media were preincubated overnight at 37 °C in a 5% CO_2 humidified atmosphere. The sperm concentration was adjusted to 20 × 10^6 spermatozoa/mL and samples were incubated for 3 h at 37 °C in a 5% CO_2 humidifying atmosphere for capacitation. During the 3 h of incubation, pH was monitored and confirmed stable at 7.4.

2.5. Sperm Thermotaxis

Sperm thermotaxis was conducted as described elsewhere [22]. Briefly, our thermotaxis selection assay is based on recovering the spermatozoa who have migrated through a capillary between two drops of the same medium (in this experiment, HTF or HTF + PEN). For selection under thermotactic conditions, a 3 °C temperature gradient was set up between both drops from 35 to 38 °C. Between 5 and 6 × 10^6 spermatozoa were loaded into the 35 °C drop and allowed to migrate for 1 h. After this time, migrated spermatozoa were recovered from the 38 °C drops and processed for tyrosine phosphorylation or DNA fragmentation analysis. As controls for random migration, two drops were placed in parallel to the thermotactic assay at the same temperature (35 to 35 °C and 38 to 38 °C, non-gradient controls). The percentage of net thermotaxis was calculated as follows: 100 × (number of spermatozoa migrating within the temperature gradient (35 to 38 °C) minus number of spermatozoa migrating within the temperature non-gradient (35 or 38 °C, selecting the temperature which resulted in higher random migration)/number of spermatozoa loaded].

2.6. Plasma Membrane Integrity

We employed propidium iodide (PI) to stain spermatozoa with damaged membranes. Sperm plasma membrane integrity was assessed using propidium iodide (PI) to stain spermatozoa with damaged membrane [25] and the fluorochrome Hoechst 33342 to stain the nuclei. Semen samples were diluted into PBS at a concentration of 2 × 10^6 spermatozoa/mL, then PI and Hoechst 33342 were added to a final concentration of 10 and 15 µM, respectively. After 5 min, the stained samples were analyzed by flow cytometry in a FASCanto II flow cytometer (BD Biosciences, San Jose, CA, USA). Spermatozoa were gathered in the forward scatter and side scatter (FSC/SSC) dot plot to exclude debris and confirmed with the violet laser (405 nm) and the blue filter (450/50 nm) to detect nuclear staining with Hoechst 33342. A total of 1 × 10^4 spermatozoa were acquired per determination. For PI, the blue laser (488 nm) and the orange filter (585/42 nm) were used. Acquired data were analyzed using FlowJo software (Becton–Dickinson, Franklin Lakes, NJ, USA) to determine the percentage of PI-stained spermatozoa per each sample.

2.7. Acrosomal Exocytosis

The method employed was based on acrosome staining using *Arachis hypogaea* (peanut) lectin conjugated with fluorescein isothiocyanate (PNA–FITC) following a standard protocol described previously [26], with minor modifications. Briefly, the spermatozoa were washed twice in fhosphate-buffered saline (PBS) by centrifugation (1 min at 500× *g*) and subsequently smeared on a microscope glass slide and air-dried on a heat plate at 37 °C. Next, the slides were immersed in absolute methanol for 30 s, air-dried, and rinsed in PBS twice for 5 min before incubation with PNA–FITC and Hoechst 33342 (15 μg/mL and 0.0065 mg/mL respectively, in H_2O) in a wet-mount box/humidified box for 30 min at room temperature. Finally, the slides were washed with distilled water for 15 min and mounted with Fluoromount ™ aqueous mounting medium. Slides were examined in a fluorescence microscope (Nikon Eclipse i50, Nikon, Tokyo, Japan) and numbers of acrosome-reacted and non-acrosome-reacted spermatozoa were counted by randomly moving across different fields of the slide (counting 200 cells per slide, 2 slides per sample).

2.8. Motility and Kinetics

Ten microliters of sperm suspension was placed in a Mackler chamber on the stage heated to 37 °C of a Nikon Eclipse E400 (Nikon, Tokyo, Japan) fitted with a digital camera, Basler acA1300-200uc (Basler AG, Ahrensburg, Germany). Three to five movies of 1.5 s were recorded at 60 frames/s using the software Pylon Viewer provided by Basler, capturing at least 100 moving spermatozoa. The motility and sperm kinetics were analyzed using the free software ImageJ 1.x [27] with the plugin CASA_bmg following instructions for analyzing stallion spermatozoa [28]. The parameters analyzed were as described by Mortimer et al. [29]: straight-line velocity (VSL; μm/s), curvilinear velocity (VCL; μm/s), average path velocity (VAP; μm/s), linearity (LIN) (defined as (VSL/VCL) × 100), straightness (STR) (defined as (VSL/ VAP) × 100), wobble (WOB) (defined as (VAP/VCL) × 100), amplitude of lateral head (ALH) displacement (μm), and beat-cross frequency (BCF; Hz). Also, we examined the percentage of spermatozoa showing more signs of hyperactivation (HYP) by determining out of all the analyzed spermatozoa (9340) the lower VCL and ALH values of the 10% of spermatozoa with the highest VCL and ALH. These values were: VCL = 150 μm/s and ALH = 5.5 μm. Thus, we defined spermatozoa showing hyperactive-like motility as those showing VCL > 150 μm/s and ALH > 5.5 μm (following the definition used in Su et al. [30]).

2.9. DNA Fragmentation

DNA fragmentation was analyzed employing the neutral version of the single cell gel electrophoresis assay (SCGE or Comet assay), as described previously [22]. Briefly, the samples were pelleted by centrifugation (600× *g*) and diluted to a maximum of 20 × 10^4 spermatozoa/mL in 0.5% low melting point agarose in PBS. Because of the low numbers obtained in the thermotaxis assay, the samples of migrated spermatozoa were used entirely. Immediately after dilution, 85 μL were placed on a slide previously coated with 1% agarose and covered with a 22 × 22 mm coverslip. The slides were then left in a wet-mount box/humidified box at 4 °C for 1 h for agarose polymerization. After removing the coverslips, slides were incubated at 37 °C for 1 h in lysis solution (2 M NaCl, 55 mM EDTA-Na_2, 8 mM Tris, 4% Triton X-100, 0.1% sodium dodecyl sulfate (SDS), 1 mM ditiotreitol (DTT), and 0.5 mg/mL of proteinase K, pH 8). Next, the slides were washed twice in neutral electrophoresis solution (90 mM Tris, 90 mM boric acid, and 2 mM EDTA, pH 8.5) and subjected to electrophoresis (25 V for 10 min). The slides were then washed with distilled water, fixed in methanol for 3 min, air-dried, and stored upon microscope examination. The samples were stained with 50 μL of 0.02 mg/mL ethidium bromide, covered with a 22 × 22 mm coverslip, and immediately observed in a fluorescence microscope Nikon Optiphot-2 (Nikon, Tokyo, Japan). Comets were digitalized with a Nikon 5100 digital camera (Nikon, Tokyo, Japan) coupled to the microscope. At least 150 comets were analyzed using the free software Casplab 1.2.3beta2 (CaspLab.com) [31].

2.10. Protein Tyrosine Phosphorylation

Protein tyrosine phosphorylation was analyzed by immunofluorescence. Spermatozoa were diluted in 500 μL of PBS to a concentration of 4×10^6 spermatozoa/mL. Due to the low numbers obtained in the thermotaxis assay, the samples of migrated spermatozoa were used entirely and undiluted. Samples were centrifuged (600× *g* for 5 min) and the resultant pellet was fixed in 2% paraformaldehyde in PBS for 10 min and stored at −20 °C for, at most, one week, until continuing with immunodetection. After defrosting at room temperature, the fixed samples were washed 3 times in PBS by centrifugation (600× *g* for 5 min) and the pellet was resuspended in 50 μL of PBS. Two drops of 25 μL were each smeared on a glass microscope slide and left to dry. Subsequently, slides were washed three times with PBS and 100 μL of PBS with 0.2% of Triton-X were placed on each slide and covered with a 20 × 60 mm coverslip, placed in a wet box, and incubated at 37 °C for 15 min. Then, slides were washed once in PBS and incubated in a wet box with 100 μL of PBS and 1% BSA (again using a coverslip) for 1 h at 37 °C. Subsequently, slides were drained, and 100 μL of the primary antibody (phosphor-tyrosine monoclonal antibody (pY20), reference 14-5001-82, ThermoFisher Scientific, Waltham, MA, USA) diluted 1:100 in PBS was added to each slide, covered with a coverslip, and incubated overnight at 4 °C. On the next day, slides were washed three times in PBS and the secondary antibody (goat anti-mouse IgG (H + L) highly cross-adsorbed secondary antibody, Alexa Fluor 488, reference A-11029, ThermoFisher Scientific, Waltham, MA, USA) was added (100 μL of a 1:100 dilution in PBS and covered with a coverslip) and incubated for 1 h in a wet box at 37 °C. Slides were then washed three times in PBS and nuclei-counterstained with 100 μL of 15 μM Hoechst 33342 by incubating for 5 min in a wet box. After an additional wash in PBS, the slides were mounted with Fluoromount ™ aqueous mounting medium and examined in a fluorescence microscope (Nikon Eclipse i50, Nikon, Tokyo, Japan). Numbers of tyrosine-phosphorylated spermatozoa were counted by randomly moving across different fields of the slide (counting 200 cells per slide, 2 slides per sample).

2.11. Statistical Analysis

Statistical analysis was carried out using the software package GraphPad Prism 8.0.2 for Windows (GraphPad Software, San Diego, CA, USA). Results are expressed as means ± standard error of the mean (SEM). Means were compared and analyzed using a one-tailed paired-sample Student's *t*-test or repeated measures one-way analysis of variance (ANOVA), followed by Tukey's post hoc test. Significance was set at $p < 0.05$.

3. Results

We employed frozen/thawed spermatozoa from 6 stallions that were prepared by density gradient centrifugation and incubated for 180 min in WHI, HTF, or HTF + PEN. To assess sample integrity, we determined percentage motility, plasma membrane integrity, and the occurrence of acrosomal exocytosis before and during incubation. The percentage of motile spermatozoa was similar between the three studied media at every time point during incubation, with a significant decrease detected in the initial 30 min (Figure 1A) ($p < 0.002$). From 30 to 180 min of incubation, motility slightly diminished, though not significantly ($p > 0.38$). Incubation with the three media for 180 min also provoked a significant reduction in the percentage of spermatozoa showing an intact plasma membrane (Figure 1B) and a significant increase in spermatozoa undergoing acrosomal exocytosis (Figure 1C) ($p < 0.0001$ and $p < 0.0019$, respectively).

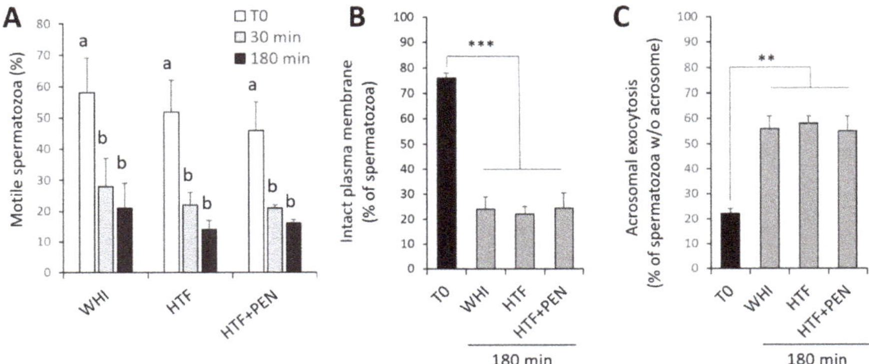

Figure 1. Sperm integrity over a 180 min period of capacitation. (**A**) Percentage of motile spermatozoa. (**B**) Percentage of spermatozoa with an intact plasma membrane as confirmed by propidium iodide staining. (**C**) Percentage of spermatozoa showing acrosomal exocytosis as determined by *Arachis hypogaea* (peanut) lectin conjugated with fluorescein isothiocyanate (PNA-FITC) staining. Spermatozoa after density gradient centrifugation (T0) and incubation for 30 or 180 min under capacitating conditions in three media: Whitten's medium (WHI), synthetic human tubal fluid (HTF), and HTF supplemented with 750 μM of penicillamine (HTF + PEN). *** $p < 0.0001$, ** $p < 0.0019$; [a,b] different letters indicate significant differences ($p < 0.05$), ($n = 6$, 12 determinations).

3.1. Effects of the Incubation Medium on Sperm Kinetics and Protein Tyrosine Phosphorylation

To determine which of the media, WHI or HTF, could potentially induce more spermatozoa to acquire the capacitation state, we conducted a comparative analysis of sperm kinetics and protein tyrosine phosphorylation over the 180 min of incubation. At the onset of incubation, VCL and ALH of swimming spermatozoa were higher in the HTF medium than WHI ($p = 0.045$ and $p = 0.04$, respectively) (Table 1). Thus, in HTF, we detected a higher fraction of spermatozoa showing a motility classified as indicating hyperactivation (relatively high VCL (>150 μm/s) and ALH (>5.5 μm), hereafter referred to as hyperactive-like motility) ($p = 0.02$) (Figure 2). Moreover, sperm kinetics in WHI did not significantly vary during the 180 min of incubation, while in HTF, sperm gradually acquired a less progressive motility type (LIN and STR reduced after 30 min ($p = 0.04$ and $p = 0.01$, respectively), as well as a decrease in WOB and beat cross frequency detected after 180 min ($p = 0.04$ and $p = 0.014$). These motion changes produced during incubation in HTF were recorded as a significant increase in the percentage of spermatozoa showing hyperactive-like motility after 30 min of incubation ($p = 0.045$) (Figure 2). However, after 180 min, hyperactive-like motility returned to the levels observed at the start of incubation.

Table 1. Kinetics of stallion spermatozoa measured at time 0 (T0), and after 30 and 180 min of incubation at 37°C in a 5% CO_2 atmosphere in Whitten's medium (WHI), synthetic human tubal fluid (HTF), and HTF supplemented with 750 μM of penicillamine (HTF + PEN).

Kinetics	T0			30 min			180 min		
	WHI	HTF	HTF + PEN	WHI	HTF	HTF + PEN	WHI	HTF	HTF + PEN
VCL (μm/s)	80 ± 6	96 ± 5 *	97 ± 2 a *	87 ± 7	109 ± 7 *	114 ± 5 b *	81 ± 7	107 ± 4 *	122 ± 3 b **
VAP (μm/s)	36 ± 1	40 ± 2	40 ± 0.8 a	37 ± 1	44 ± 3	45 ± 2 b	35 ± 2	41 ± 2	47 ± 1 b *
VSL (μm/s)	28 ± 2	30 ± 1	29 ± 1	27 ± 1	26 ± 3	27 ± 1	24 ± 2	25 ± 2	25 ± 2
LIN (%)	37 ± 4	33 ± 2 a	31 ± 2 a	34 ± 2	25 ± 3 b	25 ± 1 b	30 ± 1	24 ± 2 b *	21 ± 1 c *
STR (%)	78 ± 4	75 ± 3 a	71 ± 3 a	74 ± 2	59 ± 7 b *	61 ± 3 b *	64 ± 4	61 ± 4 b	55 ± 3 b *
WOB (%)	46 ± 2	43 ± 1 a	43 ± 1 a	44 ± 3	41 ± 1 a	40 ± 1 a b	46 ± 2	39 ± 1 b	39 ± 1 b
Beat Cross (Hz)	24.2 ± 1.4	24 ± 0.5 a	23.7 ± 0.5	25.3 ± 1	26 ± 0.5 a b	24.2 ± 0.7	26.1 ± 0.9	27 ± 1 b	24.7 ± 1.1
ALH (μm)	3 ± 0.2	3.5 ± 0.2 *	3.5 ± 0.1 a *	3.2 ± 0.2	3.9 ± 0.3 *	4.1 ± 0.2 b *	2.9 ± 0.2	3.8 ± 0.2 *	4.4 ± 0.1 b **

Curvilinear velocity (VCL), average path velocity (VAP), straight line velocity (VSL), linearity (LIN), straightness (STR), wobble (WOB), and amplitude of lateral head displacement (ALH). [a, b] Different letters indicate significant differences between time points for each medium. Asterisks indicate significant differences between media for each time point ($p < 0.05$, repeated measures one-way analysis of variance (ANOVA)).

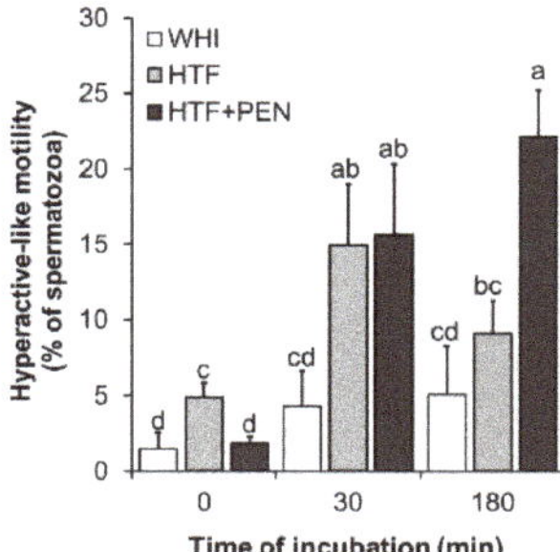

Figure 2. Percentage of spermatozoa showing hyperactive-like motility. Kinetics were examined at the time points 0, 30, and 180 min of incubation in Whitten's medium (WHI), synthetic human tubal fluid (HTF), or HTF supplemented with 750 μM of penicillamine (HTF + PEN). Spermatozoa showing hyperactive-like motility were defined as those showing VCL > 150 and ALH > 5.5. $^{a, b, c, d}$ Different letters indicate significant differences ($p < 0.05$, $n = 6$, 12 determinations).

Immunofluorescence PTP analyses revealed two staining patterns of the flagella: (i) staining showing PTP only in the midpiece (pattern I) and (ii) staining showing PTP along the whole flagellum (pattern II) (Figure 3A). The percentage of spermatozoa showing either staining pattern increased during incubation in the two media (Figure 3B,C). This increase was significantly higher for pattern I when the incubation medium was HTF rather than WHI ($p = 0.027$) (Figure 3B). No significant differences in pattern II emerged between HTF and WHI ($p = 0.1$) (Figure 3C).

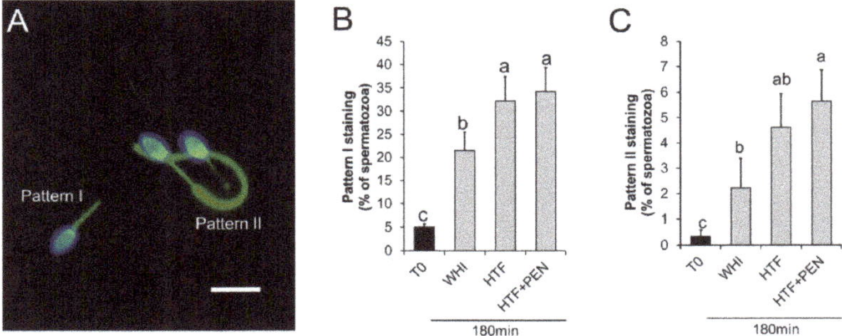

Figure 3. Protein tyrosine phosphorylation in stallion spermatozoa after 180 min of incubation for capacitation. (**A**) Micrograph of fluorescence microscopy of stallion spermatozoa after 180 min of capacitation and immune-labelled for protein tyrosine phosphorylation (PTP) (green). Nuclei were labelled with Hoechst 33342 (blue), bar = 10 μm. Pattern I: spermatozoa showing PTP at the midpiece. Pattern II: Spermatozoa showing PTP along the whole flagellum. (**B**,**C**) Percentage of spermatozoa showing pattern I (B) and pattern II (**C**) staining at time 0 and after 180 min of incubation in Whitten's medium (WHI), synthetic human tubal fluid (HTF), or HTF supplemented with 750 μM of penicillamine (HTF + PEN). $^{a, b, c}$ Different letters indicate significant differences ($p < 0.05$, $n = 6$, 12 determinations).

3.2. Effects of HTF Supplementation with Penicillamine on Sperm Kinetics and Protein Tyrosine Phosphorylation

To examine the effect of PEN on capacitated spermatozoa, it was added to HTF at a final concentration of 750 μM [20] and sperm kinetics and PTP were analyzed over the 180 min of incubation. At the start of incubation, we detected no significant differences in kinetics for HTF versus HTF + PEN (Table 1). However, at this early stage, the percentage of spermatozoa showing hyperactive-like motility was higher for HTF ($p = 0.01$). Interestingly, after 30 min of incubation, kinetics and hyperactive-like motility were similar in both media, but after 180 min, the spermatozoa incubated in HTF + PEN swam with significantly higher VCL ($p = 0.03$), VAP ($p = 0.033$), and ALH ($p = 0.04$), and lower STR ($p = 0.02$). Accordingly, the percentage of spermatozoa showing hyperactive-like motility after 180 min of incubation was higher in HTF + PEN ($p = 0.004$) (Figure 2). Our immunofluorescence analyses, nevertheless, revealed no differences in PTP for both staining patterns (Figure 3B,C). Compared to incubation in HTF alone, supplementation with PEN gave rise to a significantly higher percentage of spermatozoa showing PTP staining pattern II compared to WHI ($p = 0.04$) which, in turn, could indicate a slightly higher incidence of PTP related to incubation in HTF + PEN compared to HTF alone.

3.3. *Sperm Thermotaxis*

As thermotaxis is a capacitation-dependent process that requires spermatozoa to maintain their swimming capacity to migrate within a temperature gradient, we employed our in vitro thermotaxis assay to analyze the effect of PEN in prolonging the migration ability of capacitated spermatozoa. When incubated in HTF + PEN and not HTF alone, the number of migrated spermatozoa within the temperature gradient was significantly higher compared to those migrating in the absence of a gradient (constant temperature of 35 or 38 °C) ($p < 0.002$) (Figure 4A). These results confirm the occurrence of thermotaxis, the percentage of net thermotaxis being 1.1% ± 0.5% for HTF + PEN (percentage of spermatozoa migrated in vitro by thermotaxis referred to the loaded spermatozoa) and confirmed the protective effect of PEN supplementation. Further, as thermotaxis allows for the selection of a sperm subpopulation of high genetic integrity in humans and mice [22], we analyzed DNA fragmentation of the selected spermatozoa. Our results indicate significantly lower DNA fragmentation in selected spermatozoa compared to the unselected sample (aliquot incubated in parallel at 37 °C) (4.4% ± 0.4% and 11% ± 2% respectively, $p = 0.009$). These selected spermatozoa fractions also showed enrichment in populations with low DNA fragmentation (Figure 4B). Thus, the percentage of spermatozoa showing 0–5% DNA fragmentation (high DNA integrity) was significantly higher in the selected fraction (69% ± 4% vs 23% ± 4% respectively, $p = 0.0002$).

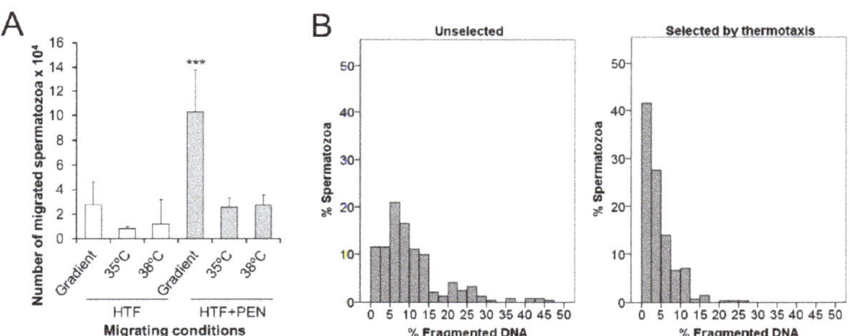

Figure 4. Sperm selection by thermotaxis. (**A**) The number of spermatozoa in vitro migrating from 35 to 38 °C or across a constant temperature (35 or 38 °C) for 60 min after 30 min of incubation in synthetic human tubal fluid (HTF) or HTF supplemented with 750 μM of penicillamine (HTF + PEN). The initial number of spermatozoa loaded in the thermotaxis system was between 5 and 6 × 10^6 per separation. (**B**) Histograms show the distributions of % fragmented DNA in individual spermatozoa unselected or selected by thermotaxis after incubation for 30 min in HTF + PEN. *** $p = 0.009$, $n = 6$.

To examine the relationship between PTP and the ability of the spermatozoa to migrate by thermotaxis, we also conducted PTP immunofluorescence analyses on selected and unselected spermatozoa. Our results revealed lower percentages of spermatozoa showing the PTP staining pattern I in the migrated spermatozoa (in all the thermotaxis and both non-gradient controls) compared to the unselected sample ($p < 0.0002$) (Figure 5A). In contrast, the percentage of spermatozoa showing PTP staining pattern II was significantly higher in the spermatozoa migrating in the non-gradient control at 38 °C and in those spermatozoa selected by thermotaxis when compared to the unselected sample ($p < 0.02$) (Figure 5B,C). In the non-migrating spermatozoa (those remaining in the drop where they were first loaded in the thermotaxis system), percentages of PTP staining patterns I and II were similar for all the conditions analyzed and in the unselected sample (Figure 5D).

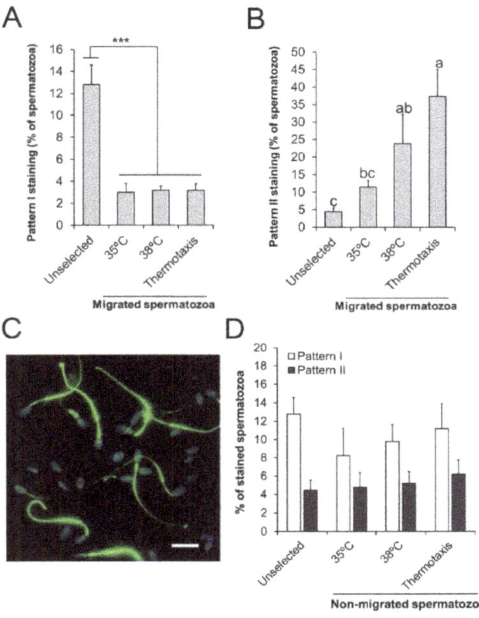

Figure 5. Protein tyrosine phosphorylation in the thermotaxis assay. (**A**,**B**) Percentage of spermatozoa showing staining patterns I or II for tyrosine phosphorylation in the unselected sample, in the spermatozoa migrating across constant temperatures (35 or 38 °C), or across a temperature gradient (from 35 to 38 °C, thermotaxis). (**C**) Representative micrograph of immunofluorescence for tyrosine phosphorylation (green) in stallion spermatozoa migrating by thermotaxis. Nuclei were labelled with Hoechst 33342 (blue). Bar = 10 µm. (**D**) Percentage of spermatozoa showing staining patterns I and II for protein tyrosine phosphorylation in the unselected sample or in those spermatozoa that did not migrate during the thermotaxis assay in the conditions described above. *** $p = 0.001$, [a, b, c] different letters indicate significant differences ($p < 0.05$, $n = 6$).

4. Discussion

Several hypotheses have been put forward to explain the unsuccessful in vitro capacitation of stallion spermatozoa [1]. The results of our study along with the literature findings detailed below suggest that the media employed should be rethought by optimizing concentrations of energy sources and adding supplements to modulate and prolong the lifespan of spermatozoa once capacitated. In our study, the HTF medium induced the greater occurrence of capacitation-related events during incubation when compared to WHI, presumably because of its higher lactate content. We also observed that supplementation with PEN prolonged the duration of hyperactive-like motility and this allowed the sperm migration by thermotaxis, suggesting a pro-survival effect on the capacitated sperm population. Interestingly, we also found that spermatozoa selected by thermotaxis showed relatively good DNA integrity, corroborating our previous results in the mouse and human [22] and opening the possibility of employing this method to improve ARTs. Another significant result that we also discuss here was the lack of a relationship found between PTP in the whole flagellum and sperm migration by thermotaxis. This might indicate the existence of a physiological PTP-independent hyperactivation response.

4.1. Capacitation-Related Events During Incubation in HTF

Incubation in HTF led to a time-dependent effect on sperm kinetics whereby hyperactive-type motility was acquired by a fraction of the spermatozoa. This effect was not observed when spermatozoa were incubated in WHI, which is a medium commonly used for stallion sperm capacitation. Further, although both media significantly increased PTP levels after 180 min of incubation, higher levels were attained with HTF. This difference was not detected by Arroyo-salvo et al. [8], who conducted a similar comparative study. In contrast, they found no time-dependent changes in sperm kinetics compatible with hyperactivation over 120 min, and PTP induction levels were similar to both media after 120 and 240 min of incubation. However, Arroyo-salvo et al. [8] employed a fresh sample washed by centrifugation, while we used frozen/thawed samples washed by DGC, which could explain the differences between both studies. Cryopreservation provokes significant changes in the sperm plasma membrane, increasing membrane peroxidation and permeability that could trigger capacitation-related events [32,33]. Further, even if capacitation is not immediately triggered, freezing/thawing may leave the spermatozoa in a poised status, making them more susceptible to capacitation than fresh sperm, as confirmed elsewhere [34]. We also detected the significant occurrence of spontaneous acrosomal exocytosis after 180 min of incubation, not detected by others employing fresh stallion semen [3,8]. This also supports the higher susceptibility of frozen/thawed spermatozoa to the destabilizing changes occurring during capacitation, as has been also shown for bull spermatozoa [35].

Differences in composition between WHI and HTF could explain the observed differences in both sperm kinetics and PTP. Both media differ in the amount of glucose and pyruvate they contain, and these are ~2 and 3 times less concentrated in HTF, respectively. However, HTF contains ~4.5 times more lactate than WHI (21.4 and 4.8 mM, respectively), which can be directly transformed to pyruvate in the sperm mitochondria by the Krebs cycle [36]. In effect, lactate and pyruvate are the main sources of energy utilized by stallion spermatozoa, and glucose may even reduce mitochondrial function [37]. Thus, the higher concentration of an energy source that can be rapidly and effectively utilized by the mitochondria could increase their activity [37], enhancing ROS production which will subsequently trigger PTP and its associated hyperactive-like motility [9]. Hence, a greater mitochondrial activity could explain the higher VCL and hyperactive-like motility observed from the onset of incubation and the higher PTP levels reported here after 180 min when using HTF rather than WHI. This could also explain the effect observed in enhancing kinetics when incubating stallion spermatozoa with follicular fluid from pre-ovulatory follicles [38]. As examined in buffalo, bull, sheep, rat, and mouse, this fluid also contains higher concentrations of lactate than glucose + pyruvate, ranging from ~7 to 27 mM depending on the species [39–41]. In humans, similar levels of glucose and lactate are reported, of around 3 mM [42]. Recently González-Fernández et al. [43] reported that in the mare's pre- and post-ovulatory oviductal fluids, concentrations of lactate were 54.66 ± 10.7 and 69.25 ± 7.3 mM, while concentrations of glucose were 0.18 ± 0.04 and 0.57 ± 0.2 mM, respectively. It is also important to point out that lactate is the most abundant source of energy within the oviduct [43,44] where capacitation, and thus hyperactivation, is triggered in vivo [45].

4.2. Effect of Penicillamine on Capacitation-Related Events During Incubation with HTF

Under our capacitating conditions, PEN was not able to rescue the time-dependent loss of motility and plasma membrane integrity. This contrasts with the results reported by Aitken et al. [20], where PEN prolonged the motility of fresh sperm incubated in BWW medium without BSA. The pro-survival effect of PEN on spermatozoa has been attributed to its ability to inactivate and slow down the production of cytotoxic aldehydes by lipid peroxidation provoked by ROS produced by the mitochondria [20]. As capacitation enhances mitochondrial function, ROS production is increased and modulates intracellular signaling for sperm capacitation, stimulating adenylyl cyclase and inhibiting tyrosine phosphatases, causing a downstream increase in PTP [9,46]. However, when oxidative stress exceeds a certain limit, the spermatozoa undergo an apoptotic-like process [18]. Thus, in our experiment, the sperm fraction that abruptly lost motility within the first 30 min of incubation and lost membrane integrity after 180 min of incubation, could have exceeded ROS production, overwhelming the protective effect of PEN. Another option is that a different deleterious process associated with sperm capacitation may have compromised sperm lifespan in a fraction of our samples. The significant changes observed in the architecture of the plasma membrane produced during capacitation, such as cholesterol removal, glycoprotein redistribution, and loss of phospholipid asymmetry [47], affects the lifespan of the spermatozoa, making them more vulnerable to damage. Thus, the plasma membrane of capacitated spermatozoa becomes more permeable to vital stains such as propidium iodide [48] and/or, as occurs with the acrosomal membrane, becomes more prone to destabilization [49]. This was likely more pronounced in our samples as we employed frozen/thawed spermatozoa, which are known to be more sensitive to the destabilizing conditions of capacitation, as the abundance of spermatozoa sustaining sublethal damage could be high. In agreement, Pommer et al. [34] showed that, in contrast to fresh sperm, frozen/thawed stallion spermatozoa incubated under capacitating conditions lost motility and membrane integrity within an hour of incubation.

We found that, unlike the case of HTF alone, supplementation with PEN led to a sustained fraction of motile spermatozoa (between 16% ± 5% and 22% ± 3%) with relatively high VCL and ALH, indicating hyperactive-like motility from 30 to 180 min of incubation. Using this medium, after 180 min, PTP along the whole flagellum reached 6% ± 1% of the total spermatozoa, from close to 0 at the onset of incubation. Assuming that only motile spermatozoa (20% ± 2% after 30 min of incubation with HTF + PEN) will acquire the capacity for PTP during incubation, then the percentage of PTP on the whole flagellum within the motile population may represent some 30%. Thus, we hypothesize that the spermatozoa that showed hyperactive-like motility could be those undergoing PTP in the whole flagellum [10,11], as each indicator was present in similar percentages of spermatozoa. No differences in PTP were observed when we compared the use of HTF + PEN to HTF alone, indicating that PEN did not induce more spermatozoa to enter a capacitated-like state, but protected those that did and also lengthened the duration of this acquired hyperactive-like motility. As commented above, physiological ROS are needed to induce hyperactivation in human sperm [50,51], but as a consequence, oxidative stress generates cytotoxic aldehydes, damaging the cell [20]. The first structure injured by this oxidative stress is the mitochondrial membrane, thus motility is the first function affected [52,53]. Accordingly, the protective effect of PEN in this setting has been shown to enhance the velocity of the motile sperm fraction in horse, rat, and human [20]. In our experiment, we also found that after 180 min of incubation in HTF + PEN, spermatozoa showed significantly higher VCL than in HTF alone. Thus, our results suggest that PEN was able to maintain the observed hyperactive-like motility, possibly prolonging the lifespan of the capacitated spermatozoa fraction. Our theory is in line with the results reported by Pavlok [21], in which PEN prolonged the fertilizing ability of frozen/thawed bovine spermatozoa.

4.3. Penicillamine Enables Sperm Selection by Thermotaxis

To carry out thermotaxis, spermatozoa must be motile and capacitated so that they can move across the temperature gradient [23,24]. Thus, for the thermotaxis experiments, spermatozoa were capacitated for 30 min in HTF or HTF + PEN, as at this time point, we had observed hyperactive-like motility with both media. However, only when incubated in HTF + PEN were spermatozoa able to migrate by thermotaxis. This observation supports the protective effect of PEN on the fraction of capacitated spermatozoa. Thus, in addition to maintaining the specific sperm kinetics needed for migration, PEN could be protecting intracellular signaling involved in the thermotactic response itself. This signaling is mediated by the phosphodiesterase and phospholipase C pathways whose thermosensors are thought to be opsins [54,55] as well as transient receptor potential cation channel subfamily V member 1 (TRPV1) [56].

In the HTF + PEN medium, thermotaxis selection yielded a net thermotaxis of 1.1% ± 0.5%, similar to the response reported in humans and mice using the same protocol [22] or employing other devices [54]. Our DNA damage assessment revealed that the fraction migrating by thermotaxis was significantly enriched in spermatozoa bearing high DNA integrity, as reported for human and mouse sperm [22]. As suggested for these two species and now also for horses, thermotaxis might be a bi-functional mechanism for the navigation and selection of high-quality capacitated spermatozoa in mammals.

Both the fraction of spermatozoa selected by thermotaxis and the fraction of spermatozoa showing random movement at a constant temperature of 38 °C showed similar strong enrichment in spermatozoa with PTP along the whole flagellum (37% ± 8% and 24% ± 8%, respectively). These percentages are similar to the percentage of PTP in the whole flagellum estimated above for the motile fraction (~30%), assuming that only motile spermatozoa can trigger PTP in the whole flagellum. This suggests that thermotaxis selects a fraction within the motile spermatozoa independently of the PTP status of the flagellum and argues against the involvement of PTP-dependent hyperactivation in the behavioral thermotaxis response of spermatozoa. Further, the lower levels of PTP detected in the sample of spermatozoa migrating in the non-gradient control at 35 °C indicates a direct PTP-inducing effect of temperature within the motile and migrating sperm fraction. This direct relationship between absolute temperature and PTP is a well-known phenomenon [57] and could indicate that PTP in the thermotactic fraction occurs during spermatozoa migration or once they have migrated. Boryshpolets et al. [57] reported that changes in the direction of swimming during the thermotactic response of human spermatozoa occur as turns that may be subtle or generated by episodes of hyperactivation. Thus, the model proposed by Boryshpolets et al. [58] for the thermotactic behavior response implies that hyperactive-like motility is transient and more frequent at lower temperatures. This contrasts with the longstanding nature of hyperactivation related to flagellum PTP and explains why in our experiment there was no significant PTP enrichment in the spermatozoa migrated by thermotaxis compared to those moving across the non-gradient control at 38 °C. We, therefore, propose the hyperactive-like motility involved in thermotaxis is PTP-independent and possibly directly linked to opsin and TRPV1 signaling for a rapid transient response. As support for this theory of PTP-independent hyperactive motility, procaine and caffeine have been shown to induce hyperactive-like motility in stallion and ram spermatozoa independently of PTP, respectively [59,60]. Further work is needed to elucidate the full transduction signaling pathway coupled to the sperm temperature sensing machinery along with the behavior changes involving the transient acquisition of hyperactive-like motility.

5. Conclusions

In this study, we observed a protective effect of penicillamine used for the in vitro capacitation of stallion spermatozoa in prolonging the duration of hyperactive-like motility of a fraction of the sperm sample and in allowing sperm migration by thermotaxis, a process that is capacitation-dependent. In addition, we report here that thermotaxis selects a sperm fraction enriched in PTP also showing high DNA integrity, thus supporting its potential use for sperm preparation before assisted reproductive techniques in the horse. The results reported here also point to a relevant role of lactate in the capacitation of stallion spermatozoa and also identify no relationship between protein tyrosine phosphorylation in the sperm flagellum and migration by thermotaxis.

Author Contributions: S.P.-C., I.O.-L., and S.R.-D. conceived the study, conducted the experiments, and wrote the paper; D.D.L.C.-D., C.S., and B.G.-L. conducted the experiments; M.J.S.-C. and A.G.-A. conceived the study and wrote the paper. All authors have read and agreed to the published version of the manuscript.

Funding: This work is supported by Grants RTI2018-096736-A-I00 and RTI2018-093548-B-I00 from the Spanish Ministry of Science, Innovation, and Universities. I.O.L. and S.R.D. are supported by CONACYT fellowship of the Mexican government (283833) and "Doctorados Industriales 2018" fellowship of Comunidad de Madrid (IND2018/BIO-9610), respectively. S.P.C. is supported by a Ramón y Cajal contract from the Spanish Ministry of Science, Innovation, and Universities (RYC-2016-20147).

Acknowledgments: We thank L. González-Fernández and B. Macías-García from the University of Extremadura for providing the cryopreserved semen samples. This research is especially dedicated to the memory of Serafín Pérez-Cerezales, an invaluable person, mentor, and friend.

Conflicts of Interest: The authors declare that there is no conflict of interest that could be perceived as prejudicing the impartiality of the research reported.

References

1. Leemans, B.; Gadella, B.M.; Stout, T.A.E.; De Schauwer, C.; Nelis, H.; Hoogewijs, M.; Van Soom, A. Why doesn't conventional IVF work in the horse? The equine oviduct as a microenvironment for capacitation/fertilization. *Reproduction* **2016**, *152*, 233–245. [CrossRef] [PubMed]
2. Morris, L.H.; Maclellan, L.J. Update on advanced semen-processing technologies and their application for in vitro embryo production in horses. *Reprod. Fertil. Dev.* **2019**, *31*, 1771–1777. [CrossRef] [PubMed]
3. Bromfield, E.G.; Aitken, R.J.; Gibb, Z.; Lambourne, S.R.; Nixon, B. Capacitation in the presence of methyl-β-cyclodextrin results in enhanced zona pellucida-binding ability of stallion spermatozoa. *Reproduction* **2014**, *147*, 153–166. [CrossRef]
4. Loux, S.C.; Crawford, K.R.; Ing, N.H.; González-Fernández, L.; Macías-García, B.; Love, C.C.; Varner, D.D.; Velez, I.C.; Choi, Y.H.; Hinrichs, K. CatSper and the Relationship of hyperactivated motility to intracellular calcium andpH kinetics in equine sperm. *Biol. Reprod.* **2013**, *89*, 1–15. [CrossRef] [PubMed]
5. Macías-García, B.; Gonzalez-Fernandez, L.; Loux, S.C.; Rocha, A.M.; Guimarães, T.; Pena, F.J.; Varner, D.D.; Hinrichs, K. Effect of calcium, bicarbonate, and albumin on capacitation-related events in equine sperm. *Reproduction* **2015**, *149*, 87–99. [CrossRef]
6. McPartlin, L.A.; Suarez, S.S.; Czaya, C.A.; Hinrichs, K.; Bedford-Guaus, S.J. Hyperactivation of Stallion Sperm Is Required for Successful In Vitro Fertilization of Equine Oocytes. *Biol. Reprod.* **2009**, *81*, 199–206. [CrossRef]
7. Ortgies, F.; Klewitz, J.; Görgens, A.; Martinsson, G.; Sieme, H. Effect of procaine, pentoxifylline and trolox on capacitation and hyperactivation of stallion spermatozoa. *Andrologia* **2012**, *44*, 130–138. [CrossRef]
8. Arroyo-Salvo, C.; Sanhueza, F.; Fuentes, F.; Treulén, F.; Arias, M.E.; Cabrera, P.; Silva, M.; Felmer, R. Effect of human tubal fluid medium and hyperactivation inducers on stallion sperm capacitation and hyperactivation. *Reprod. Domest. Anim.* **2018**, *54*, 184–194. [CrossRef]
9. Aitken, R.J.; Nixon, B. Sperm capacitation: A distant landscape glimpsed but unexplored. *Mol. Hum. Reprod.* **2013**, *19*, 785–793. [CrossRef]
10. Si, Y.; Okuno, M. Role of Tyrosine Phosphorylation of Flagellar Proteins in Hamster Sperm Hyperactivation1. *Biol. Reprod.* **1999**, *61*, 240–246. [CrossRef]
11. Nassar, A.; Mahony, M.; Morshedi, M.; Lin, M.H.; Srisombut, C.; Oehninger, S. Modulation of sperm tail protein tyrosine phosphorylation by pentoxifylline and its correlation with hyperactivated motility. *Fertil. Steril.* **1999**, *71*, 919–923. [CrossRef]
12. Leemans, B.; Stout, T.A.E.; De Schauwer, C.; Heras, S.; Nelis, H.; Hoogewijs, M.; Van Soom, A.; Gadella, B.M. Update on mammalian sperm capacitation: How much does the horse differ from other species? *Reproduction* **2019**, *157*, 181–197. [CrossRef] [PubMed]
13. Romero-Aguirregomezcorta, J.; Sugrue, E.; Martínez-Fresneda, L.; Newport, D.; Fair, S. Hyperactivated stallion spermatozoa fail to exhibit a rheotaxis-like behaviour, unlike other species. *Sci. Rep.* **2018**, *8*, 16897. [CrossRef] [PubMed]
14. Cohen-Dayag, A.; Tur-Kaspa, I.; Dor, J.; Mashiach, S.; Eisenbach, M. Sperm capacitation in humans is transient and correlates with chemotactic responsiveness to follicular factors. *PNAs* **1995**, *92*, 11039–11043. [CrossRef]
15. Eisenbach, M. Mammalian sperm chemotaxis and its association with capacitation. *Dev. Genet.* **1999**, *25*, 87–94. [CrossRef]
16. Giojalas, L.C.; Rovasio, R.A.; Fabro, G.; Gakamsky, A.; Eisenbach, M. Timing of sperm capacitation appears to be programmed according to egg availability in the female genital tract. *Fertil. Steril.* **2004**, *82*, 247–249. [CrossRef]
17. Harrison, R.A.P. Capacitation mechanisms, and the role of capacitation as seen in eutherian mammals. *Reprod. Fertil. Dev.* **1996**, *8*, 581–594. [CrossRef]

18. Aitken, R.J.; Baker, M.A.; Nixon, B. Are sperm capacitation and apoptosis the opposite ends of a continuum driven by oxidative stress? *Asian J. Androl.* **2015**, *17*, 633–639. [CrossRef]
19. Petrunkina, A.M.; Waberski, D.; Günzel-Apel, A.R.; Töpfer-Petersen, E. Determinants of sperm quality and fertility in domestic species. *Reproduction* **2007**, *134*, 3–17. [CrossRef]
20. Aitken, R.J.; Gibb, Z.; Mitchell, L.A.; Lambourne, S.R.; Connaughton, H.S.; De Iuliis, G.N. Sperm Motility Is Lost In Vitro as a Consequence of Mitochondrial Free Radical Production and the Generation of Electrophilic Aldehydes but Can Be Significantly Rescued by the Presence of Nucleophilic Thiols. *Biol. Reprod.* **2012**, *87*, 110. [CrossRef]
21. Pavlok, A. D-penicillamine and granulosa cells can effectively extend the fertile life span of bovine frozen-thawed spermatozoa in vitro: Effect on fertilization and polyspermy. *Theriogenology* **2000**, *53*, 1135–1146. [CrossRef]
22. Pérez-Cerezales, S.; Laguna-Barraza, R.; De Castro, A.C.; Sánchez-Calabuig, M.J.; Cano-Oliva, E.; De Castro-Pita, F.J.; Montoro-Buils, L.; Pericuesta, E.; Fernández-González, R.; Gutiérrez-Adán, A. Sperm selection by thermotaxis improves ICSI outcome in mice. *Sci. Rep.* **2018**, *8*, 2902. [CrossRef] [PubMed]
23. Bahat, A.; Tur-Kaspa, I.; Gakamsky, A.; Giojalas, L.C.; Breitbart, H.; Eisenbach, M. Thermotaxis of mammalian sperm cells: A potential navigation mechanism in the female genital tract. *Nat. Med.* **2003**, *9*, 149–150. [CrossRef] [PubMed]
24. Bahat, A.; Caplan, S.R.; Eisenbach, M. Thermotaxis of human sperm cells in extraordinarily shallow temperature gradients over a wide range. *PLoS ONE* **2012**, *7*, e41915. [CrossRef]
25. Garner, D.L.; Johnson, L.A. Viability Assessment of Mammalian Sperm Using SYBR-14 and Propidium Iodide1. *Biol. Reprod.* **1995**, *53*, 276–284. [CrossRef]
26. Lybaert, P.; Danguy, A.; Leleux, F.; Meuris, S.; Lebrun, P. Improved methodology for the detection and quantification of the acrosome reaction in mouse spermatozoa. *Histol. Histopathol.* **2009**, *24*, 999–1007.
27. Schneider, C.A.; Rasband, W.S.; Eliceiri, K.W. NIH Image to ImageJ: 25 years of image analysis. *Nat. Methods* **2012**, *9*, 671–675. [CrossRef]
28. Giaretta, E.; Munerato, M.; Yeste, M.; Galeati, G.; Spinaci, M.; Tamanini, C.; Mari, G.; Bucci, D. Implementing an open-access CASA software for the assessment of stallion sperm motility: Relationship with other sperm quality parameters. *Anim. Reprod. Sci.* **2017**, *176*, 11–19. [CrossRef]
29. Mortimer, D.; Serres, C.; Mortimer, S.T.; Jouannet, P. Influence of image sampling frequency on the perceived movement characteristics of progressively motile human spermatozoa. *Gamete Res.* **1988**, *20*, 313–327. [CrossRef]
30. Su, T.W.; Choi, I.; Feng, J.; Huang, K.; McLeod, E.; Ozcan, A. Sperm trajectories form chiral ribbons. *Sci. Rep.* **2013**, *3*, 1–8. [CrossRef]
31. Końca, K.; Lankoff, A.; Banasik, A.; Lisowska, H.; Kuszewski, T.; Góźdź, S.; Koza, Z.; Wojcik, A. A cross-platform public domain PC image-analysis program for the comet assay. *Mutat. Res. Genet. Toxicol. Environ. Mutagen.* **2003**, *534*, 15–20. [CrossRef]
32. De Andrade, A.F.C.; Zaffalon, F.G.; Celeghini, E.C.C.; Nascimento, J.; Bressan, F.F.; Martins, S.M.M.K.; de Arruda, R.P. Post-thaw addition of seminal plasma reduces tyrosine phosphorylation on the surface of cryopreserved equine sperm, but does not reduce lipid peroxidation. *Theriogenology* **2012**, *77*, 1866–1872. [CrossRef] [PubMed]
33. Naresh, S.; Atreja, S.K. The protein tyrosine phosphorylation during in vitro capacitation and cryopreservation of mammalian spermatozoa. *Cryobiology* **2015**, *70*, 211–216. [CrossRef] [PubMed]
34. Pommer, A.C.; Rutllant, J.; Meyers, S.A. Phosphorylation of Protein Tyrosine Residues in Fresh and Cryopreserved Stallion Spermatozoa under Capacitating Conditions. *Biol. Reprod.* **2003**, *68*, 1208–1214. [CrossRef] [PubMed]
35. Fukuda, M.; Sakase, M.; Fukushima, M.; Harayama, H. Changes of IZUMO1 in bull spermatozoa during the maturation, acrosome reaction, and cryopreservation. *Theriogenology* **2016**, *86*, 2179–2188.e3. [CrossRef] [PubMed]
36. Ferramosca, A.; Zara, V. Bioenergetics of mammalian sperm capacitation. *BioMed Res. Int.* **2014**, *2014*, 902953. [CrossRef]
37. Darr, C.R.; Varner, D.D.; Teague, S.; Cortopassi, G.A.; Datta, S.; Meyers, S.A. Lactate and Pyruvate Are Major Sources of Energy for Stallion Sperm with Dose Effects on Mitochondrial Function, Motility, and ROS Production. *Biol. Reprod.* **2016**, *95*, 34. [CrossRef]

38. Leemans, B.; Gadella, B.M.; Stout, T.A.E.; Nelis, H.; Hoogewijs, M.; Van Soom, A. An alkaline follicular fluid fraction induces capacitation and limited release of oviduct epithelium-bound stallion sperm. *Reproduction* **2015**, *150*, 193–208. [CrossRef]
39. Orsi, N.M.; Gopichandran, N.; Leese, H.J.; Picton, H.M.; Harris, S.E. Fluctuations in bovine ovarian follicular fluid composition throughout the oestrous cycle. *Reproduction* **2005**, *129*, 219–228. [CrossRef]
40. Zeilmaker, G.H.; Verhamme, C.M.P.M. Lactate concentrations in pre-ovulatory follicles of pro-oestrous rats before and after onset of oocyte maturation. *Acta Endocrinol.* **1977**, *86*, 380–383. [CrossRef]
41. Nandi, S.; Girish Kumar, V.; Manjunatha, B.M.; Ramesh, H.S.; Gupta, P.S.P. Follicular fluid concentrations of glucose, lactate and pyruvate in buffalo and sheep, and their effects on cultured oocytes, granulosa and cumulus cells. *Theriogenology* **2008**, *69*, 186–196. [CrossRef] [PubMed]
42. Gull, I.; Geva, E.; Lerner-Geva, L.; Lessing, J.B.; Wolman, I.; Amit, A. Anaerobic glycolysis. The metabolism of the preovulatory human oocyte. *Eur. J. Obstet. Gynecol. Reprod. Biol.* **1999**, *85*, 225–228. [CrossRef]
43. González-Fernández, L.; Sánchez-Calabuig, M.J.; Calle-Guisado, V.; García-Marín, L.J.; Bragado, M.J.; Fernández-Hernández, P.; Gutiérrez-Adán, A.; Macías-García, B. Stage-specific metabolomic changes in equine oviductal fluid: New insights into the equine fertilization environment. *Theriogenology* **2020**, *143*, 35–43. [CrossRef] [PubMed]
44. Ruiz-Pesini, E.; Díez-Sánchez, C.; López-Pérez, M.J.; Enríquez, J.A. The Role of the Mitochondrion in Sperm Function: Is There a Place for Oxidative Phosphorylation or Is This a Purely Glycolytic Process? *Curr. Top. Dev. Biol.* **2007**, *77*, 3–19.
45. Pérez-Cerezales, S.; Ramos-Ibeas, P.; Acuña, O.S.; Avilés, M.; Coy, P.; Rizos, D.; Gutiérrez-Adán, A. The oviduct: From sperm selection to the epigenetic landscape of the embryo. *Biol. Reprod.* **2018**, *98*, 262–276. [CrossRef]
46. Leclerc, P.; De Lamirande, E.; Gagnon, C. Regulation of protein-tyrosine phosphorylation and human sperm capacitation by reactive oxygen derivatives. *Free Radic. Biol. Med.* **1997**, *22*, 643–656. [CrossRef]
47. Gadella, B.M.; Harrison, R.A.P. The capacitating agent bicarbonate induces protein kinase A-dependent changes in phospholipid transbilayer behavior in the sperm plasma membrane. *Development* **2000**, *127*, 2407–2420.
48. Kerns, K.; Sharif, M.; Zigo, M.; Xu, W.; Hamilton, L.E.; Sutovsky, M.; Ellersieck, M.; Drobnis, E.Z.; Bovin, N.; Oko, R.; et al. Sperm cohort-specific zinc signature acquisition and capacitation-induced zinc flux regulate sperm-oviduct and sperm-zona pellucida interactions. *Int. J. Mol. Sci.* **2020**, *21*, 2121. [CrossRef]
49. Ruiz-Díaz, S.; Grande-Pérez, S.; Arce-López, S.; Tamargo, C.; Hidalgo, C.O.; Pérez-Cerezales, S. Changes in the Cellular Distribution of Tyrosine Phosphorylation and Its Relationship with the Acrosomal Exocytosis and Plasma Membrane Integrity during In Vitro Capacitation of Frozen/Thawed Bull Spermatozoa. *Int. J. Mol. Sci.* **2020**, *21*, 2725. [CrossRef]
50. De Lamirande, E.; Gagnon, C. Impact of reactive oxygen species on spermatozoa: A balancing act between beneficial and detrimental effects. *Hum. Reprod.* **1995**, *10*, 15–24. [CrossRef]
51. De Lamirande, E.; Gagnon, C. A positive role for the superoxide anion in triggering hyperactivation and capacitation of human spermatozoa. *Int. J. Androl.* **1993**, *16*, 21–25. [CrossRef] [PubMed]
52. Bardaweel, S.K.; Gul, M.; Alzweiri, M.; Ishaqat, A.; Alsalamat, H.A.; Bashatwah, R.M. Reactive oxygen species: The dual role in physiological and pathological conditions of the human body. *Eurasian J. Med.* **2018**, *50*, 193–201. [CrossRef] [PubMed]
53. Jones, R.; Mann, T.; Sherins, R. Peroxidative breakdown of phospholipids in human spermatozoa, spermicidal properties of fatty acid peroxides, and protective action of seminal plasma. *Fertil. Steril.* **1979**, *31*, 531–537. [CrossRef]
54. Pérez-Cerezales, S.; Boryshpolets, S.; Afanzar, O.; Brandis, A.; Nevo, R.; Kiss, V.; Eisenbach, M. Involvement of opsins in mammalian sperm thermotaxis. *Sci. Rep.* **2015**, *5*, 16146. [CrossRef]
55. Roy, D.; Levi, K.; Kiss, V.; Nevo, R.; Eisenbach, M. Rhodopsin and melanopsin coexist in mammalian sperm cells and activate different signaling pathways for thermotaxis. *Sci. Rep.* **2020**, *10*, 112. [CrossRef]
56. De Toni, L.; Garolla, A.; Menegazzo, M.; Magagna, S.; Di Nisio, A.; Šabović, I.; Rocca, M.S.; Scattolini, V.; Filippi, A.; Foresta, C. Heat Sensing Receptor TRPV1 Is a Mediator of Thermotaxis in Human Spermatozoa. *PLoS ONE* **2016**, *11*, e0167622. [CrossRef]
57. Si, Y. Hyperactivation of hamster sperm motility by temperature-dependent tyrosine phosphorylation of an 80-kDa protein. *Biol. Reprod.* **1999**, *61*, 247–252. [CrossRef]

58. Boryshpolets, S.; Pérez-Cerezales, S.; Eisenbach, M. Behavioral mechanism of human sperm in thermotaxis: A role for hyperactivation. *Hum. Reprod.* **2015**, *30*, 884–892. [CrossRef]
59. Leemans, B.; Stout, T.A.E.; Van Soom, A.; Gadella, B.M. pH-dependent effects of procaine on equine gamete activation. *Biol. Reprod.* **2019**, *101*, 1056–1074. [CrossRef]
60. Colás, C.; Cebrián-Pérez, J.A.; Muiño-Blanco, T. Caffeine induces ram sperm hyperactivation independent of cAMP-dependent protein kinase. *Int. J. Androl.* **2010**, *33*, 187–197. [CrossRef]

MDPI

Article

Effects of In Vitro Interactions of Oviduct Epithelial Cells with Frozen–Thawed Stallion Spermatozoa on Their Motility, Viability and Capacitation Status

Brenda Florencia Gimeno [1,†], María Victoria Bariani [1,†], Lucía Laiz-Quiroga [1], Eduardo Martínez-León [2], Micaela Von-Meyeren [1], Osvaldo Rey [2], Adrián Ángel Mutto [1,*] and Claudia Elena Osycka-Salut [1,*]

1 Laboratorio de Biotecnologías Reproductivas y Mejoramiento Genético Animal, Instituto de Investigaciones Biotecnológicas, Universidad Nacional de San Martín (UNSAM), Campus Miguelete, Avenida 25 de Mayo y Francia, San Martín, Buenos Aires, CP 1650, Argentina; bre.gimeno@gmail.com (B.F.G.); vickiba86@gmail.com (M.V.B.); lucia.laizquiroga@gmail.com (L.L.-Q.); micaelavonmeyeren@gmail.com (M.V.-M.)

2 Signaling and Cancer Laboratory, Consejo Nacional de Investigaciones Científicas y Técnicas, Instituto de Inmunología, Genética y Metabolismo, Facultad de Farmacia y Bioquímica, Hospital de Clínicas "José de San Martín", Universidad de Buenos Aires, Ciudad Autónoma de Buenos Aires (CABA), CP 1120, Argentina; eduartinez@gmail.com (E.M.-L.); osrey@ucla.edu (O.R.)

* Correspondence: amutto@iib.unsam.edu.ar (A.Á.M.); claudia.osycka@gmail.com (C.E.O.-S.)

† Contributed equally to this study.

Simple Summary: The use of assisted reproductive techniques, which involve the manipulation of sperm and oocytes in the laboratory, support owner production of valuable animals' offspring. However, several limitations remain underlining the need to further optimize existing protocols as well as to develop new strategies. For example, the required conditions to make equine spermatozoa competent to fertilize an oocyte in vitro (IVF) have not been established. Therefore, our initial goal was to optimize different conditions associated with frozen equine sperm manipulations in order to improve their quality. We observed that simple factors such as sample concentration, incubation period and centrifugation time affect the sperm motility. Since in vivo fertilization involves the interaction between spermatozoa and epithelial cells in the mare's oviductal tract, our next goal was to mimic this environment by establishing primary cultures of oviductal cells. Using this in vitro system, we were able to select a sperm population capable of fertilization. In short, this study provides a novel protocol that improves the yield of fertilization-capable sperm obtained from equine frozen spermatozoa.

Abstract: Cryopreservation by negatively affecting sperm quality decreases the efficiency of assisted reproduction techniques (ARTs). Thus, we first evaluated sperm motility at different conditions for the manipulation of equine cryopreserved spermatozoa. Higher motility was observed when spermatozoa were incubated for 30 min at 30×10^6/mL compared to lower concentrations ($p < 0.05$) and when a short centrifugation at 200× g was performed ($p < 0.05$). Moreover, because sperm suitable for oocyte fertilization is released from oviduct epithelial cells (OECs), in response to the capacitation process, we established an in vitro OEC culture model to select a sperm population with potential fertilizing capacity in this species. We demonstrated E-cadherin and cytokeratin expression in cultures of OECs obtained. When sperm–OEC cocultures were performed, the attached spermatozoa were motile and presented an intact acrosome, suggesting a selection by the oviductal model. When co-cultures were incubated in capacitating conditions a greater number of alive ($p < 0.05$), capacitated ($p < 0.05$), with progressive motility ($p < 0.05$) and with the intact acrosome sperm population was observed ($p < 0.05$) suggesting that the sperm population released from OECs in vitro presents potential fertilizing capacity. Improvements in handling and selection of cryopreserved sperm would improve efficiencies in ARTs allowing the use of a population of higher-quality sperm.

Citation: Gimeno, B.F.; Bariani, M.V.; Laiz-Quiroga, L.; Martínez-León, E.; Von-Meyeren, M.; Rey, O.; Mutto, A.Á.; Osycka-Salut, C.E. Effects of In Vitro Interactions of Oviduct Epithelial Cells with Frozen–Thawed Stallion Spermatozoa on Their Motility, Viability and Capacitation Status. *Animals* **2021**, *11*, 74. https://doi.org/10.3390/ani11010074

Received: 25 November 2020
Accepted: 28 December 2020
Published: 3 January 2021

Keywords: cryopreserved sperm; sperm–oviduct interaction; sperm selection; ARTs; equines

1. Introduction

The equine industry is a very strong, economically diverse and productive business worth US$300 billion worldwide [1]. Moreover, horses represent an enormous value as sports and companion animals as well as a valuable model for the study of several human pathologies [2–5]. Nevertheless, the low success rate of assisted reproduction techniques (ARTs) available for in vitro equine embryo production is insufficient to satisfy the current needs.

The development of protocols for gametes and embryo cryopreservation has facilitated the implementation of ARTs. For example, cryopreserved sperm are widely employed in domestic animal production since cryopreservation facilitates the transport and storage of the samples for later use in different reproductive biotechnologies [6]. Nevertheless, cryopreservation procedures can negatively affect spermatozoa quality, causing changes at the structural and molecular levels, compromising sperm function [7]. Membranes are thought to be the primary site of cryopreservation injury, an injury associated with an increase in the intracellular concentration of reactive oxygen species (ROS) [8,9]. Generally, the key for cryopreservation of stallion semen is the individual stallion itself [10,11]. However, success in cryopreserving has been variable [10,12,13] as revealed by limited pregnancy rates [10] very likely due to differences in sperm freezing efficiencies [11]. For example, 20% of stallions produce semen that freezes well (high cryosurvival rates and post-thaw sperm progressive motility and total motility figures of 40–60% and >70%), 60% freezes acceptably (post-thaw sperm progressive motilities figures of 25–35%) and 20% freezes poorly (low tolerance to cryopreservation and post-thaw sperm progressive motilities as low as 10–15%) [11,14]. Therefore, the optimization of semen cryopreservation is critical for a successful ART.

The conditions to induce in vitro capacitation of stallion spermatozoa have not yet been described, a limitation that specially affects the efforts to perform equine in vitro fertilization (IVF), a widely used ART [15,16]. Thus, intracytoplasmic sperm injection (ICSI) is a technique widely employed within the equine breeding industry despite its modest yield [17,18]. This technique employs mature oocytes and fresh, cryopreserved, and/or low-quality stallion semen selected by sperm motility and morphology [19]. On the contrary, high fertilization rates in equines are obtained via artificial insemination (AI) [14,20], very likely due to the promoting effects of the oviductal environment upon sperm capacitation [21]. Therefore, the use of an in vitro system that mimics the in vivo conditions should significantly improve spermatozoa selection, fertilization and embryonic development employing ARTs protocols [22].

The oviductal epithelial cells (OECs), which form the oviductal sperm reservoir [23], are specialized cells associated with the selection of sperm suitable for oocyte fertilization [24]. The sperm–OEC interaction is highly specific [25–27] and takes place between the sperm plasma membrane at the acrosome region and the ciliated cells of the oviductal epithelium [28]. Several studies showed that spermatozoa attached to the oviductal cells have an intact acrosome and chromatin [29,30] and are morphologically normal [31] and motile [31,32]. Furthermore, OECs preferentially bind non-capacitated sperm [33–35] indicating that in vivo sperm capacitation is associated with its release from OECs [23,26,36,37] by a mechanism that involves sperm plasma membrane remodeling and molecular processes including an increase in cyclic adenosine monophosphate (cAMP) levels, increase in calcium influx, changes in protein phosphorylation and protein kinases activity, and an increment in the production of reactive oxygen species, such as nitric oxide [38–40]. These modifications lead to hyperactivated motility and prepare the spermatozoa to undergo the acrosome reaction to fertilize the oocyte [41,42].

Considering the high demand for in vitro developed equine embryos, first, we aimed to evaluate the effect of different conditions associated with the manipulation of stallion cryopreserved sperm. Furthermore, the second aim of this work was to establish an in vitro OEC primary culture to select a sperm population from cryopreserved samples with potential fertilizing capacity.

2. Materials and Methods

2.1. Chemicals

Fetal bovine serum (FBS), gentamicin and fungizone were purchased from GIBCO (Thermo Fisher Scientific, Waltham, MA, USA). Dulbecco's Modified Eagle Medium-high glucose (DMEM-HG) medium, *Pisum sativum* agglutinin-FITC staining (PSA-FITC), Hoechst 33258 (viability studies), Hoechst, L-glutamine, penicillin–streptomycin and bovine serum albumin (BSA; V fraction) were obtained from Sigma Chemicals (St. Louis, MI, USA). Glass wool columns for sperm selection were acquired from MicroFiber Manville. Salts used to prepare sperm, Whitten's medium (WM), were purchased from MERK (Darmstadt, Germany). All PCR reagents were obtained from Biodynamics and Genbiotech (Buenos Aires, Argentina). Cytokeratin-7 antibody (catalogue # ab9021; lot # Gr3225265-2) and E-cadherin antibody (catalogue # ab76055; lot # Gr317373-14) were purchased from Abcam Inc. (Cambridge, MA, USA). Antibodies against protein kinase A (PKA) phosphorylated Ser/Thr containing-substrates (clone 100G7E catalogue #9624S, lot # 21) were purchased from Cell Signaling Technology (Danvers, MA, USA). Anti-phosphorylated tyrosine antibody (clone 4G10 catalogue #05-321; lot # 3272262) was purchased from EMD Millipore (Burlington, MA, USA). Secondaries antibodies, Alexa 488-conjugated chicken anti-mouse IgGs and Alexa 568-conjugated goat anti-rabbit IgGs, were obtained from Invitrogen (Carlsbad, CA, USA). All the other chemicals were of analytical grade and obtained from standard sources.

2.2. Culture Media

DMEM-HG medium supplemented with 10% FBS, 0.25 mg/mL gentamicin, 1 µg/mL fungizone, 2 mM L-glutamine and 50 U/mL penicillin–streptomycin (complete DMEM-HG) was employed during oviduct handling and monolayer cultures establishment. Sperm handling and co-culture experiments were performed with non-capacitating modified Whitten's base medium (NCWM) (BSA and bicarbonate free: 100 mM NaCl; 4.7 mM KCl; 1.2 mM $MgCl_2 * 6H_2O$, 22 mM HEPES acid-free; 4.8 mM L-lactic acid hemicalcium; 5.5 mM D-glucose; 1 mM pyruvate; pH = 7.4; Osm = 300 mOsm). Capacitating medium (CWM) was prepared by adding 25 mM $NaHCO_3$ and 7 mg/mL BSA to the NCWM (pH = 7.4; Osm = 300 mOsm) [43,44].

2.3. Cryopreserved Stallion Sperm Handling and Processing

For each set of experiments, four cryopreserved semen straws (ejaculates) from four different stallions (200×10^6 spermatozoa/0.5 mL straw) were used. These samples were obtained from GeneTec by Ativet (Pilar, Buenos Aires, Argentina; website: www.genetec.com.ar) and Los Pingos del Taita (Rio Cuarto, Córdoba, Argentina; website: www.lospingosdeltaita.com). Straws were thawed in a water bath at 37 °C for 30 s. Spermatozoa were selected using glass wool columns and washed by centrifugation at 600× *g* for 5 min at room temperature (RT). Pellets were resuspended in 100 µL of NCWM and sperm concentration and motility were examined using a hemocytometer mounted on a bright-field microscope stage heated at 38.5 °C at 100× magnification (Nikon Instruments Inc., Tokyo, Japan).

2.3.1. Effect of Incubation Time, Sample Concentration and Centrifugation Time on Sperm Motility

After glass wool column selection, sperm were incubated in NCWM at two different concentrations, 12×10^6 sperm/mL and 30×10^6 sperm/mL, at different times (0, 30, 60 and 120 min) and sperm motility was measured by Computer-Assisted Sperm Analysis (CASA system: IVOS II™ Animal—Hamilton Thorne). These concentrations were chosen based on the CASA system optimal-concentration working range (10–50 × 10^6 sperm/mL). To study the effect of centrifugation time on sperm motility, sperm were incubated at 30 × 10^6 sperm/mL and centrifugated for 0 (control), 1 min or 2 min at 200× *g* at RT. We did centrifugations studies with 30 × 10^6 sperm/mL because we observed that sperm motility was not affected for short times. Motility was studied using the CASA system. Results are expressed as % of total motile sperm.

2.3.2. Sperm Motility Assay

Sperm motility was analyzed with CASA-system (IVOS II™ Animal—Hamilton Thorne). Fourteen randomly selected microscopic fields were scanned at 60 Fr/s with ~45 sperm per field (*n* = at least 3 independent replicates). Moreover, sperm total motility was subjectively evaluated using a field microscope stage heated at 38.5 °C at 100× magnification.

2.4. Oviducts Collection

Mare oviducts were obtained from the Lamar S.A. slaughterhouse (Mercedes, Buenos Aires, Argentina). Oviducts were collected at the time of slaughter and transported to the laboratory at 4 °C in saline solution with 50 µg/mL of gentamycin.

2.4.1. Oviductal Cell Collection and Cultures

The oviducts were cleaned of surrounding tissues, and the oviductal content was collected by flushing the oviducts with PBS and squeezing (applying pressure) the entire oviduct with tweezers within a laminar flow hood. The ampulla and isthmus OECs from 6 different animals were collected, pooled, resuspended in 10 mL of PBS and pelleted by centrifugation at 1500× *g* for 5 min at RT. The pelleted cells were resuspended in complete DMEM-HG, plated in 24-well tissue culture plates, and maintained at 38.5 °C in a 5% CO_2 atmosphere. After 24 h, the explants were collected by micropipette-aspiration, subjected to the same clarification procedure, plated again and further incubated with complete DMEM-HG at 38.5 °C in a 5% CO_2 atmosphere until the cultures reached confluence (10–13 days). The medium was changed every 48 h. The epithelial phenotype of the cultured cells was confirmed by immunocytochemical analysis and RT-PCR. To perform co-cultures with sperm, confluent cell monolayers were washed three times with NCWM medium and maintained in the same medium for 1 h before sperm addition.

2.4.2. E-Cadherin and Cytokeratin-7 MRNA Expression in OEC Primary Cultures

The expression of mRNAs encoding for the epithelial markers, cytokeratin-7 (e-KRT7) and E-cadherin (e-CDH1), was examined by RT-PCR. Specifically, total RNA was isolated from OECs using Trizol reagent (Invitrogen, Carlsbad, CA, USA) according to the manufacturer's recommendations. Only samples with a 260 nm/280 nm ratio greater than 1.7 were used for further analysis. cDNA was synthesized from 1 µg of total mRNA by using SuperScriptIII enzyme (Invitrogen TM, Carlsbad, CA, USA) and random primers (Invitrogen TM, Carlsbad, CA, USA), according to the manufacturer's instructions in the presence of recombinant RNAase inhibitor (Invitrogen TM, Carlsbad, CA, USA). After first-strand synthesis, PCR was performed with the following oligonucleotide primers: e-KRT7 (cytokeratin-7): 5′-GTGGTGAATTCTTCTGGCGG-3′ (sense), 5′-AATAGGCTTTGAGGACCCCC-3′ (antisense); e-CDH1 (E-cadherin): 5′-TCACCACAGAC CCAGTAACC-3′ (sense), 5′-CGTTCACATCCATCACGTCC-3′ (antisense); equine GAPDH: 5′-CATCATCCCTGCTTCTACTGG-3′ (sense), 5′-TCCACGACTGACACGTTAGG-3′ (antisense). Amplifications were performed using Taq DNA polymerase enzyme (Invitrogen, Carlsbad, California, USA). PCR was performed as follows: 95 °C for 5 min (initial denaturation) and 35 cycles at 95 °C for 30 s, 56 °C for 30 s, 72 °C for 1 min and finally 72 °C for 10 min. Negative controls were performed without cDNA template. PCR products were separated on a 2% (*w*/*v*) agarose gel, stained with ethidium bromide, and recorded under UV light with an Olympus C5060 digital camera (Olympus Corp., Japan).

2.4.3. Cytokeratin and E-Cadherin Distribution in OEC Primary Cultures

OECs were fixed for 10 min at RT in 4% *w*/*v* paraformaldehyde and permeabilized with PBS-Tritón X-100 0.4%. Non-specific binding sites were blocked (60 min, PBS-1% gelatin IRA grade, Bio-Rad Laboratories, Hercules, CA, USA) and samples incubated with anti-cytokeratin antibody (1:50) or anti E-cadherin antibody (1:50) diluted in PBS-Tween 0.05% for 18 h at 4 °C. After three washes in PBS-Tween 0.05% at RT, the samples were incubated with anti-mouse antibody (1:1000) diluted in PBS-Tween 0.05%. After three washes with PBS-Tween 0.05% at RT, DNA was stained for 7 min with Hoechst 33352 (1 μg/mL). The specificity of the immunodetection was assessed by a) omitting the primary antibody and b) replacing the primary antibody with serum from non-immunized rabbits at the same concentration as the corresponding primary antibody (IgG control). Samples were examined with a Nikon Eclipse Ti-E microscope (Nikon Instruments Inc., Tokyo, Japan) and fluorescence images were captured with an Andor Neo 5.5 sCMOS camera (Oxford Instruments, Abingdon, United Kingdom) driven by NIS-Elements AR v 4.30.01 software (Nikon Instruments Inc., Tokyo, Japan). Not less than 20 fields per experiment were analyzed, and both markers were studied at least on three different pools of OECs (n = 3). Results are shown with one representative image.

2.5. Co-Cultures of OECs and Spermatozoa

To perform co-cultures with sperm, confluent OEC monolayers were washed three times with NCWM medium and maintained in the same medium for 1 h before sperm addition. OEC monolayers were co-cultured with sperm suspensions (7×10^6 sperm/mL of NCWM/well) for 60 min at 38.5 °C in a 5% CO_2 atmosphere. Sperm concentration used was selected to recover enough numbers of sperm released from OECs to perform the different assays, where 7×10^6 sperm/mL was the optimal sperm concentration. When lower concentrations were used ($<7 \times 10^6$ sperm/mL), we did not recover enough sperm to perform the assays, whereas higher concentrations ($>7 \times 10^6$ sperm/mL) showed a similar number of sperm released and attached to the OECs to that obtained with 7×10^6 sperm/mL, likely due to the saturation of the sperm-binding sites on the OECs. The sperm's viability and motility were not affected by sperm concentration because the gametes were selected and immediately co-incubated with oviductal cells. Unbound spermatozoa were removed by washing the monolayers three times with NCWM. At this point, we evaluated the acrosome status and motility of sperm bound to OECs (Figure 1a).

Evaluation of Acrosome Status of Sperm Bound to OEC Primary Cultures

To evaluate the acrosome status of spermatozoa bound to OECs (Figure 1a), co-cultures were washed and fixed in 0.4% *w*/*v* paraformaldehyde in PBS for 1 h at RT and permeabilized with methanol for 5 min at 4 °C. The fixed co-cultures were incubated with PSA-FITC (10 μg/mL) for 1 h at RT. After three washes with PBS, DNA was stained with Hoechst 33352 (1 μg/mL) for 10 min at RT. Samples were mounted with Fluor Save (Merck Millipore, Burlington, MA, USA) and examined with a fluorescence Nikon 80i microscope (Nikon Instruments Inc., Tokyo, Japan) coupled to a digital camera. Not less than 20 fields per experiment were analyzed and acrosome status was studied at least on three different sperm–OEC co-cultures (n = 3). Results are shown with one representative image.

2.6. Retrieval of the Released and Bound Sperm Population after the Incubation of the Co-Cultures under Capacitating Conditions

The OEC/sperm co-cultures were washed three times with NCWM to remove unbound sperm. Then, were incubated with CWM or NCWM (control) for 15 min and washed again three times with NCWM to recover released sperm (Figure 1b) and the population that remained bound to the OECs (Figure 1c). The following analyses were performed in the released sperm population retrieved: the number of sperm, viability, acrosome integrity, motility and capacitation status. Additionally, we evaluated the number of sperm bound to the OECs. A replicate (*n*) in these experiments was defined as the co-culture of an OEC monolayer co-cultured with sperm from one stallion (total 4 stallions). All the treatments (including the control) were performed for each replicate.

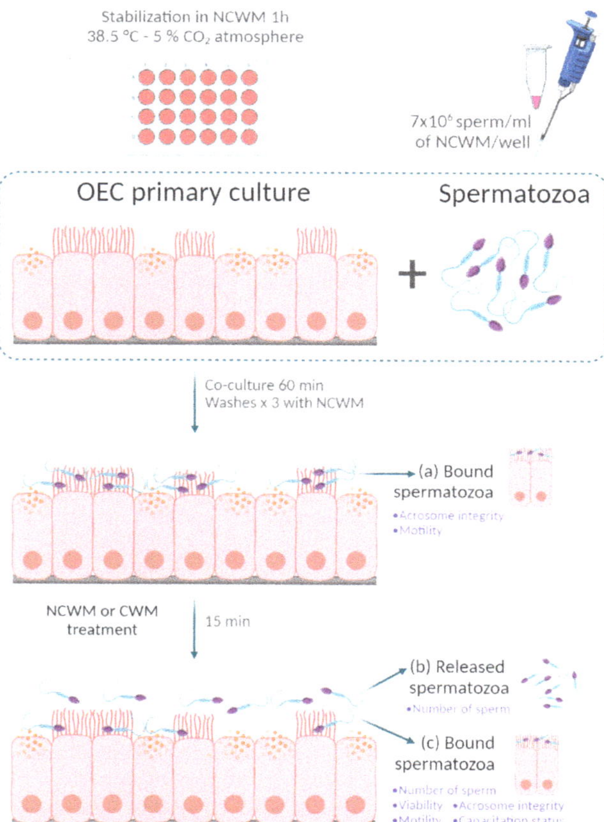

Figure 1. Oviductal epithelial cells and sperm co-culture experimental design. Confluent oviductal epithelial cell (OEC) monolayers were stabilized with non-capacitating modified Whitten's base medium (NCWM) 1 h before sperm addition. Scheme 7 × 10^6 sperm/mL of NCWM) were added to each well and were co-cultured with OECs for 60 min at 38.5 °C in a 5% CO_2 atmosphere. Unbound spermatozoa were removed by washing the monolayers three times with NCWM. Acrosome status and motility were evaluated on (**a**) sperm that remained bound to OECs. Next, the sperm–OEC co-cultures were incubated with NCWM or capacitating medium (CWM) for 15 min. After that time, monolayers were washed three times with NCWM to analyze (**b**) the number of released sperm and (**c**) the number, viability, motility, acrosome integrity and capacitation status of the sperm that remained bound to the OECs after treatments. Created with BioRender.com.

2.6.1. Evaluation of the Number of Sperm Bound to the OECs after Capacitating Treatment

Co-cultures were fixed in glutaraldehyde 2.5% *v*/*v* for 60 min at RT and washed three times with PBS. The number of sperm that remained attached (Figure 1c) was determined by examining 20 fields under a phase-contrast microscope (Olympus, Tokyo, Japan). Results were expressed as the mean of the average number of bound spermatozoa in a 0.11 mm^2 area per replicate.

2.6.2. Evaluation of the Number of Sperm Released from the OECs under Capacitating Conditions

The released sperm population recovered (Figure 1b) was fixed (0.2% *w*/*v* paraformaldehyde) for 30 min at RT and the number of sperm was determined using a hemocytometer (results are shown as the total number of released spermatozoa).

2.6.3. Evaluation of Viability of the Released Sperm Population from the OECs under Capacitating Conditions

To assess viability, the released sperm population recovered was incubated with Hoechst 33258 (2 µg/mL) for 5 min, fixed (1% *w*/*v* paraformaldehyde) for 8 min at RT, washed with PBS and aliquots were air-dried onto glass slides. Hoechst 33258 is a fluorescent DNA-binding supravital stain with limited membrane permeability [45–47]. At least 200 stained cells/treatment were scored in an epifluorescence Nikon 80i microscope (Nikon Instruments Inc., Tokyo, Japan). Results are shown as the percentage of live spermatozoa.

2.6.4. Evaluation of Acrosome Status of the Released Sperm Population from the OECs under Capacitating Conditions

To assess acrosome status, spermatozoa were incubated with Hoechst 33258 as described before (see Section 2.6.3.). Aliquots were air-dried onto glass slides and permeabilized in methanol for 10 min at 4 °C. Slides were incubated with *Pisum sativum* agglutinin-FITC (PSA-FITC, 50 mg/mL) for 1 h at RT. At least 200 stained cells per treatment were evaluated using an epifluorescence Nikon 80i microscope (Nikon Instruments Inc., Tokyo, Japan). Results are shown as the percentage of live spermatozoa with intact acrosomes.

2.6.5. Evaluation of Total Motility of the Released Sperm Population from the OECs under Capacitating Conditions

Sperm released were briefly concentrated by a 3 s centrifugation and total sperm motility was analyzed using CASA-system (IVOS II^{TM} Animal—Hamilton Thorne) as described before (see Section 2.3.2). Moreover, sperm total motility was subjectively evaluated using a field microscope stage heated at 38.5 °C at 100× magnification with similar results. Results are shown as the % of total motile sperm.

2.6.6. Evaluation of Capacitation Status of the Released Sperm Population from the OECs under Capacitating Conditions

The capacitation status of released sperm was analyzed by examining the phosphorylation of protein kinase A substrates (pPKAs) and tyrosine-phosphorylation of proteins (pY) using immunofluorescence ([48] with some modifications). Specifically, we first determined sperm viability with Hoechst 33258 as described before (see Section 2.6.3.). Then, spermatozoa were fixed (20 min, at RT with 0.2% *w*/*v* paraformaldehyde), immobilized on slides and permeabilized with TPBS-Triton X100 0.5% for 20 min at RT. Non-specific binding sites were blocked (60 min, at RT with 3% *w*/*v* BSA TPBS) and incubated with pPKA (1:500) and pY (1:500) antibodies diluted in PBS for 18 h at 4 °C. Samples were washed and further incubated with Alexa 555-conjugated goat anti-rabbit IgG (red, 1:500) and Alexa 488 chicken anti-mouse IgG (green, 1:500) diluted in PBS for 1 h at RT. The specificity of the immunodetection was assessed by omitting the first antibody. Sperm cells were mounted and examined under a fluorescence Nikon 80i microscope (Nikon Instruments Inc., Tokyo, Japan). The proportion of spermatozoa with green and red fluorescent tails among the live sperm population (without Hoechst 33258 fluorescent heads) was determined by randomly scoring 200 spermatozoa. Results are shown as the percentage of live spermatozoa with both stains: positive pPKA and positive pY.

2.7. Statistical Analysis

The effect of the incubation time and sample concentration on % total motility was analyzed by two-way ANOVA in a completely randomized design with repeated measures. Comparisons were made with Tukey's post hoc test ($p < 0.05$). The statistical analysis of

% total motility between the times of centrifugation was analyzed by one-way ANOVA (time) in a completely randomized design. Comparisons between the mean of each time of centrifugation and control (0 min) were made with Dunnett's post hoc test ($p < 0.05$). For the evaluation of the number of sperm that remained attached to OECs after treatments (Section 2.6.1., number of bound sperm), comparison between groups was performed with a one-way ANOVA in blocks, where each pool of OECs with spermatozoa was considered a block, and all treatments were applied to it (randomized blocks). When the ANOVA tests were significant ($p < 0.05$), multiple comparisons were performed by Tukey´s post hoc test ($p < 0.05$). The effect of dependent variables measured in the released sperm population from OEC primary cultures (i.e., number of released sperm, % viability, % acrosome-intact sperm, % motile sperm and % capacitated sperm) were analyzed by *t*-test ($p < 0.05$). The assumptions of normality (ANOVA and *t*-test) and homogeneity of variances (ANOVA) were assessed prior to performing the statistical analysis by Shapiro–Wilks test and Levene test, respectively. In the case of the two-way ANOVA, as we chose not to accept the assumption of the sphericity, we used the method of Geisser and Greenhouse to correct for violations of the assumption. All values represent the mean $\pm$ S.E.M. Statistical analyses were performed using the Prism 7 software package (GraphPad, La Jolla, CA, USA).

3. Results

3.1. Effect of Different Processing Conditions on Sperm Motility

To determine whether different processing conditions of cryopreserved samples affect sperm motility, we evaluated the effect of incubation time, centrifugation time and sample concentration on this parameter. A repeated-measures two-way ANOVA was applied to study the interaction between time of incubation and sample concentration on the % of sperm total motility. The results showed a statistically significant interaction between the two independent variables on the dependent variable ($p < 0.05$), i.e., that the changes in the motility during the incubation depended on the sample concentration. Specifically, we observed that samples incubated at 12×10^6 sperm/mL did not show changes in motility at different times of incubation, while samples incubated at 30×10^6 sperm/mL showed a decrease in motility after 30 and 120 min of incubation (Figure 2A). It is important to highlight that these equine cryopreserved sperm samples present poor motility after the thawing process (0 min; total motility <35%) (Figure 2A). However, the motility of the lower-concentrated samples was always lower than 10% and we found that at time 0 samples incubated at 30×10^6 sperm/mL presented higher motility compared to 12×10^6 sperm/mL samples, suggesting that it is convenient to use more concentrated samples (Figure 2A). After more than 60 min of incubation, samples at 30×10^6 sperm/mL did not show differences in motility compared to lower-concentrated samples (Figure 2A). Centrifugation steps could be necessary for some ART protocols and in the study of sperm physiology. In this sense, we evaluated how two different centrifugation (at $200\times g$) times may affect sperm motility when sperm were incubated at concentrations of 30×10^6 sperm/mL. This sperm concentration was used because sperm motility was not affected for short times (Figure 2A). We did not find differences in this parameter after centrifugations for 1 min in comparison to samples without centrifugation (control) (Figure 2B). However, we observed a significant decrease in sperm motility when the samples were centrifuged for 2 min compared with control (Figure 2B).

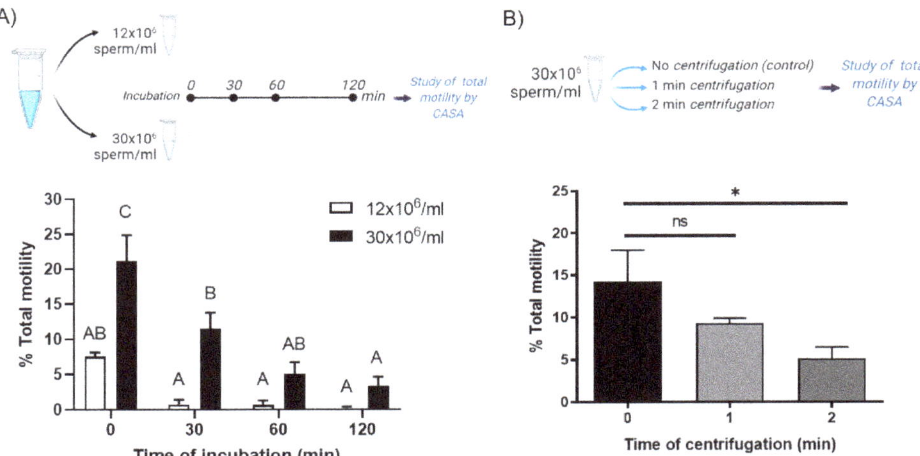

Figure 2. Parameters that affect cryopreserved equine sperm motility. (**A**) Time of incubation and sample concentration. Spermatozoa were incubated in NCWM at 12 × 10^6/mL or at 30 × 10^6/mL and motility was determined at different times (0, 30, 60 and 120 min). Two-way ANOVA with repeated measures ($p < 0.05$). $n = 4$. Means with different letters are significantly different (Tuckey post hoc test). (**B**) Time of centrifugation. Spermatozoa were incubated at 30 × 10^6/mL centrifuged for 0 min (control), 1 min or 2 min at 200× *g*. One-way ANOVA ($p < 0.05$). ns: not statistically different. * Indicates statistically significant differences (Dunnett's post hoc test between the mean of each time of centrifugation and control). $n = 4$. Bars represent the mean ± SEM of total motile sperm. Percentage of total motility was assessed by Computer Assisted Sperm Analysis (CASA).

3.2. OEC Primary Culture Characterization

To characterize the obtained OEC primary cultures, we examined the expression of the epithelial markers E-cadherin and cytokeratin-7 using RT-PCR and immunofluorescence. As Figure 3A shows, OECs cultured in vitro expressed mRNAs for both epithelial cell markers whereas equine fibroblasts did not (data not shown). In agreement with these observations, E-cadherin and cytokeratin-7 protein expression was detected in OECs, with E-cadherin mainly present on cell membranes and cytokeratin-7 in the cytoplasm (Figure 3B).

3.3. Characterization of Cryopreserved Semen Bound to OEC Primary Cultures

As shown in Figure 4A, sperm not only adhered to the OECs immediately after insemination (0 min), but it remained bound even after 1 h of co-culture and several washes with NCWM. Moreover, we observed that the majority of the OEC attached sperm were motile (Supplementary Materials, Video S1) and with an intact acrosome (Figure 4B).

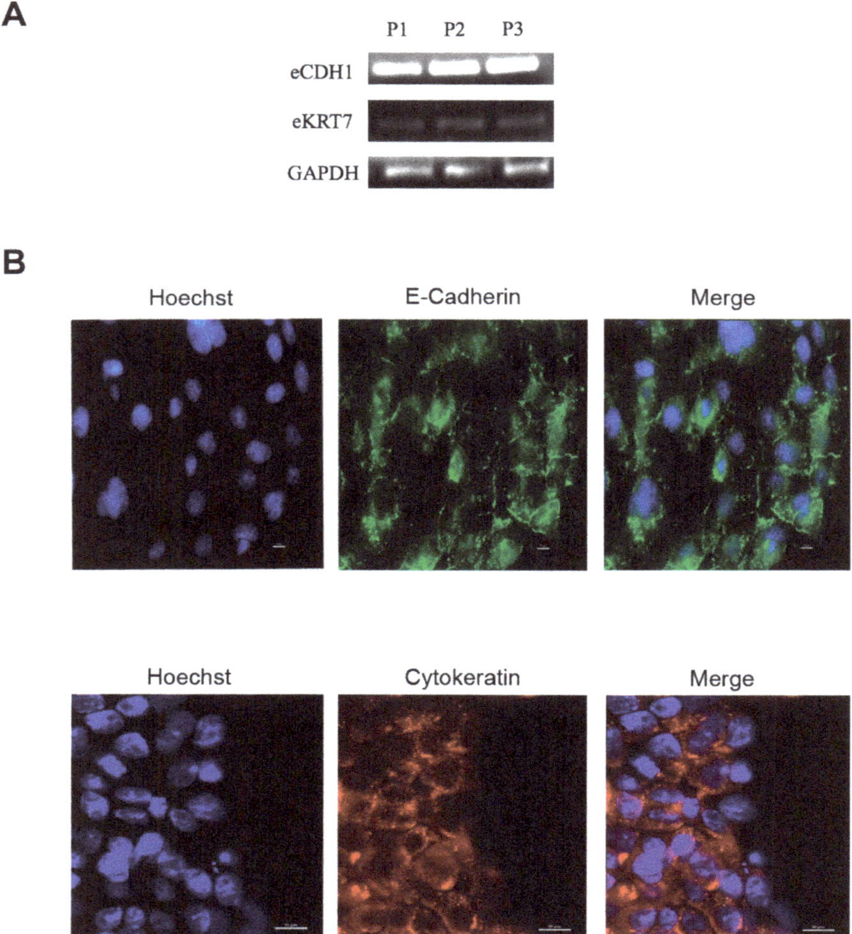

Figure 3. Characterization of OEC primary cultures derived from equine oviduct explants. (**A**) Expression of epithelial markers in equine OECs was determined by RT-PCR employing equine specific primers for eCDH1 (E-cadherin), eKRT7 (cytokeratin-7) and GAPDH. Representative results of three different pools of OECs are shown (P1, P2, P3; $n = 3$); (**B**) Intracellular distribution of E-cadherin (green) and cytokeratin (red) were examined using immunofluorescence (see Section 2.4.3., Materials and Methods). DNA was labeled with Hoechst 33352 (blue), E-cadherin: 200× magnification, cytokeratin: 400× magnification ($n = 3$). Bar = 10 µm.

3.4. Characterization of Cryopreserved Sperm Released after Co-Culture with OECs under Capacitating Conditions

Previous studies indicated that spermatozoa release from oviductal cells in vivo and in vitro is associated with sperm capacitation. Based on that, our results showed that cryopreserved stallion sperm's attachment to OECs in vitro did not impact their motility or acrosomes integrity, we examined whether released spermatozoa from OECs under capacitating condition (CWM) affected sperm viability, acrosome status and progressive motility. We also assessed PKA phosphorylated substrates (pPKA) and protein tyrosine phosphorylation (pY), two widely employed sperm capacitation indicators.

A

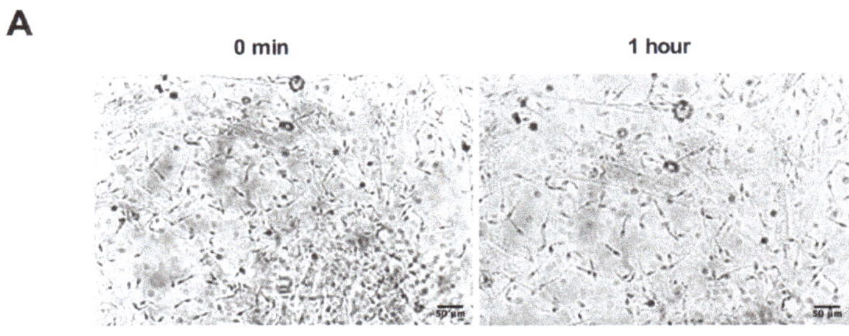

B

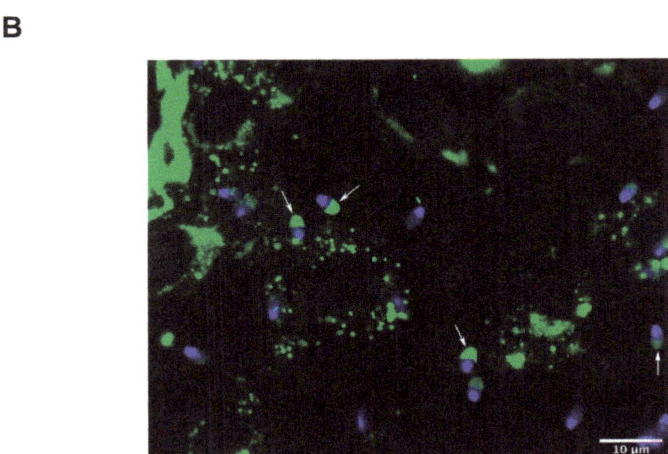

Figure 4. Characterization of cryopreserved semen bound to OEC primary cultures (**A**) Phase-contrast representative images of cryopreserved equine sperm co-cultured with OECs, 400× magnification, $n = 6$. Bar = 50 µm. (**B**) Representative images of intact acrosomes (white arrows) examined by immunofluorescence 1 h after co-culture with OECs stained with PSA-FITC (green, acrosome) and Hoechst 33352 (blue, DNA). 1000× magnification, $n = 3$. Bar = 10 µm.

As Figure 5A,B shows, the treatment with CWM reduced the number of OEC-bound spermatozoa ($p < 0.05$) concurrently increasing the number of released ones ($p < 0.05$). The released sperm in response to CWM treatment also showed a significantly higher percentage of alive spermatozoa ($p < 0.05$) with intact acrosomes ($p < 0.05$) and total motility ($p < 0.05$) (Figure 5C–E). In contrast, this effect on motility was not observed during the incubation of cryopreserved spermatozoa at the same conditions (7×10^6 sperm/mL, 60 min) in OECs' absence where total motility was around 0% (subjectively evaluated using a field microscope, data not shown). Moreover, spermatozoa released in response to CWM showed an increase in PKA activity and protein tyrosine phosphorylation (pY) ($p < 0.05$) (Figure 5F).

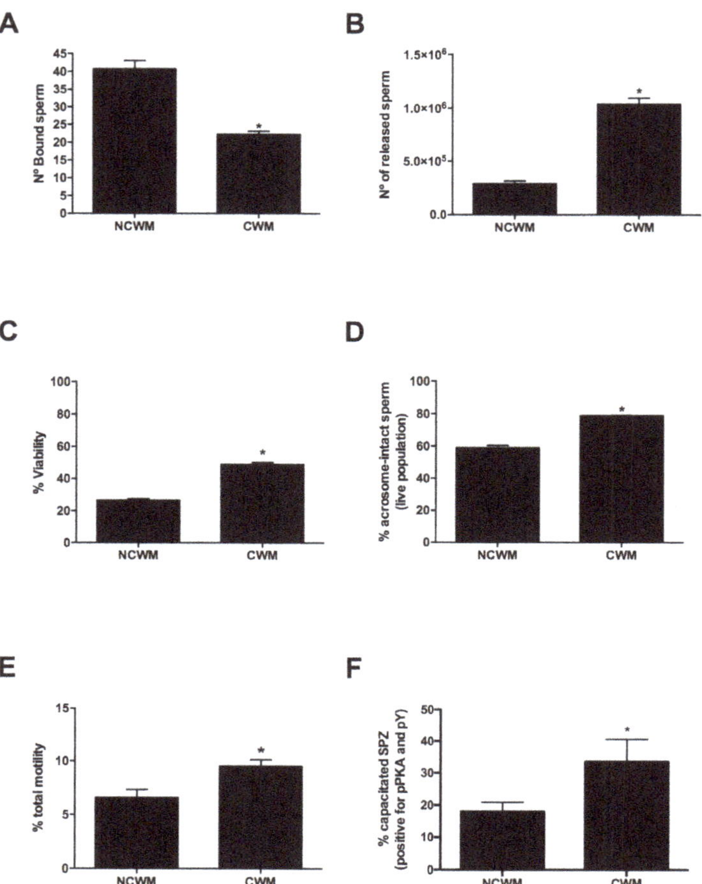

Figure 5. Characterization of sperm released from OECs primary cultures. OEC and sperm co-cultures were incubated for 15 min at 37 °C in the presence of NCWM (no capacitating condition) or CWM (capacitating condition) and the spermatozoa bound (**A**) to OECs were quantified using bright field microscopy while (**B**) the released sperm were recovered and quantified with a hemocytometer. Bars represent the mean ± SEM of bound spermatozoa/0.11 mm^2 monolayer (**A**) and (**B**) the mean ± SEM of the number of released sperm. * $p < 0.05$, $n = 6$. (**C**) Sperm released from the co-cultures were incubated with Hoechst 33258, fixed and the percentage of live sperm was determined using fluorescence microscopy. Bars represent the mean ± SEM of live spermatozoa. * $p < 0.05$, $n = 6$. (**D**) Sperm viability and acrosome status were evaluated as described in Section 2. Bars represent the mean ± SEM of live spermatozoa with intact acrosomes. * $p < 0.05$, $n = 6$. (**E**) The total motility of released sperm was determined by CASA. Bars represent the mean ± SEM of total motile spermatozoa. * $p < 0.05$, $n = 6$. (**F**) OEC-released sperm positive for pPKA and pY were evaluated using immunofluorescence. Bars represent the mean ± SEM of live sperm. * $p < 0.05$, $n = 6$.

4. Discussion

The use of stallion frozen semen minimizes the spread of diseases, eliminates geographic barriers, and preserves the genetic material of the animal for an unlimited time [12]. The efficiency of equine semen cryopreservation depends mainly on the animal's characteristics such as genetics and age [49]. Additionally, stallions that are satisfactorily fertile under normal field conditions can produce semen that after freezing and thawing results in very low pregnancy rates [10]. Several factors influence the cryo-survival of stallion sperm including freezing regimes [12,50,51], oxidative and osmotic stress, ice crystal formation, toxicity of the cryoprotectants [8,52,53], sample processing [54] and variability among stallions [11]. Consequently, the attachment of cryopreserved equine spermatozoa to equine OECs or zona pellucida in vitro is reduced compared to that of fresh spermatozoa [55]. This limitation is very likely due to a reduction in post-thaw motility associated with changes in the integrity of the sperm membrane, suggesting a possible mechanism to explain the reduced fertility achieved with cryopreserved samples versus fresh spermatozoa in horses. These negatives effects of cryopreservation on sperm function result in low ART success rates. Given the characteristics of cryopreserved semen samples, it is important to know how to manipulate them, not only for their application in ART but also for their use in research studies.

Previously, Hayden et al. showed that raw stallion semen dilution with commercial extenders decreased the total motility [56]. In agreement with these observations, our results indicate that sperm total motility—measured shortly after thawing—was concentration-dependent, i.e., more concentrated samples displayed higher motility. Moreover, after more than 60 min of incubation, samples at 30×10^6 sperm/mL did not show differences in motility compared to lower-concentrated samples. We studied these sperm concentration and sperm total motility from 0 to 120 min because there are different protocols described for post-thawed sperm incubation times for research studies, such as sperm capacitation a process with high relevance in ARTs applied in equines. Those concentrations range between 10×10^6 sperm/mL to 200×10^6 sperm/mL [57–62] and times ranged between 10 and 120 min [58,63,64]. Sperm concentrations used were chosen based on the CASA system optimal-concentration working range (10–50 $\times 10^6$ sperm/mL). Moreover, several sperm selection protocols used during ARTs require multiple centrifugations. It was previously described that different conditions of centrifugation alter motility and oxidative status, and consequently increase the DNA damage of cryopreserved stallion sperm [65,66]. We tested a $200\times g$ force, not previously described for cryopreserved stallion sperm and its effect on motility. As previously reported, we found that total sperm motility was negatively affected by centrifugation times longer than 2 min, even at lower g-forces [65]. In this work, we have used samples that were frozen poorly (<35% motility post-thawing) [67], suggesting that a better outcome could be possible using samples with better post-thaw motility.

The mammalian oviduct plays an essential role during the selection of competent sperm subpopulations, and it is involved in the maintenance of fertilization capacity during sperm storage (sperm reservoir) [23,68]. Accordingly, we developed an in vitro OEC culture model to select sperm populations from cryopreserved samples with fertilizing potential.

In view of the role of OECs in sperm selection under physiological conditions, we speculated whether the co-culture of equine cryopreserved semen and equine-derived OEC primary cultures would replicate parameters observed in the intact reproductive tract including OEC–sperm binding, motility and acrosome integrity [23,69,70].

Thomas et al. have shown that a subpopulation of morphologically normal and motile spermatozoa attach to equine OEC monolayers using fresh stallion semen [31]. In agreement with these results, we observed that cryopreserved equine sperm, processed under our optimized conditions, not only attached to the OEC culture model but also maintained their motility and presented an intact acrosome. These spermatozoa maintained their motility after 1 h of in vitro co-culture with OECs. In contrast, this effect on motility was not observed during the incubation of cryopreserved spermatozoa at the same conditions in OECs' absence. Thus, the in vitro equine OEC monolayer culture established in this work would be a useful tool for the selection of a sperm population with fertilization potential from cryopreserved samples.

Previous studies indicated that spermatozoa release from oviductal cells in vivo and in vitro is associated with sperm capacitation and hyperactivation [23,34,36,37,69,71,72]. These processes stimulate sperm release from the oviductal epithelium, to come into contact and fertilize an oocyte [23,73]. Based on that, our results showed that cryopreserved stallion sperm's attachment to OECs in vitro did not impact the motility or the acrosome's integrity, we examined whether spermatozoa released from OECs under capacitating conditions (CWM) presented affected sperm viability, acrosome status and progressive motility. We also assessed PKA phosphorylated substrates (pPKA) and protein tyrosine phosphorylation (pY), two widely utilized sperm capacitation indicators [74].

In previous works, it has been described that the in vitro incubation of equine sperm in CWM medium increases protein phosphorylation in tyrosine residues, progressive motility and the induction of the acrosomal reaction, all events associated with sperm capacitation [44]. In agreement with these observations, our results showed that the incubation of sperm–OEC cultures in the presence of CWM promoted the release of a greater number of sperm. The released sperm were alive, motile and presented an intact acrosome and an increase in molecular markers, i.e., PKA activity and Tyr phosphorylation, associated with sperm capacitation. Thus, these results indicate that the frozen equine sperm–OEC co-culture model provides a useful system to enrich spermatozoa populations with fertilization potential.

ICSI is a low-embryo-yield technique within the equine breeding industry that can achieve similar embryo development using frozen or fresh equine spermatozoa [17,75]. Different methods are employed to select stallion sperm prior to ICSI (swim-up procedure, density gradient centrifugation or microfluidics) in order to increase the probability of selecting sperm that when used will result in optimal fertility [76–78]. Within this context, the probability that sperm-injected oocytes develop into an embryo (morula or blastocyst) improves when frozen–thawed stallion sperm show high membrane integrity [79]. In this regard, we speculated that sperm population released from OEC co-culture under capacitating conditions could be used to enhance ICSI efficiency and embryo quality in equines due to their potential fertilizing capacity.

Further studies to achieve a better understanding of the molecular mechanisms that regulate the acquisition of spermatozoa fertilization capacity during their transit through the female reproductive tract will favor the development of new sperm selection methods to be incorporated into ARTs for equines and other animal species.

5. Conclusions

Our results show that the total motility of previously frozen equine sperm samples is dependent on its concentration, the incubation time and the centrifugation duration applied during processing. We also found that cryopreserved spermatozoa interacted with OEC cultures in vitro and that this equine sperm–OEC co-culture model could be a useful tool to select a sperm population with potential fertilizing capacity under capacitating conditions.

In conclusion, this work contributed to the existing knowledge on the effect of different conditions associated with the manipulation of stallion cryopreserved sperm. Although further analyses are needed, we speculated that the selection of higher-quality male gametes using the in vitro OEC primary culture established in this study would improve, in future, the efficiency of ARTs as well as the quality of the obtained embryos in equines.

Supplementary Materials: The following are available online at https://www.mdpi.com/2076-2615/11/1/74/s1, Video S1: Cryopreserved equine sperm attached to OECs in vitro after 1 h of co-culture.

Author Contributions: Conceptualization, C.E.O.-S.; methodology, B.F.G., M.V.B., L.L.-Q., E.M.-L., M.V.-M. and C.E.O.-S.; formal analysis, B.F.G., L.L.-Q. and M.V.B.; investigation, B.F.G., L.L.-Q., M.V.B. and C.E.O.-S.; resources, O.R. and A.Á.M.; writing—original draft preparation, C.E.O.-S.; writing—review and editing, M.V.B., E.M.-L., O.R., A.Á.M. and C.E.O.-S.; supervision, A.Á.M. and C.E.O.-S.; funding acquisition, A.Á.M. and C.E.O.-S. All authors have read and agreed to the published version of the manuscript.

Funding: This study was supported by grants from Agencia Nacional de Promoción de la Investigación, el Desarrollo Tecnológico y la Innovación of Argentina PICT 2016-3138 to C.E.O.-S and PICT 2015-1548 to A.A.M.

Institutional Review Board Statement: Not applicable.

Informed Consent Statement: Not applicable.

Data Availability Statement: The data presented in this study are available within the article or supplementary material.

Acknowledgments: We thank to the Equine Breeding Centers (GeneTec by Ativet and Los Pingos del Taita) and Lamar S.A. slaughterhouse for providing semen and oviducts respectively. Authors thank to Francisco Guaimas for his technical support.

Conflicts of Interest: The authors declare no conflict of interest. The funders had no role in the design of the study; in the collection, analyses, or interpretation of data; in the writing of the manuscript; or in the decision to publish the results.

References

1. Tecirlioglu, R.T.; Trounson, A.O. Embryonic stem cells in companion animals (horses, dogs and cats): Present status and future prospects. *Reprod. Fertil. Dev.* **2007**, *19*, 740–747. [CrossRef] [PubMed]
2. McCoy, A.M. Animal Models of Osteoarthritis: Comparisons and Key Considerations. *Vet. Pathol.* **2015**, *52*, 803–818. [CrossRef] [PubMed]
3. Bond, S.; Léguillette, R.; Richard, E.A.; Couetil, L.; Lavoie, J.P.; Martin, J.G.; Pirie, R.S. Equine asthma: Integrative biologic relevance of a recently proposed nomenclature. *J. Vet. Intern. Med.* **2018**, *32*, 2088–2098. [CrossRef] [PubMed]
4. Cong, X.; Zhang, S.M.; Ellis, M.W.; Luo, J. Large Animal Models for the Clinical Application of Human Induced Pluripotent Stem Cells. *Stem Cells Dev.* **2019**, *28*, 1288–1298. [CrossRef] [PubMed]
5. Koch, T.G.; Betts, D.H. Stem cell therapy for joint problems using the horse as a clinically relevant animal model. *Expert Opin. Biol. Ther.* **2007**, *7*, 1621–1626. [CrossRef] [PubMed]
6. Moore, S.G.G.; Hasler, J.F.F. A 100-Year Review: Reproductive technologies in dairy science. *J. Dairy Sci.* **2017**, *100*, 10314–10331. [CrossRef]
7. Hezavehei, M.; Sharafi, M.; Kouchesfahani, H.M.; Henkel, R.; Agarwal, A.; Esmaeili, V.; Shahverdi, A. Sperm cryopreservation: A review on current molecular cryobiology and advanced approaches. *Reprod. Biomed. Online* **2018**, *37*, 327–339. [CrossRef]
8. Treulen, F.; Aguila, L.; Arias, M.E.; Jofré, I.; Felmer, R. Impact of post-thaw supplementation of semen extender with antioxidants on the quality and function variables of stallion spermatozoa. *Anim. Reprod. Sci.* **2019**, *201*, 71–83. [CrossRef]
9. Ball, B.A.; Vo, A.T.; Baumber, J. Generation of reactive oxygen species by equine spermatozoa. *Am. J. Vet. Res.* **2001**, *62*, 508–515. [CrossRef]
10. Sieme, H.; Harrison, R.A.P.; Petrunkina, A.M. Cryobiological determinants of frozen semen quality, with special reference to stallion. *Anim. Reprod. Sci.* **2008**, *107*, 276–292. [CrossRef]
11. Vidament, M.; Dupere, A.M.; Julienne, P.; Evain, A.; Noue, P.; Palmer, E. Equine frozen semen: Freezability and fertility field results. *Theriogenology* **1997**, *48*, 907–917. [CrossRef]
12. Alvarenga, M.A.; Papa, F.O.; Ramires Neto, C. Advances in Stallion Semen Cryopreservation. *Vet. Clin. N. Am. Equine Pract.* **2016**, *32*, 521–530. [CrossRef] [PubMed]
13. Neuhauser, S.; Bollwein, H.; Siuda, M.; Handler, J. Effects of Different Freezing Protocols on Motility, Viability, Mitochondrial Membrane Potential, Intracellular Calcium Level, and DNA Integrity of Cryopreserved Equine Epididymal Sperm. *J. Equine Vet. Sci.* **2019**, *82*. [CrossRef] [PubMed]
14. Allen, W. The Development and Application of the Modern Reproductive Technologies to Horse Breeding. *Reprod. Domest. Anim.* **2005**, *40*, 310–329. [CrossRef] [PubMed]
15. Leemans, B.; Gadella, B.M.; Stout, T.A.E.; De Schauwer, C.; Nelis, H.; Hoogewijs, M.; Van Soom, A. Why doesn't conventional IVF work in the horse? The equine oviduct as a microenvironment for capacitation/fertilization. *Reproduction* **2016**, *152*, R233–R245. [CrossRef]
16. Leemans, B.; Stout, T.A.E.; De Schauwer, C.; Heras, S.; Nelis, H.; Hoogewijs, M.; Van Soom, A.; Gadella, B.M. Update on mammalian sperm capacitation: How much does the horse differ from other species? *Reproduction* **2019**, *157*, R181–R197. [CrossRef]
17. Hinrichs, K. Assisted reproductive techniques in mares. *Reprod. Domest. Anim.* **2018**, *53*, 4–13. [CrossRef]
18. Bedford, S.J.; Kurokawa, M.; Hinrichs, K.; Fissore, R.A. Patterns of Intracellular Calcium Oscillations in Horse Oocytes Fertilized by Intracytoplasmic Sperm Injection: Possible Explanations for the Low Success of This Assisted Reproduction Technique in the Horse. *Biol. Reprod.* **2004**, *70*, 936–944. [CrossRef]
19. Morris, L.H.A. The development of in vitro embryo production in the horse. *Equine Vet. J.* **2018**, *50*, 712–720. [CrossRef]
20. Squires, E.L. Integration of future biotechnologies into the equine industry. *Anim. Reprod. Sci.* **2005**, *89*. [CrossRef]
21. Holt, W.V.; Elliott, R.M.; Fazeli, A.; Sostaric, E.; Georgiou, A.S.; Satake, N.; Prathalingam, N.; Watson, P.F. Harnessing the biology of the oviduct for the benefit of artificial insemination. *Soc. Reprod. Fertil. Suppl.* **2006**, *62*, 247–259. [PubMed]
22. Li, X.; Morris, L.H.A.; Allen, W.R. Influence of co-culture during maturation on the developmental potential of equine oocytes fertilized by intracytoplasmic sperm injection (ICSI). *Reproduction* **2001**, *121*, 925–932. [CrossRef] [PubMed]

23. Suarez, S.S. Regulation of sperm storage and movement in the mammalian oviduct. *Int. J. Dev. Biol.* **2008**, *52*, 455–462. [CrossRef] [PubMed]
24. Ellington, J.E.; Ignotz, G.G.; Varner, D.D.; Marcucio, R.S.; Mathison, P.; Ball, B.A. In vitro interaction between oviduct epithelial and equine sperm. *Arch. Androl.* **1993**, *31*, 79–86. [CrossRef] [PubMed]
25. Aldarmahi, A.; Elliott, S.; Russell, J.; Klonisch, T.; Hombach-Klonisch, S.; Fazeli, A. Characterisation of an in vitro system to study maternal communication with spermatozoa. *Reprod. Fertil. Dev.* **2012**, *24*, 988–998. [CrossRef] [PubMed]
26. Dobrinski, I.; Ignotz, G.G.; Thomas, P.G.; Ball, B.A. Role of carbohydrates in the attachment of equine spermatozoa to uterine tubal (oviductal) epithelial cells in vitro. *Am. J. Vet. Res.* **1996**, *57*, 1635–1639.
27. Dobrinski, I.; Suarez, S.S.; Ball, B.A. Intracellular calcium concentration in equine spermatozoa attached to oviductal epithelial cells in vitro. *Biol. Reprod.* **1996**, *54*, 783–788. [CrossRef]
28. Suarez, S.; Redfern, K.; Raynor, P.; Martin, F.; Phillips, D.M. Attachment of boar sperm to mucosal explants of oviduct in vitro: Possible role in formation of a sperm reservoir. *Biol. Reprod.* **1991**, *44*, 998–1004. [CrossRef]
29. Gualtieri, R.; Talevi, R. In vitro-cultured bovine oviductal cells bind acrosome-intact sperm and retain this ability upon sperm release. *Biol. Reprod.* **2000**, *62*, 1754–1762. [CrossRef]
30. Ardón, F.; Helms, D.; Sahin, E.; Bollwein, H.; Töpfer-Petersen, E.; Waberski, D. Chromatin-unstable boar spermatozoa have little chance of reaching oocytes in vivo. *Reproduction* **2008**, *135*, 461–470. [CrossRef]
31. Thomas, P.G.A.; Ball, B.A.; Miller, P.G.; Brinsko, S.P.; Southwood, L. A Subpopulation of Morphologically Normal, Motile Spermatozoa Attach to Equine Oviductal Epithelial Cell Monolayers. *Biol. Reprod.* **1994**, *51*, 303–309. [CrossRef] [PubMed]
32. Ellington, J.E.; Samper, J.C.; Jones, A.E.; Oliver, S.A.; Burnett, K.M.; Wright, R.W. In vitro interactions of cryopreserved stallion spermatozoa and oviduct (uterine tube) epithelial cells or their secretory products. *Anim. Reprod. Sci.* **1999**, *56*, 51–65. [CrossRef]
33. Thomas, P.G.A.; Ball, B.A.; Brinsko, S.P. Interaction of Equine Spermatozoa with Oviduct Epithelial Cell Explants Is Affected by Estrous Cycle and Anatomic Origin of Explant. *Biol. Reprod.* **1994**, *228*, 222–228. [CrossRef] [PubMed]
34. Lefebvre, R.; Suarez, S.S. Effect of capacitation on bull sperm binding to homologous oviductal epithelium. *Biol. Reprod.* **1996**, *54*, 575–582. [CrossRef] [PubMed]
35. Fazeli, A.; Duncan, A.E.; Watson, P.F.; Holt, W. V Sperm-oviduct interaction: Induction of capacitation and preferential binding of uncapacitated spermatozoa to oviductal epithelial cells in porcine species. *Biol. Reprod.* **1999**, *60*, 879–886. [CrossRef] [PubMed]
36. Osycka-Salut, C.E.; Castellano, L.; Fornes, D.; Beltrame, J.S.; Alonso, C.A.I.; Jawerbaum, A.; Franchi, A.; Díaz, E.S.; Perez Martinez, S. Fibronectin From Oviductal Cells Fluctuates During the Estrous Cycle and Contributes to Sperm–Oviduct Interaction in Cattle. *J. Cell. Biochem.* **2017**, *118*, 4095–4108. [CrossRef]
37. Gervasi, M.G.; Rapanelli, M.; Ribeiro, M.L.; Farina, M.; Billi, S.; Franchi, A.M.; Perez Martinez, S. The endocannabinoid system in bull sperm and bovine oviductal epithelium: Role of anandamide in sperm-oviduct interaction. *Reproduction* **2009**, *137*, 403–414. [CrossRef]
38. de Lamirande, E.; O'Flaherty, C. Sperm activation: Role of reactive oxygen species and kinases. *Biochim. Biophys. Acta* **2008**, *1784*, 106–115. [CrossRef]
39. Visconti, P.E.; Westbrook, V.A.; Chertihin, O.; Demarco, I.; Sleight, S.; Diekman, A.B. Novel signaling pathways involved in sperm acquisition of fertilizing capacity. *J. Reprod. Immunol.* **2002**, *53*, 133–150. [CrossRef]
40. Buffone, M.G.; Wertheimer, E.V.; Visconti, P.E.; Krapf, D. Central role of soluble adenylyl cyclase and cAMP in sperm physiology. *Biochim. Biophys. Acta* **2014**, *1842*, 2610–2620. [CrossRef]
41. Hirohashi, N.; Yanagimachi, R. Sperm acrosome reaction: Its site and role in fertilization. *Biol. Reprod.* **2018**, *99*, 127–133. [CrossRef] [PubMed]
42. Yanagimachi, R. Mammalian fertilization. In *The Physiology of Reproduction*; Academic Press: Cambridge, MA, USA, 1994; pp. 189–317.
43. McPartlin, L.; Suarez, S.S.; Czaya, C.; Hinrichs, K.; Bedford-Guaus, S.J. Hyperactivation of stallion sperm is required for successful in vitro fertilization of equine oocytes. *Biol. Reprod.* **2009**, *81*, 199–206. [CrossRef] [PubMed]
44. McPartlin, L.A.; Littell, J.; Mark, E.; Nelson, J.L.; Travis, A.J.; Bedford-Guaus, S.J. A defined medium supports changes consistent with capacitation in stallion sperm, as evidenced by increases in protein tyrosine phosphorylation and high rates of acrosomal exocytosis. *Theriogenology* **2008**, *69*, 639–650. [CrossRef] [PubMed]
45. Galantino-homer, H.L.; Visconti, P.E.; Kopf, G.S. Regulation of Protein Tyrosine Phosphorylation during Bovine Sperm Capacitation by a Cyclic Adenosine 3′, 5′ -Monophosphate-Dependent Pathway. *Biol. Reprod.* **1997**, *719*, 707–719. [CrossRef]
46. Osycka-Salut, C.; Gervasi, M.G.; Pereyra, E.; Cella, M.; Ribeiro, M.L.; Franchi, A.M.; Perez-Martinez, S. Anandamide induces sperm release from oviductal epithelia through nitric oxide pathway in bovines. *PLoS ONE* **2012**, *7*, e30671. [CrossRef]
47. Kinger, S.; Rajalakshmi, M. Assessment of the vitality and acrosomal status of human spermatozoa using fluorescent probes. *Int. J. Androl.* **1995**, *18*, 12–18. [CrossRef]
48. Leemans, B.; Stout, T.A.E.; Van Soom, A.; Gadella, B.M. pH dependent effects of procaine on equine gametes. *Biol. Reprod.* **2019**, *101*, 1056–1074. [CrossRef]
49. Aurich, J.; Kuhl, J.; Tichy, A.; Aurich, C. Efficiency of Semen Cryopreservation in Stallions. *Animals* **2020**, *10*, 1033. [CrossRef]
50. Höfner, L.; Luther, A.M.; Waberski, D. The role of seminal plasma in the liquid storage of spermatozoa. *Anim. Reprod. Sci.* **2020**, *2020*, 106290. [CrossRef]

51. Salazar, J.L.; Teague, S.R.; Love, C.C.; Brinsko, S.P.; Blanchard, T.L.; Varner, D.D. Effect of cryopreservation protocol on postthaw characteristics of stallion sperm. *Theriogenology* **2011**, *76*, 409–418. [CrossRef]
52. Yeste, M. Sperm cryopreservation update: Cryodamage, markers, and factors affecting the sperm freezability in pigs. *Theriogenology* **2016**, *85*, 47–64. [CrossRef] [PubMed]
53. Brum, A.M.; Sabeur, K.; Ball, B.A. Apoptotic-like changes in equine spermatozoa separated by density-gradient centrifugation or after cryopreservation. *Theriogenology* **2008**, *69*, 1041–1055. [CrossRef] [PubMed]
54. Ellerbrock, R.E.; Honorato, J.; Curcio, B.R.; Stewart, J.L.; Souza, J.A.T.; Love, C.C.; Lima, F.S.; Canisso, I.F. Effect of urine contamination on stallion semen freezing ability. *Theriogenology* **2018**, *117*, 1–6. [CrossRef] [PubMed]
55. Dobrinski, I.; Thomas, P.G.A.; Ball, B.A. Cryopreservation Reduces the Ability of Equine Spermatozoa to Attach to Oviductal Epithelial Cells and Zonae Pellucidae In Vitro. *J. Androl.* **1995**, *16*, 536–542. [CrossRef] [PubMed]
56. Hayden, S.S.; Blanchard, T.L.; Brinsko, S.P.; Varner, D.D.; Hinrichs, K.; Love, C.C. The "dilution effect" in stallion sperm. *Theriogenology* **2015**, *83*, 772–777. [CrossRef] [PubMed]
57. Moreno-Irusta, A.; Dominguez, E.M.; Marín-Briggiler, C.I.; Matamoros-Volante, A.; Lucchesi, O.; Tomes, C.N.; Treviño, C.L.; Buffone, M.G.; Lascano, R.; Losinno, L.; et al. Reactive oxygen species are involved in the signaling of equine sperm chemotaxis. *Reproduction* **2020**, *159*, 423–436. [CrossRef]
58. Martins, H.S.; Filho, O.A.M.; Araujo, M.S.S.; Martins, N.R.; De Albuquerque Lagares, M. Evaluation of fertilizing ability of frozen equine sperm using a bovine zona pellucida binding assay. *Cryo-Letters* **2018**, *39*. 298–305.
59. Al-Essawe, E.M.; Wallgren, M.; Wulf, M.; Aurich, C.; Macías-García, B.; Sjunnesson, Y.; Morrell, J.M. Seminal plasma influences the fertilizing potential of cryopreserved stallion sperm. *Theriogenology* **2018**, *115*, 99–107. [CrossRef]
60. Macías García, B.; Morrell, J.M.; Ortega-Ferrusola, C.; González-Fernández, L.; Tapia, J.A.; Rodriguez-Martínez, H.; Peña, F.J. Centrifugation on a single layer of colloid selects improved quality spermatozoa from frozen-thawed stallion semen. *Anim. Reprod. Sci.* **2009**, *114*, 193–202. [CrossRef]
61. Ruiz-Díaz, S.; Oseguera-López, I.; De La Cuesta-Díaz, D.; García-López, B.; Serres, C.; Sanchez-Calabuig, M.J.; Gutiérrez-Adán, A.; Perez-Cerezales, S. The presence of d-penicillamine during the in vitro capacitation of stallion spermatozoa prolongs hyperactive-like motility and allows for sperm selection by thermotaxis. *Animals* **2020**, *10*, 1467. [CrossRef]
62. Choi, Y.H.; Landim-Alvarenga, F.C.; Seidel, G.E.; Squires, E.L. Effect of capacitation of stallion sperm with polyvinylalcohol or bovine serum albumin on penetration of bovine zona-free or partially zona-removed equine oocytes. *J. Anim. Sci.* **2003**, *81*, 2080–2087. [CrossRef] [PubMed]
63. Pommer, A.C.; Rutllant, J.; Meyers, S.A. Phosphorylation of protein tyrosine residues in fresh and cryopreserved stallion spermatozoa under capacitating conditions. *Biol. Reprod.* **2003**, *68*, 1208–1214. [CrossRef] [PubMed]
64. Rota, A.; Sabatini, C.; Przybył, A.; Ciaramelli, A.; Panzani, D.; Camillo, F. Post-thaw Addition of Caffeine and/or Pentoxifylline Affect Differently Motility of Horse and Donkey-Cryopreserved Spermatozoa. *J. Equine Vet. Sci.* **2019**, *75*, 41–47. [CrossRef] [PubMed]
65. Marzano, G.; Moscatelli, N.; Di Giacomo, M.; Martino, N.A.; Lacalandra, G.M.; Dell'aquila, M.E.; Maruccio, G.; Primiceri, E.; Chiriacò, M.S.; Zara, V.; et al. Centrifugation force and time alter CASA parameters and oxidative status of cryopreserved stallion sperm. *Biology* **2020**, *9*, 22. [CrossRef] [PubMed]
66. Rappa, K.L.; Rodriguez, H.F.; Hakkarainen, G.C.; Anchan, R.M.; Mutter, G.L.; Asghar, W. Sperm processing for advanced reproductive technologies: Where are we today? *Biotechnol. Adv.* **2016**, *34*, 578–587. [CrossRef]
67. Katila, T. In vitro evaluation of frozen-thawed stallion semen: A review. *Acta Vet. Scand.* **2001**, *42*, 199–217. [CrossRef]
68. Miller, D.J. Review: The epic journey of sperm through the female reproductive tract. *Animals* **2018**, 1–11. [CrossRef]
69. Suarez, S.S. Formation of a reservoir of sperm in the oviduct. *Reprod. Domest. Anim.* **2002**, *37*, 140–143. [CrossRef]
70. Miller, D.J. Regulation of Sperm Function by Oviduct Fluid and the Epithelium: Insight into the Role of Glycans. *Reprod. Domest. Anim.* **2015**, *50*, 31–39. [CrossRef]
71. Martínez-León, E.; Osycka-Salut, C.; Signorelli, J.; Pozo, P.; Perez, B.; Kong, M.; Morales, P.; Perez-Martinez, S.; Diaz, E.S. Fibronectin stimulates human sperm capacitation through the cyclic AMP/protein kinase A pathway. *Hum Reprod* **2015**, *30*, 2138–2151. [CrossRef]
72. Talevi, R.; Gualtieri, R. Molecules involved in sperm-oviduct adhesion and release. *Theriogenology* **2010**, *73*, 796–801. [CrossRef] [PubMed]
73. Suarez, S.S. Mammalian sperm interactions with the female reproductive tract. *Cell Tissue Res.* **2015**. [CrossRef] [PubMed]
74. Gervasi, M.G.; Visconti, P.E. Chang's meaning of capacitation: A molecular perspective. *Mol. Reprod. Dev.* **2016**, *83*, 860–874. [CrossRef]
75. Choi, Y.H.; Love, C.C.; Love, L.B.; Varner, D.D.; Brinsko, S.; Hinrichs, K. Developmental competence in vivo and in vitro of in vitro-matured equine oocytes fertilized by intracytoplasmic sperm injection with fresh or frozen-thawed spermatozoa. *Reproduction* **2002**, *123*, 455–465. [CrossRef] [PubMed]
76. Colleoni, S.; Lagutina, I.; Lazzari, G.; Rodriguez-Martinez, H.; Galli, C.; Morrell, J.M. New Methods for Selecting Stallion Spermatozoa for Assisted Reproduction. *J. Equine Vet. Sci.* **2011**, *31*, 536–541. [CrossRef]
77. Choi, Y.H.; Velez, I.C.; Macías-García, B.; Riera, F.L.; Ballard, C.S.; Hinrichs, K. Effect of clinically-related factors on in vitro blastocyst development after equine ICSI. *Theriogenology* **2016**, *85*, 1289–1296. [CrossRef]

78. Gonzalez-Castro, R.A.; Carnevale, E.M. Use of microfluidics to sort stallion sperm for intracytoplasmic sperm injection. *Anim. Reprod. Sci.* **2019**, *202*, 1–9. [CrossRef]
79. Gonzalez-Castro, R.A.; Carnevale, E.M. Association of equine sperm population parameters with outcome of intracytoplasmic sperm injections. *Theriogenology* **2018**, *119*, 114–120. [CrossRef]

Article

The Effects of Red Light on Mammalian Sperm Rely upon the Color of the Straw and the Medium Used

Jaime Catalán [1,2,3], Iván Yánez-Ortiz [1], Sabrina Gacem [1], Marion Papas [1], Sergi Bonet [2,3], Joan E. Rodríguez-Gil [1], Marc Yeste [2,3,*,†] and Jordi Miró [1,*,†]

1 Equine Reproduction Service, Department of Animal Medicine and Surgery, Faculty of Veterinary Sciences, Autonomous University of Barcelona, E-08193 Bellaterra (Cerdanyola del Vallès), Spain; dr.jcatalan@gmail.com (J.C.); ivan.yanez22@gmail.com (I.Y.-O.); swp.sabrina.gacem@gmail.com (S.G.); papas.marion@gmail.com (M.P.); JuanEnrique.Rodriguez@uab.cat (J.E.R.-G.)
2 Biotechnology of Animal and Human Reproduction (TechnoSperm), Institute of Food and Agricultural Technology, University of Girona, E-17003 Girona, Spain; sergi.bonet@udg.edu
3 Unit of Cell Biology, Department of Biology, Faculty of Sciences, University of Girona, E-17003 Girona, Spain
* Correspondence: marc.yeste@udg.edu (M.Y.); jordi.miro@uab.cat (J.M.); Tel.: +34-972-419514 (M.Y.); +34-93-5814293 (J.M.)
† These authors contributed equally to this work.

Simple Summary: Several studies have shown that the exposure of semen to red light improves sperm quality and fertilizing ability, which could improve the efficiency of assisted reproductive techniques with irradiated semen. However, despite being considered as possible sources of variation, the effects of the color of the container (straws) or the medium have not yet been evaluated. In this study, 13 ejaculates from different stallions were split into equal fractions, diluted either with Kenney or Equiplus extender, and subsequently packed into straws of five different colors. After storage at 4 °C for 24 h, the sperm were irradiated and different variables, including sperm motility, plasma membrane integrity, and mitochondrial membrane potential, were evaluated. Our results confirm that irradiation increases some motion characteristics and mitochondrial membrane potential without affecting sperm viability and demonstrate that the effects depend on the color of the straw and the extender used.

Abstract: Previous research has determined that irradiation of mammalian sperm with red light increases motility, mitochondrial activity, and fertilization capacity. In spite of this, no study has considered the potential influence of the color of the straw and the extender used. Therefore, this study tests the hypothesis that the response of mammalian sperm to red light is influenced by the color of the straw and the turbidity/composition of the extender. Using the horse as a model, 13 ejaculates from 13 stallions were split into two equal fractions, diluted with Kenney or Equiplus extender, and stored at 4 °C for 24 h. Thereafter, each diluted fraction was split into five equal aliquots and subsequently packed into 0.5-mL straws of red, blue, yellow, white, or transparent color. Straws were either nonirradiated (control) or irradiated with a light–dark–light pattern of 3–3–3 (i.e., light: 3 min, dark: 3 min; light: 3 min) prior to evaluating sperm motility, acrosome and plasma membrane integrity, mitochondrial membrane potential, and intracellular ROS and calcium levels. Our results showed that irradiation increased some motion variables, mitochondrial membrane potential, and intracellular ROS without affecting the integrities of the plasma membrane and acrosome. Remarkably, the extent of those changes varied with the color of the straw and the extender used; the effects of irradiation were more apparent when sperm were diluted with Equiplus extender and packed into red-colored straws or when samples were diluted with Kenney extender and packed into transparent straws. As the increase in sperm motility and intracellular ROS levels was parallel to that of mitochondrial activity, we suggest that the impact of red light on sperm function relies upon the specific rates of energy provided to the mitochondria, which, in turn, vary with the color of the straw and the turbidity/composition of the extender.

Citation: Catalán, J.; Yánez-Ortiz, I.; Gacem, S.; Papas, M.; Bonet, S.; Rodríguez-Gil, J.E.; Yeste, M.; Miró, J. The Effects of Red Light on Mammalian Sperm Rely upon the Color of the Straw and the Medium Used. *Animals* **2021**, *11*, 122. https://doi.org/10.3390/ani11010122

Received: 16 October 2020
Accepted: 5 January 2021
Published: 8 January 2021

Keywords: horse; sperm; red light irradiation; extender; straw

1. Introduction

Artificial insemination (AI) is a tool widely used today for horse breeding, especially when looking for genetic improvement [1]. The increasing use of this technology, both in the horse and other species, has augmented the interest for semen processing techniques and their optimization, aimed at maximizing their survival and fertilization capacity [2,3]. Unfortunately, semen quality often deviates from expectations and leads to unsatisfactory pregnancy rates [1]. In this context, any protocol or procedure that optimizes its use and helps increase reproductive performance should be considered; for this reason, several approaches have been undertaken in recent years [3,4]. One of these approaches is sperm irradiation; in effect, previous research has demonstrated that red light stimulation, either with low-level lasers or light-emitting diodes (LEDs), increases the motility, ability to elicit in vitro capacitation, fertilizing ability, and lifespan of fresh, liquid-stored, and frozen-thawed sperm in fish [5], birds [6], humans [7–12], pigs [13–15], sheep [16], dogs [17,18], buffalos [19] donkeys [3], and horses [20,21]. In addition to this, recent studies have shown that the increase in sperm motility in response to LED-based red light is concomitant with that of mitochondrial activity in pigs, donkeys, and horses [3,13,20,22].

The mechanisms through which light exerts its effects are not entirely clear. Three potential mechanisms have been surmised to explain the response of mammalian sperm to red light (reviewed in Yeste et al. [23]). The first of these hypotheses is related to the possible influence of light on transient receptor proteins (TRPs) [24–26], which reside in sperm plasmalemma and have been purported to participate in the modulation of thermotaxis [27,28]. The second hypothesis is related to the presence of opsins in mammalian spermatozoa, which absorb light from different spectra [23]; despite being mainly related to the response to thermotaxis [28], they could also be involved in the sperm response to light. Finally, mounting evidence supports the third hypothesis, which confers a crucial role on endogenous cellular photosensitizers, especially those present in the mitochondria [22,23,29]. These photosensitizers absorb light from electromagnetic radiation and then ionize and transfer the absorbed energy into adjacent molecules [30]. This increased energy induces a rise in electrochemical mitochondria potential, which may result in an augmentation of ATP and Ca^{2+} levels [23]. In spite of this, it cannot be ruled out that more than one of the mechanisms proposed by these hypotheses are involved in the sperm response to red light [22,29].

Previous studies carried out with low-level laser therapy devices and light-emitting diodes (LEDs) have reported an increase in ATP production via the mitochondrial electron chain [31,32] without damaging the irradiated cells [32,33] or the integrity of their DNA [32,34]. Therefore, it has been suggested that light stimulation can have a safe and positive effect on sperm motility and fertilizing ability both in vivo and in vitro [32]. Nevertheless, the sperm response to irradiation has been reported to depend on different factors, including the type (i.e., fresh, cooled-stored or frozen-thawed) and state of the sample [3], the irradiation of the light beam used [32], the time or pattern of exposure [13], and the species [5]. Given the properties of light emission/absorption, other factors such as the color of the straw and choice of extender could also affect the sperm response to red light. However, to the best of our knowledge, no previous study has examined this possibility despite the wide variety of extenders and colors of commercial straws.

Taking the results obtained in the aforementioned studies (especially those conducted with fresh, cooled-stored, and frozen-thawed horse sperm) [20,21] into account, this study aims at determining whether the color of the straw and the extender used affect the response of cooled-stored sperm to LED-based red light (620–630 nm). Our hypothesis is that the effects of red light on horse sperm depend on the color of the straw and the extender.

2. Materials and Methods

2.1. Suppliers

All reagents used were of analytical grade and were purchased from Boehringer-Mannheim (Mannheim, Germany), Merck (Darmstadt, Germany), and Sigma-Aldrich (Saint Louis, MO, USA). As far as fluorochromes are concerned, unless otherwise stated, all were purchased from Molecular Probes (Thermo Fisher Scientific; Waltham, MA, USA) and were previously prepared with dimethyl sulfoxide (Sigma-Aldrich). Plastic materials were provided by Nunc (Roskilde, Denmark), and empty straws of different colors (transparent, red, white, blue, and yellow) were purchased from Minitüb GmbH (Tiefenbach, Germany).

2.2. Animals and Ejaculates

This study included 13 ejaculates from 13 different adult stallions (age: 5–8 years old) with proven fertility. Animals were housed at the Equine Reproduction Service, Autonomous University of Barcelona (Bellaterra, Cerdanyola del Vallès, Spain), which is an EU-approved semen collection center (Authorization code: ES09RS01E) that operates under strict protocols of animal welfare and health control. All animals were semen donors and were collected under CEE health conditions (free of equine arteritis, infectious anemia, and contagious metritis). As indicated in Catalán et al. [21], this Service runs under the rules of the Regional Government of Catalonia, Spain, and no manipulation of the animals other than semen collection was carried out. The study was approved by the Ethics Committee, Autonomous University of Barcelona (Code: CEEAH 1424).

Ejaculates were collected through a Hannover artificial vagina (Minitüb GmbH, Tiefenbach, Germany), and an in-line nylon mesh filter was used to remove the gel fraction. Upon collection, gel-free semen was split into two fractions of equal volume and immediately diluted 1:5 (*v:v*) either in Kenney [35] or Equiplus extender (Minitüb GmbH; Tiefenbach, Germany), which were selected for their different turbidity. The absorbance of these two extenders was evaluated at 625 nm with a spectrophotometer (Biochrom WPA, Lightwave II; Cambridge, UK) and sterilized; ultrafiltered Milli-Q water was used as blank. Absolute absorbance values of the Equiplus and Kenney extenders were 0.090 and >2.5, respectively. Both extenders were preheated to 37 °C, and sperm concentration was adjusted in all cases to 30×10^6 sperm/mL with a Neubauer chamber (Paul Marienfeld GmbH and Co. KG; Lauda-Königshofen, Germany). Following this, sperm motility (Computer Assisted Semen Analysis, CASA), morphology (eosin–nigrosin staining), and plasma membrane integrity (SYBR14/PI) of each sample were evaluated. All samples were confirmed to fulfill the standard thresholds: ≥60% SYBR14$^+$/PI$^-$ spermatozoa and ≥70% morphologically normal spermatozoa. Thereafter, semen samples were stored in a refrigerator at 4 °C for 24 h.

2.3. Experimental Design

After 24 h of storage, samples extended in either Kenney or Equiplus were split and packed into 0.5-mL straws (Minitüb GmbH) of five different colors (blue, red, yellow, white, and transparent); the sperm concentration was maintained at 30×10^6 sperm/mL at all experimental points. Straws were placed within a programmable photoactivation system (MaxiCow; IUL, SA, Barcelona, Spain). In this device, each straw is in contact with a triple-LED configuration system that emits red light (wavelength window: 620 to 630 nm). The apparatus is equipped with software (IUL, SA) that allows the regulation of intensity and time of exposure. In all cases, the intensity was set at 100%.

Straws of different colors containing sperm diluted by both extenders were irradiated with a light–dark–light interval pattern of 3–3–3 min. Nonirradiated samples (control) were also packed into 0.5-mL straws and left for 9 min in the dark, which was the same time used to irradiate the samples. Upon light stimulation, irradiated and nonirradiated samples were transferred into 1.5-mL tubes. Sperm motility was evaluated with a computer-assisted sperm analysis (CASA) system, and plasma membrane and acrosome integrity, mitochondrial membrane potential, intracellular ROS (peroxides and superoxides), and calcium levels were determined through flow cytometry.

2.4. Analysis of Sperm Motility

Sperm motility was evaluated using a computer-assisted sperm analysis (CASA) system (Integrated Sperm Analysis System V1.0; Proiser S.L.; Valencia, Spain). In brief, samples were incubated at 38 °C in a water bath for 5 min, and 5 μL of each sperm sample was placed onto a Makler chamber (Sefi Medical Instruments; Haifa, Israel), previously warmed at 38 °C. Samples were then analyzed under a 10× negative phase-contrast objective (Olympus BX41 microscope; Olympus, Tokyo, Japan). A minimum of 1000 sperm cells was counted per analysis. In each evaluation, percentages of total motility (TMOT, %) and progressively motile spermatozoa (PMOT, %) were recorded together with the following kinetic measures: curvilinear velocity (VCL, μm/s), which is the mean path velocity of the sperm head along its actual trajectory; straight-line velocity (VSL, μm/s), which is the mean path velocity of the sperm head along a straight line from its first to its last position; average path velocity (VAP, μm/s), which is the mean velocity of the sperm head along its average trajectory; percentage of linearity (LIN, %), which is the quotient between VSL and VCL multiplied by 100; percentage of straightness (STR, %), which is the quotient between VSL and VAP multiplied by 100; percentage of oscillation (WOB, %), which is the quotient between VAP and VCL multiplied by 100; mean amplitude of lateral head displacement (ALH, μm), which is the mean value of the extreme side-to-side movement of the sperm head in each beat cycle; frequency of head displacement (BCF, Hz), which is the frequency at which the actual sperm trajectory crosses the average path trajectory.

CASA settings were those recommended by the manufacturer, i.e., frames: 25 images captured per second; particle area >4 and <75 μm^2; connectivity = 6; minimum number of images to calculate ALH: 10. The cut-off value for motile spermatozoa was VAP $\geq$ 10 μm/s; for progressively motile spermatozoa, the cut-off value was STR $\geq$ 75%.

2.5. Flow Cytometry

The integrity of sperm plasma membrane (SYBR14/PI), acrosome integrity (PNA-FITC/PI), mitochondrial membrane potential (JC1), and intracellular levels of peroxides (H_2DCFDA/PI), superoxides (HE/YO-PRO-1), and calcium (Fluo3/PI) were determined through flow cytometry. Samples were stained and evaluated following the protocol described by Prieto-Martínez et al. [36] and adjusted to horse spermatozoa.

Management of the flow cytometer and analysis of the samples were carried out in accordance with the recommendations of the International Society of Cytometry [37]. The flow cytometer used in this study was a Cell Lab Quanta SC™ (Beckman Coulter, Fullerton, CA, USA), and particles were excited with an argon laser (488 nm) at a power of 22 mW. Prior to staining, sperm concentration was adjusted to 1×10^6 sperm/mL. Every day, the electronic volume (EV) channel was calibrated with 10-μm diameter fluorescent beads (Beckman Coulter), following the manufacturer's instructions. The flow rate was set at 4.17 μL/min, and the analyzer threshold was established to exclude cell aggregates (particles with a diameter >12 μm) and debris (particles with a diameter < 7 μm). Sperm cells were gated on the basis of EV and side scatter (SS) distributions. Three different optical filters were used (FL1 for the analysis of SYBR14, PNA, H_2DCFDA, Fluo3, and JC1 monomers, detection width: 505–545 nm; FL2 for the analysis of JC1 aggregates, detection width: 560–590 nm; FL3 for the analysis of PI and HE, detection width: 655–685 nm).

Dot plots were examined using Cell Lab Quanta SC™ MPL Analysis Software (version 1.0; Beckman Coulter) and data from PNA/PI, JC1, H_2DCFDA/PI, HE/YO-PRO-1; Fluo3/PI were corrected using the percentage of nonstained debris particles found in SYBR14/PI staining, as recommended by Petrunkina et al. [38].

2.5.1. Analysis of Plasma Membrane Integrity

Sperm viability (plasma membrane integrity) was assessed using the LIVE/DEAD® Sperm Viability Kit (SYBR14/PI; Molecular Probes, Thermo Fisher Scientific; Waltham, MA, USA), according to the protocol described by Garner and Johnson [39], and adapted to horse spermatozoa. In brief, samples were first incubated with SYBR14 (final concentration: 100 nM) at 38 °C for 10 min, and then with PI (final concentration: 12 μM) at 38 °C for 5 min. Three sperm populations were distinguished: (i) viable spermatozoa emitting green fluorescence ($SYBR14^+/PI^-$), which appeared on the right side of the lower half of the FL1/FL3 dot plots; (ii) nonviable spermatozoa emitting red fluorescence ($SYBR14^-/PI^+$), which appeared on the left side of the upper half of the FL1/FL3 dot plots; (iii) nonviable spermatozoa emitting both green and red fluorescence ($SYBR14^+/PI^+$), which appeared on the right side of the upper half of the FL1/FL3 dot plots. Nonstained particles ($SYBR14^-/PI^-$), which appeared on the left side of the lower half of the FL1/FL3 dot plots, showed EV/SS distributions similar to spermatozoa and were considered non-DNA debris particles. Percentages of nonstained particles were used to correct the percentages of double-negative sperm populations in the other assessments. Spill-over of FL1 into the FL3 channel was compensated (2.45%).

2.5.2. Analysis of Acrosome Integrity

Plasma membrane integrity was evaluated through PNA/PI costaining, following the procedure described for horse spermatozoa by Rathi et al. [40]. With this purpose, spermatozoa were stained with PNA conjugated with FITC (final concentration: 5 μg/mL) and PI (final concentration: 12 μm) and incubated at 38 °C for 10 min in the dark. Green fluorescence from PNA was collected through FL1, whereas red fluorescence from PI was collected through FL3. As spermatozoa were not previously permeabilized, they were identified and placed in one of the four following populations: (i) spermatozoa with intact plasma membranes (PNA^-/PI^-); (ii) spermatozoa with damaged plasma membranes that presented an acrosome membrane that could not be fully intact (PNA^+/PI^+); (iii) spermatozoa with damaged plasma membranes and lost outer acrosome membranes (PNA^-/PI^+); (iv) spermatozoa with damaged plasma membranes (PNA^+/PI^-). Therefore, after PNA/PI staining, two main categories were detected: (i) spermatozoa with an intact plasma membrane (PNA^-/PI^-) and (ii) spermatozoa that had damaged their plasma membrane and/or their acrosome membrane (these were represented by the other three categories: PNA^+/PI^-, PNA^+/PI^+, PNA^-/PI^+). Unstained and single-stained samples were used for setting the EV gain, FL1 and FL3 PMT voltages, and for compensation of PNA spill over into the PI channel (2.45%).

2.5.3. Analysis of Mitochondrial Membrane Potential

Mitochondrial membrane potential (MMP) was determined through incubation with JC1 (5,5′,6,6′-tetrachloro-1,1′,3,3′tetraethyl-benzimidazolylcarbocyanine iodide; final concentration: 0.3 μM) at 38 °C for 30 min in the dark. When MMP is low, JC1 forms monomers emitting green fluorescence ($JC1_{mon}$), which are collected through FL1. When mitochondrial membrane potential is high, JC1 forms aggregates emitting orange fluorescence ($JC1_{agg}$), which are detected through FL2. Three sperm populations were distinguished: (i) spermatozoa with green-stained mitochondria (low MMP), (ii) spermatozoa with orange-stained mitochondria (high MMP), and (iii) spermatozoa with heterogeneous mitochondria, stained both green and orange in the same cell (intermediate MMP). Ratios between FL2 ($JC1_{agg}$) and FL1 fluorescence ($JC1_{mon}$) for each of these sperm populations were also evaluated. Spill-over of FL1 into the FL2 channel was compensated (68.5%). Percentages of debris particles found in SYBR14/PI staining ($SYBR14^-/PI^-$) were subtracted from those of spermatozoa with low MMP, and the percentages of all sperm populations were recalculated.

2.5.4. Analysis of Intracellular ROS Levels: H_2O_2 and O_2^-

Intracellular ROS levels were determined through two oxidation sensitive fluorescent probes, 2′,7′-dichlorodihydrofluorescein diacetate (H_2DCFDA) and hydroethidine (HE), which detect hydrogen peroxides (H_2O_2) and superoxide anions ($\cdot O_2^-$), respectively [41]. Following a modified procedure from Guthrie and Welch [42], a simultaneous differentiation of viable and nonviable sperm was performed using PI (H_2DCFDA) or YO-PRO-1 (HE).

In the case of peroxides, spermatozoa were incubated with H_2DCFDA (final concentration: 200 μM) and PI (final concentration: 12 μM) at room temperature for 30 min in the dark. H_2DCFDA is a stable, cell-permeable, nonfluorescent probe that is converted into 2′,7′-dichlorofluorescein (DCF) in the presence of H_2O_2 [42]. Fluorescence of DCF^+ was measured through FL1 and that of PI was detected through FL3. Four sperm populations were distinguished: (i) viable spermatozoa with low levels of peroxides (DCF^-/PI^-), (ii) viable spermatozoa with high levels of peroxides (DCF^+/PI^-), (iii) nonviable spermatozoa with low levels of peroxides (DCF^-/PI^+), and (iv) nonviable spermatozoa with high levels of peroxides (DCF^+/PI^+). Percentages of debris particles found in SYBR14/PI staining ($SYBR14^-/PI^-$) were subtracted from those of viable spermatozoa with low levels of peroxides (DCF^-/PI^-) and the percentages of all sperm populations were recalculated. Spill-over of FL1 into the FL3 channel was compensated (2.45%). Data are shown as corrected percentages of viable spermatozoa with high levels of peroxides (DCF^+/PI^-) and the geometric mean of DCF^+-fluorescence intensity in the DCF^+/PI^- sperm population.

Regarding superoxide anions, samples were incubated with HE (final concentration: 4 μM) and YO-PRO-1 (final concentration: 25 nM) at room temperature for 30 min in the dark [42]. Hydroethidine diffuses freely through the plasma membrane and converts into ethidium (E^+) in the presence of superoxide anions (O_2^-) [43]. Fluorescence of ethidium (E^+) was detected through FL3 and that of YO-PRO-1 was detected through FL1. Four sperm populations were distinguished: (i) viable spermatozoa with low levels of superoxides (E^-/YO-PRO-1^-), (ii) viable spermatozoa with high levels of superoxides (E^+/YO-PRO-1^-), (iii) nonviable spermatozoa with low levels of superoxides (E^-/YO-PRO-1^+), and (iv) nonviable spermatozoa with high levels of superoxides (E^+/YO-PRO-1^+). Percentages of debris particles found in the SYBR14/PI test ($SYBR14^-/PI^-$) were subtracted from those of viable spermatozoa with low levels of superoxides (E^-/YO-PRO-1^-) and the percentages of all sperm populations were recalculated. Spill-over of FL3 into the FL1 channel was compensated (5.06%). Data are shown as corrected percentages of viable spermatozoa with high levels of superoxides (E^+/YO-PRO-1^-) and the geometric mean of E^+-fluorescence intensity in the E^+/YO-PRO-1^- sperm population.

2.5.5. Intracellular Calcium Levels

Previous studies found that Fluo3 mainly stains mitochondrial calcium in mammalian sperm [44]. For this reason, we combined this fluorochrome with propidium iodide (Fluo3/PI), as described by Kadirvel et al. [45]; the following four populations were identified: (i) viable spermatozoa with low levels of intracellular calcium ($Fluo3^-/PI^-$), (ii) viable spermatozoa with high levels of intracellular calcium ($Fluo3^+/PI^-$), (iii) nonviable spermatozoa with low levels of intracellular calcium ($Fluo3^-/PI^+$), and (iv) nonviable spermatozoa with high levels of intracellular calcium ($Fluo3^+/PI^+$). FL1 spill-over into the FL3 channel (2.45%) and FL3 spill-over into the FL1 channel (28.72%) were compensated.

2.6. Statistical Analyses

Statistical analyses were conducted using a statistical package (SPSS® Ver. 25.0 for Windows; IBM Corp., Armonk, NY, USA). Data were first tested for normal distribution (Shapiro–Wilk test) and homogeneity of variances (Levene test), and, if required, they were transformed with arcsin $\sqrt{x}$. The effects of the color of the straw and the extender on the response of horse sperm to red light were tested with a two-way analysis of variance (ANOVA), followed by a post hoc Sidak test. Sperm motility measures; percentages of spermatozoa with an intact plasma membrane ($SYBR14^+/PI^-$), acrosome-intact spermatozoa (PNA-FITC$^-$/PI$^-$), spermatozoa with high and intermediate mitochondrial membrane potential, viable spermatozoa with high intracellular calcium levels (Fluo3$^+$/PI$^-$), viable spermatozoa with high superoxide levels (E$^+$/YO-PRO-1$^-$), and viable spermatozoa with high peroxide levels (DCF$^+$/PI$^-$); and geometric mean fluorescence intensities (GMFI) of JC1$_{agg}$, Fluo3$^+$, E$^+$, and DCF$^+$ were analyzed.

Motile sperm subpopulations were determined through the protocol described in Luna et al. [46]. In brief, individual kinematic variables (VCL, VSL, VAP, LIN, STR, WOB, ALH, and BCF) recorded for each spermatozoon were used as independent variables in a principal component analysis (PCA). Kinematic measures were sorted into PCA components, and the obtained matrix was subsequently rotated using the Varimax method with Kaiser normalization. As a result, each sperm cell was assigned a regression score for each of the new PCA components, and these values were subsequently used to run a two-step cluster analysis based on the log-likelihood distance and Schwarz's Bayesian criterion. Four sperm subpopulations were identified, and each individual spermatozoon was assigned to one of these subpopulations (SP1, SP2, SP3, or SP4). Following this, percentages of spermatozoa belonging to each subpopulation were calculated per sample and used to determine the effects of the color of the straw and the extender on the response of horse sperm to red light through two-way ANOVA and Sidak's post hoc test.

In all analyses, the level of significance was set at $p \leq 0.05$. Data are shown as mean $\pm$ standard error of the mean (SEM).

3. Results

As expected, no differences in variables were observed between the straws of different colors in the nonirradiated group. For this reason, and in order to simplify the presentation of data, all these results have been grouped and identified as "control" (nonirradiated samples).

3.1. Plasma Membrane Integrity

Percentages of membrane-intact spermatozoa (Figure S1a) did not differ between nonirradiated and irradiated samples. In addition, neither the color of the straw nor the type of extender had any effect on the percentages of membrane-intact spermatozoa in irradiated and nonirradiated samples (e.g., nonirradiated sperm in Equiplus extender: 46.2% $\pm$ 3.9% vs. sperm diluted in Equiplus, packed into blue straws, and irradiated: 47.9% $\pm$ 3.6% vs. sperm diluted in Kenney, packed into blue straws, and irradiated: 38.6% $\pm$ 3.0%).

3.2. Acrosomal Integrity

In a similar fashion to that observed for plasma membrane integrity, percentages of acrosomal-intact spermatozoa (Figure S1b) did not differ between nonirradiated and irradiated samples, regardless of the color of the straw or the extender (e.g., nonirradiated sperm in Kenney extender: 39.2% $\pm$ 3.1% vs. sperm diluted in Kenney, packed into red straws, and irradiated: 39.7% $\pm$ 3.2% vs. sperm diluted in Equiplus, packed into red straws, and irradiated: 49.8% $\pm$ 3.6%).

3.3. Sperm Motility

No significant differences were observed in the percentages of sperm with total (Figure S2a) and progressive motility (Figure S2b) between nonirradiated and irradiated samples when compared within each extender. In addition, neither the color of the straw nor the type of extender had any effect on the percentages of sperm with total and progressive motility when the two extenders were compared within each of the colored straws (e.g., total motility: nonirradiated sperm in Equiplus: 56.0% $\pm$ 4.5% vs. sperm diluted in Equiplus, packed into yellow straws, and irradiated: 56.5% $\pm$ 3.8% vs. sperm diluted in Kenney, packed into yellow straws, and irradiated: 56.3% $\pm$ 3.4%; progressive motility: nonirradiated sperm diluted in Kenney: 27.1% $\pm$ 2.4% vs. sperm diluted in Kenney, packed into transparent straws, and irradiated: 31.9% ± 2.9% vs. sperm diluted in Equiplus, packed into transparent straws, and irradiated: 28.4% ± 2.6%).

Regarding sperm kinetic variables (Table 1), VCL, VSL, and VAP were significantly ($p < 0.05$) higher in samples diluted in Equiplus and packed into red straws than in their respective control (nonirradiated samples). In addition, VCL and VAP were significantly ($p < 0.05$) higher in sperm diluted in Kenney extender and packed into transparent straws than in the nonirradiated control. In addition to this, STR in samples packed into blue straws and irradiated was significantly higher ($p < 0.05$) in sperm diluted in Kenney than in those diluted in Equiplus extender.

As shown in Table 2, four different motile sperm subpopulations were identified (SP1, SP2, SP3, and SP4); SP1 was characterized as the fastest subpopulation since it exhibited the highest values in VCL, VSL, and VAP. SP2 was the slowest sperm subpopulation. SP3, although characterized by intermediate speed values (but lower than SP1 and SP4) and LIN and STR values similar to SP1, was the one that showed the highest BCF. Finally, SP4 was characterized by intermediate speed values, which were higher than in SP3, but it was the least linear.

Figure 1a shows the percentages of sperm belonging to SP1. Compared to their respective controls, these percentages were significantly ($p < 0.05$) higher in irradiated samples diluted in Equiplus extender and packed into blue, yellow, or red straws and in irradiated samples diluted in Kenney extender and packed into transparent straws. Percentages of sperm belonging to SP2 were significantly ($p < 0.05$) higher in the control than in irradiated samples diluted in Equiplus extender and packed into yellow, red, or transparent straws (Figure 1b). On the contrary, no significant differences ($p > 0.05$) between nonirradiated and irradiated samples were observed for SP3 and SP4 (Figure 1c,d).

3.4. Mitochondrial Membrane Potential

As shown in Figure 2a and Figure S3, percentages of sperm with high MMP were significantly ($p < 0.05$) higher in samples diluted in Equiplus and packed into yellow, red, and transparent straws and in those diluted in Kenney and packed into transparent straws than in their respective controls. In contrast, no significant differences between extenders were observed when nonirradiated and irradiated samples packed into straws of different color were compared. With regard to the percentages of sperm with intermediate MMP, no significant differences between nonirradiated and irradiated samples were observed, regardless of the color of the straw and the extender (Figure 2b).

No significant differences in the geometric mean of $JC1_{agg}$ intensity (orange, FL2) of sperm populations with high (Figure 2c) and intermediate MMP (Figure 2d) were observed between nonirradiated and irradiated samples, regardless of the color of the straw and the extender used. However, the geometric mean of $JC1_{agg}$ intensity (orange, FL2) of the sperm population with a high MMP (Figure 2c) was significantly ($p < 0.05$) higher in samples diluted in Kenney extender and nonirradiated (control) or packed into blue, yellow, red, white, or transparent straws than in their counterparts diluted in Equiplus extender. In addition, the geometric mean of $JC1_{agg}$ intensity (orange, FL2) of the sperm population, with an intermediate MMP in nonirradiated samples (control), was significantly ($p < 0.05$) higher when they were diluted in Kenney than when they were diluted in Equiplus extenders (Figure 2d).

Finally, we also evaluated $JC1_{agg}/JC1_{mon}$ ratios of sperm populations with high (Figure 2e) and intermediate MMP (Figure 2f). No significant differences ($p > 0.05$) were observed when comparing nonirradiated and irradiated samples, regardless of the color of the straw and extender used, either within the same diluent or when comparing the two extenders.

Table 1. Effects of the color of the straw (blue, yellow, red, white, and transparent), extender, and light stimulation on sperm kinetic variables in nonirradiated (control) and irradiated samples.

Extender	Treatment (Straw Color)	Kinetic Variables (Mean ± SEM)							
		VCL (μm/s)	VSL (μm/s)	VAP (μm/s)	LIN (%)	STR (%)	WOB (%)	ALH (μm)	BCF (Hz)
Equiplus	Control	91.2 ± 3.1 [1]	54.0 ± 2.9	74.2 ± 3.2 [1]	59.3 ± 1.4	73.5 ± 1.9	80.8 ± 1.1	3.0 ± 0.2	8.9 ± 0.2
	Blue	98.5 ± 4.4	60.4 ± 3.6	82.6 ± 4.3	61.2 ± 214	72.9 ± 1.3 [a]	83.0 ± 1.4	3.0 ± 0.2	8.9 ± 0.4
	Yellow	102.0 ± 4.7	62.1 ± 4.1	82.8 ± 3.4	61.0 ± 1.7	75.6 ± 1.8	80.7 ± 1.4	3.3 ± 0.2	9.2 ± 0.1
	Red	106.5 ± 3.2 [2]	68.8 ± 2.6 [2]	89.3 ± 3.0 [2]	64.7 ± 2.1	77.2 ± 2.9	83.8 ± 0.9	3.3 ± 0.2	8.8 ± 0.2
	White	96.7 ± 4.0	61.3 ± 3.9	81.2 ± 4.3	63.1 ± 1.4	75.5 ± 1.6	83.5 ± 0.8	2.9 ± 0.2	8.7 ± 0.2
	Transparent	102.9 ± 4.9	62.9 ± 3.1	83.6 ± 4.2	62.8 ± 1.8	77.8 ± 2.1	82.7 ± 0.6	3.2 ± 0.2	8.9 ± 0.2
Kenney	Control	93.8 ± 2.8 [1]	56.7 ± 3.6	72.3 ± 3.1 [1]	59.9 ± 2.6	78.7 ± 2.2	76.1 ± 2.7	2.9 ± 0.1	10.1 ± 0.4
	Blue	97.5 ± 4.3	63.3 ± 2.8	77.7 ± 4.5	63.8 ± 2.2	82.4 ± 1.6 [b]	81.0 ± 2.2	2.8 ± 0.1	9.2 ± 0.4
	Yellow	100.9 ± 4.1	63.4 ± 2.5	80.4 ± 3.9	63.3 ± 2.0	81.2 ± 2.5	78.6 ± 2.3	3.0 ± 0.1	9.9 ± 0.4
	Red	96.1 ± 3.6	62.6 ± 2.5	77.7 ± 4.3	63.7 ± 1.8	81.8 ± 1.8	78.6 ± 2.3	3.0 ± 0.1	9.8 ± 0.5
	White	92.1 ± 3.2	56.0 ± 2.5	70.8 ± 3.7	60.7 ± 2.1	80.1 ± 2.0	76.2 ± 2.5	3.0 ± 0.1	10.4 ± 0.4
	Transparent	107.4 ± 3.0 [2]	66.3 ± 2.8	86.0 ± 3.0 [2]	62.0 ± 2.2	78.8 ± 2.4	79.1 ± 2.5	3.1 ± 0.1	9.8 ± 0.5

SEM: standard error of the mean; VCL (μm/s): curvilinear velocity; VSL (μm/s): straight line velocity; VAP (μm/s): average path velocity; LIN (%): linearity; STR (%): straightness; WOB (%): wobble; ALH (μm): amplitude of lateral head displacement; BCF (Hz): beat-cross frequency. Different numbers ([1, 2]) indicate significant differences ($p < 0.05$) between nonirradiated and irradiated samples packed into straws of different color within the same extender (i.e., Kenney and Equiplus). Different letters ([a, b]) indicate significant differences ($p < 0.05$) between extenders within nonirradiated or samples packed into straws of different color. The absence of numbers indicates the lack of statistical differences ($p > 0.05$) between irradiated and nonirradiated samples within the same diluent. The absence of letters means the lack of significant differences between irradiated/nonirradiated sperm between the two extenders. Data are shown as mean ± SEM of 13 separate experiments.

Table 2. Descriptive parameters (mean ± SEM; range) of the four sperm subpopulations (SP1, SP2, SP3, and SP4) identified in nonirradiated and irradiated samples diluted in the two extenders and packed into straws of different color.

	SP1		SP2		SP3		SP4	
N	10,647		12,622		14,674		8748	
Parameter	Mean ± SEM	Range	Mean ± SEM	Range	Mean ± SEM	Range	Mean ± SEM	Range
VCL (μm/s)	146.7 ± 0.4	108.2–247.3	61.9 ± 0.2	10.0–126.6	94.5 ± 0.1	54.9–149.2	126.3 ± 0.3	71.9–248.3
VSL (μm/s)	108.5 ± 0.2	63.4–198.3	33.4 ± 0.1	4.0–54.3	71.4 ± 0.1	42.6–123.6	40.0 ± 0.2	4.2–112.6
VAP (μm/s)	126.6 ± 0.2	10.2–121.0	45.4 ± 0.1	7.6–80.4	82.5 ± 0.1	48.9–120.0	97.4 ± 0.3	25.1–201.5
LIN (%)	75.0 ± 0.1	10.0–99.5	56.4 ± 0.2	8.3–96.6	76.6 ± 0.1	42.0–99.3	32.0 ± 0.2	2.7–59.8
STR (%)	85.9 ± 0.1	5.3–96.9	74.6 ± 0.2	4.0–95.2	87.1 ± 0.08	52.1–99.7	42.4 ± 0.2	1.5–97.0
WOB (%)	87.2 ± 0.1	37.8–100.0	75.1 ± 0.2	23.6–98.6	88.0 ± 0.09	45.9–100.0	78.0 ± 0.2	19.3–99.2
ALH (μm)	3.9 ± 0.02	0.5–11.6	2.2 ± 0.01	0.2–6.8	2.6 ± 0.01	0.3–7.4	3.9 ± 0.02	1.0–10.3
BCF (Hz)	8.2 ± 0.03	0.0–21.0	8.7 ± 0.03	0.0–22.0	9.1 ± 0.03	0.0–22.0	7.7 ± 0.05	0.0–21.6

SEM: standard error of the mean; VCL (μm/s): curvilinear velocity; VSL (μm/s): straight line velocity; VAP (μm/s): average path velocity; LIN (%): linearity; STR (%): straightness; WOB (%): wobble; ALH (μm): amplitude of lateral head displacement; BCF (Hz): beat-cross frequency.

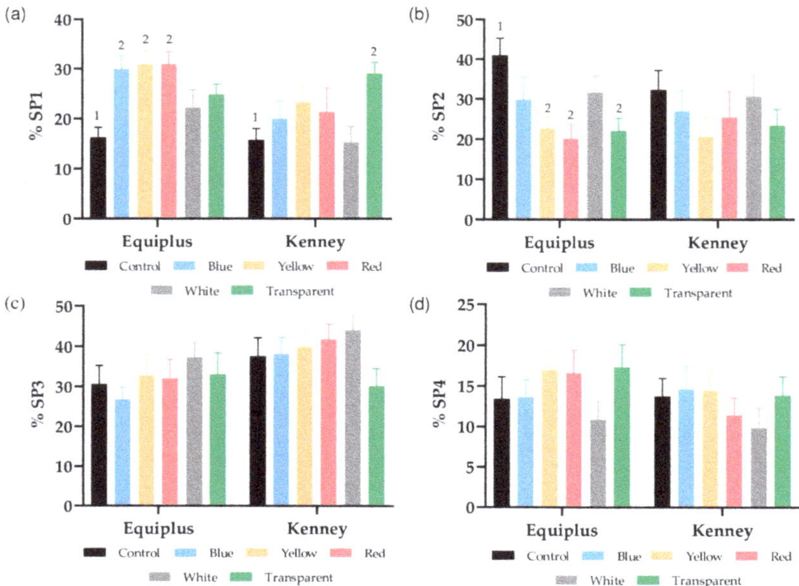

Figure 1. Effects of the color of the straw, extender, and light stimulation on the structure of motile sperm subpopulations in control (nonirradiated) and irradiated samples packed into straws of different color and extended either with Equiplus or Kenney extender. (**a**) Subpopulation 1 (SP1, which was the fastest subpopulation for VCL, VSL and VAP); (**b**) Subpopulation 2 (SP2, the slowest); (**c**) Subpopulation 3 (SP3); (**d**) Subpopulation 4 (SP4). The different numbers (1, 2) indicate significant differences ($p < 0.05$) between irradiated and nonirradiated samples packed into different colored straws within the same diluent. The absence of numbers indicates the lack of statistical differences between irradiated and nonirradiated samples packed into different colored straws within the same diluent. On the other hand, no significant differences between nonirradiated and irradiated samples were observed when the two extenders were compared within the same treatment. Data are shown as mean ±SEM of 13 separate experiments.

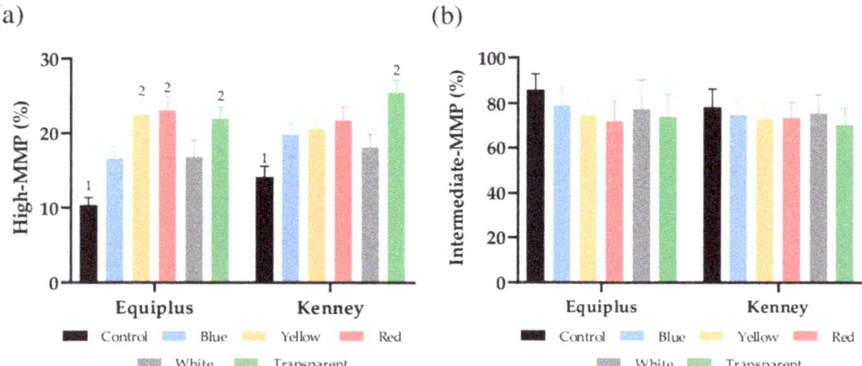

Figure 2. *Cont.*

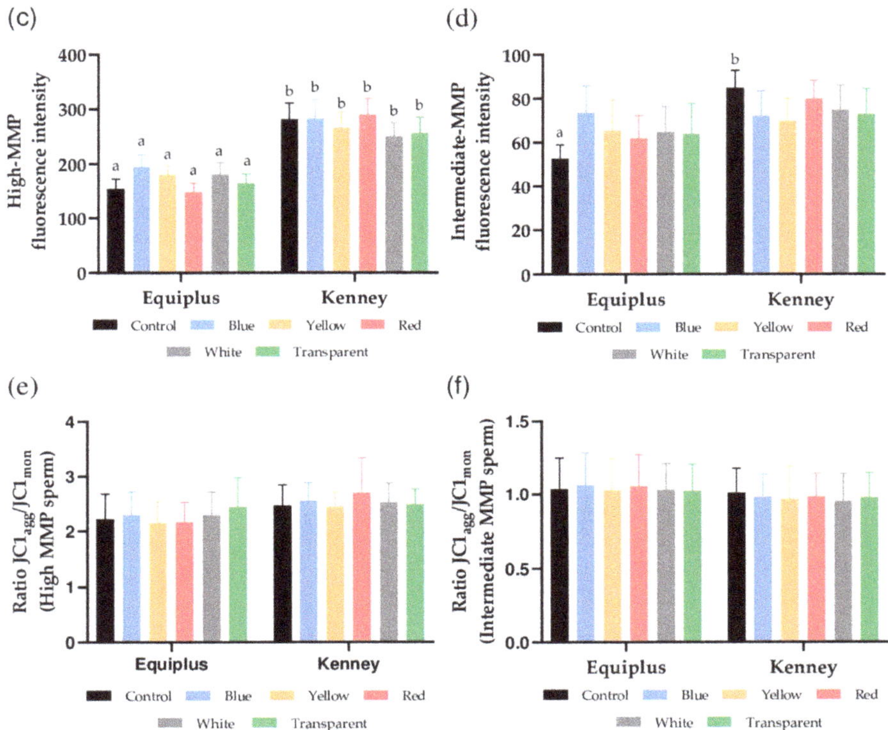

Figure 2. Effects of the color of the straw, extender, and light stimulation on mitochondrial membrane potential in control (nonirradiated) and irradiated samples packed into straws of different color and extended either with Equiplus or Kenney extender. The results are presented as percentages of sperm with high mitochondrial membrane potential (MMP; **a**) and with intermediate mitochondrial membrane potential (MMP; **b**), geometric mean of fluorescence intensity of $JC1_{agg}$ (GMFI, FL2) in sperm populations with high (**c**) and intermediate MMP (**d**), and $JC1_{agg}/JC1_{mon}$ ratios (GMFI FL2/GMFI FL1) in sperm populations with high (**e**) and intermediate MMP (**f**) in nonirradiated (control) and irradiated samples. Different numbers (1, 2) indicate significant differences ($p < 0.05$) between nonirradiated and irradiated samples packed into straws of different color within the same diluent. Different letters (a, b) indicate significant differences ($p < 0.05$) between the two extenders within nonirradiated or samples irradiated and packed into straws of different color. The absence of numbers indicates the lack of statistical differences between irradiated and nonirradiated samples within the same diluent, and the absence of letters indicates the lack of differences when comparing a given treatment between both diluents. Data are shown as mean ± SEM of 13 separate experiments.

3.5. Intracellular Peroxide and Superoxide Levels

Figure 3a shows the percentage of viable sperm with high peroxide levels. No significant differences between nonirradiated and irradiated samples were observed, regardless of the color of the straw or the extender used. However, as Figure 3b shows, GMFI of DCF^+ in the population of viable sperm with high levels of peroxides (DCF^+/PI^-) was significantly higher ($p < 0.05$) in transparent, irradiated straws diluted in Equiplus extender than in their respective control (i.e., nonirradiated samples diluted in Equiplus) and transparent, irradiated straws diluted in Kenney extender.

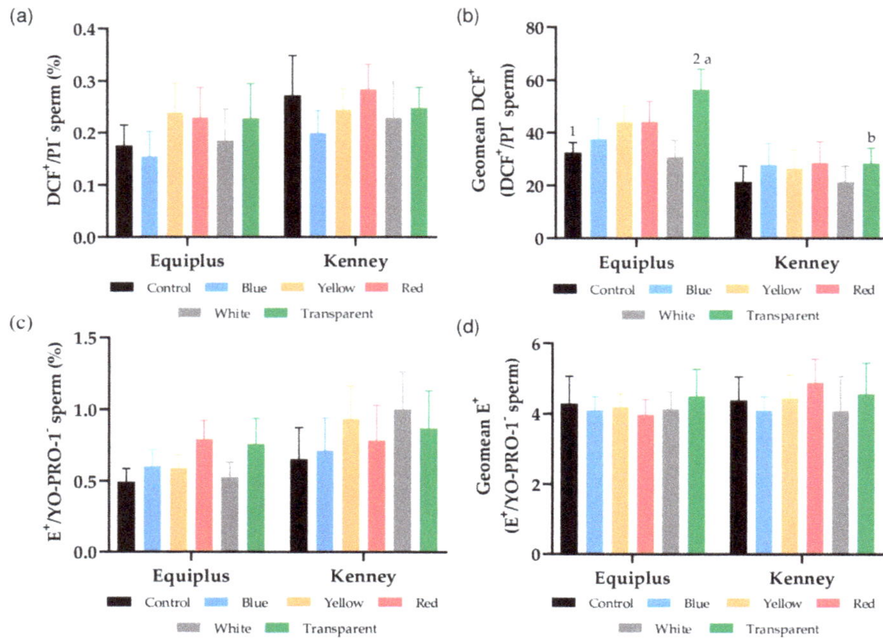

Figure 3. Effects of the color of the straw, extender, and light stimulation on intracellular ROS levels in control (nonirradiated) and irradiated samples packed into straws of different color and extended either with Equiplus or Kenney extender. Data are shown as (**a**) percentages of viable spermatozoa with high peroxide levels (DCF^+/PI^-); (**b**) geometric mean of DCF^+-intensity (GMFI, FL1 channel) in the population of viable spermatozoa with high peroxide levels (DCF^+/PI^-); (**c**) percentages of viable spermatozoa with high superoxide levels (E^+/YO-PRO-1^-); (**d**) geometric mean of E^+-intensity (GMFI, FL3 channel) in the population of viable spermatozoa with high superoxide levels (E^+/YO-PRO-1^-). Different numbers (1, 2) indicate significant differences ($p < 0.05$) between nonirradiated and irradiated samples packed into straws of different color within the same diluent. Different letters (a, b) indicate significant differences ($p < 0.05$) between the two extenders within nonirradiated or samples irradiated packed into straws of different color. The absence of numbers or letters indicates the lack of statistical difference between irradiated and nonirradiated samples within the same diluent or when comparing a given treatment between Kenny and Equiplus extenders. Data are shown as mean ± SEM of 13 separate experiments.

As shown in Figure 3c, percentages of viable spermatozoa with high levels of superoxides (E^+/YO-PRO-1^-) and GMFI of E^+ in the population of viable spermatozoa with high levels of superoxide (Figure 3d) did not differ ($p > 0.05$) between irradiated and nonirradiated samples, regardless of the color of the straw and the extender used.

3.6. Intracellular Calcium Levels

Percentages of viable sperm with high intracellular calcium levels ($Fluo3^+/PI^-$; Figure S4a) did not differ between nonirradiated and irradiated samples, regardless of the color of the straw and the extender used (e.g., nonirradiated samples diluted in Equiplus: 0.5% ± 0.1% vs. sperm diluted in Equiplus, packed into red straws, and irradiated: 0.8% ± 0.1% vs. samples diluted in Kenney, packed into red straws, and irradiated: 0.8% ± 0.2%). Similar results were observed for the GMFI of $Fluo3^+$ in the viable sperm population with high intracellular calcium levels (Figure S4b; e.g., nonirradiated sperm diluted in Kenney: 4.4 ± 0.2 vs. sperm diluted in Kenney, packed into white straws, and irradiated: 4.1 ± 0.2 vs. sperm diluted in Equiplus, packed into white straws, and irradiated: 4.1 ± 0.1).

4. Discussion

The results of this study agree with previous research, as irradiation with LED-based red light was found to modify some sperm motion variables and increase mitochondrial membrane potential and intracellular ROS of horse sperm without affecting the integrity of the plasma membrane and acrosome. The most remarkable and novel finding, however, was that these effects varied with the color of the straw used to pack sperm before irradiation and with the turbidity of the extender.

Regarding the effects on sperm motility, red light stimulation did not affect TMOT or PMOT, regardless of the color of the straw or the type of diluent used, which agrees with the data reported for dogs [17,18], bulls [47], and horses [20,21]. However, other studies found that irradiation of sperm with red light increases total and progressive motility in humans [7–11], buffaloes [19], sheep [4], pigs [13], and donkeys [3]. In evaluating the presence of motile subpopulations of sperm in horse ejaculates, we identified four separate subpopulations. These results are similar to those previously reported for this species [21,48]. In addition to this, we observed that the percentages of sperm belonging to SP1, which was the fastest subpopulation according to VCL, VSL, and VAP, were significantly higher in irradiated samples that were either diluted with Equiplus extender and packed into blue, yellow, and red straws or diluted with Kenney extender and packed into transparent straws. Furthermore, samples diluted in Equiplus extender and packed into red, yellow, and transparent straws showed significantly lower percentages of sperm belonging to SP2 (the slowest subpopulation) than the control. Therefore, our data confirm the results obtained in previous studies, where irradiation with red light was found to modify the structure of motile sperm subpopulations by decreasing the percentages of the slowest sperm subpopulation [21] and increasing those of the fastest one [18,21,22]. Moreover, we observed that light-stimulation increased some kinetic measures, which agrees with the data reported for other species such as humans [9,34], dogs [17,18], cattle [47], buffaloes [19], pigs [13], donkeys [3,22], and horses [20,21]. These observed differences reinforce the hypothesis that the effects of red light on spermatozoa depend on the specific irradiation pattern [3,13,22,29] and also differ between species [3,5,20–22]. At this point, the increase of VCL, VSL, and VAP observed in sperm diluted in Equiplus extender, packed into red straws, and irradiated and the increase of VCL and VAP found in semen diluted in Kenney extender, packed into transparent straws, and irradiated should be emphasized. Moreover, STR also increased in sperm diluted in Kenney extender, packed into blue straws, and irradiated. All these data suggest that the effects of red light on sperm depend on the color of the straw and the medium used. Based on these results, it is reasonable to surmise that the color of the straw and the turbidity of the extender modify the amount of light/energy that reaches the sperm cells.

At present, there is no clear explanation of how irradiation affects these sperm motion measures as the exact mechanism(s) through which red light stimulates sperm still remains unknown. However, one of the established hypotheses postulates that red light may boost mitochondrial activity, which could be relevant to explain the effects observed in sperm kinetics. Related to this, our data on the analysis of mitochondrial membrane potential (JC1) would agree with this possibility because there was an increase in the percentages of sperm with high mitochondrial membrane potential in samples packed into transparent and red straws and irradiated, regardless of the extender used (Equiplus or Kenney). This matches with Siqueira et al. [47], who found that irradiation of bovine sperm with a He-Ne laser at a wavelength of 633 nm increases the percentage of sperm cells with high mitochondrial membrane potential, and with Yeste et al. [13], who observed that irradiation with red LED light at a wavelength between 620–630 nm augments the percentages of pig sperm with high mitochondrial membrane potential. All these data suggest that red light stimulation could increase mitochondrial activity through photosensitizers present in the electronic chain, such as cytochrome C [13,22,29,49], which would underlie the increase observed in sperm motility.

In addition to the aforementioned, because ROS are mainly generated in the mitochondria as a byproduct of the electronic chain and following the previously established hypothesis, which points out that one of the first effects of light on sperm is the production of ROS [5,50], the generation of intracellular peroxide and superoxide levels was also evaluated. While sperm irradiation did not affect superoxide generation, we found an increase in the levels of peroxides in those irradiated after dilution in Equiplus and packing into transparent straws. This rise in intracellular ROS levels agrees with Zan-Bar et al. [5], Catalán et al. [20], and Cohen et al. [50], who suggested that ROS formation would be mediated through specific endogenous cellular photosensitizers such as mitochondrial cytochromes. In this sense, it has been reported that although an excess of ROS production produced by irradiation with light could be detrimental to sperm cells [5,50], low ROS levels are beneficial for sperm motility and fertilizing ability [5]. Whilst more studies are needed to set a relationship between fertilization ability and high mitochondrial membrane potential, intracellular ROS, and sperm motility, H_2O_2 has been suggested to be the active molecule involved in the light-mediated changes of sperm fertilizing capacity [50], which is consistent with Zan-Bar et al. [5] and de Lamirande et al. [51], who indicated that low concentrations of ROS participate in the signaling transduction pathways related to sperm capacitation and acrosomal reaction. Therefore, ROS can have both harmful and beneficial effects on sperm, and the delicate balance between the amounts of ROS produced and ROS scavengers at any time point determines whether a particular sperm function parameter is compromised or boosted [50]. In this sense, the extent of increase in intracellular ROS levels observed in this study was not enough to negatively affect sperm motility and viability, which is similar to that reported by Catalán et al. [3] in a study conducted with fresh and cooled-stored donkey semen. This increase in intracellular ROS (peroxides), observed herein after irradiation, together with the variation seen due to the color of the straw and the extender used, was concomitant with a rise in mitochondrial activity. These findings reinforce the conjecture that ROS formation caused by light would be mediated by specific endogenous cellular photosensitizers such as mitochondrial cytochromes. Furthermore, cytochrome complexes are also known to be implicated in the intrinsic apoptotic pathway [52], and both ROS generation and modulation of apoptotic-like changes are crucial to cause and control sperm capacitation [53]. Therefore, red light-induced changes in cytochrome C complex activity could ultimately affect sperm capacitation and survival. Surprisingly, however, our results did not show an increase in intracellular calcium levels, which is a crucial secondary messenger involved in the modulation of sperm motility and capacitation [54,55]. Related to this, it is worth noting that our data differ from those reported in previous studies, where light-stimulation was found to increase intracellular levels of calcium [30,50]. This could be explained by different conditions of time and intensity of radiation between the current study and the others, as previous research indicates that sperm irradiation can have stimulatory or inhibitory effects on calcium transport, depending on the intensity of the light used [56].

Regarding the effects of irradiation on the integrity of the plasma membrane, no significant differences were found between irradiated and nonirradiated samples. These results were similar to those reported by Yeste et al. [13] and Pezo et al. [14] in pigs and by Catalán et al. in horses and donkeys [3,20,22]. Similarly, no negative impact of irradiation on acrosomal integrity was observed, which concurs with previous studies in rabbits [16], pigs [13,57], and donkeys [22]. This supports the idea that under the conditions tested herein, stimulation of sperm with red light is safe and can have a positive effect on sperm motility and mitochondrial membrane potential, in agreement with Gabel et al. [32].

Finally, the differences observed in this study between straw colors and extenders with regard to mitochondrial activity, intracellular levels of peroxides, and motility suggest that these two factors also influence the sperm response to light. In fact, the impact of red light on mammalian sperm has been previously reported to rely on the precise rhythm and intensity of light [13] and the functional status of the cell [3]. In agreement with this and with the hypothesis that light acts on endogenous cellular photosensitizers of mitochondria, it is reasonable to suggest that the energy supplied to the mitochondrial electron chain by red light is proportional to the exposure time and the intensity of the light used. The final consequence of this phenomenon would be that the color of the straw and the opacity/turbidity of the medium influence the intensity of the light that feeds mitochondria, which would generate a different effect on sperm cells.

5. Conclusions

Our results confirm that LED-based red light irradiation increases some sperm motion variables, mitochondrial membrane potential, and intracellular ROS without affecting the integrity of the sperm membrane and acrosome. However, these effects vary with the color of the straw and the extender/medium used. Given that increased motility and intracellular ROS levels are concomitant with a rise in mitochondrial activity, we suggest that the impact of irradiation on sperm depends on the precise rates of energy provided by the light that feeds the mitochondria. Remarkably, such an energy rate, sensed by mitochondrial photosensitizers, varies with the color of the straw and the extender/medium used, so that these two aspects have to be taken into consideration when sperm are irradiated. In effect, as could be observed in this study, the greatest effects were obtained in samples diluted in Equiplus extender, packed into red straws, and irradiated and samples diluted in Kenney extender, packed into transparent straws, and irradiated.

Supplementary Materials: The following are available online at https://www.mdpi.com/2076-2615/11/1/122/s1: Figures S1–S3.

Author Contributions: Conceptualization, S.B., J.E.R.-G., J.M., and M.Y.; methodology, J.C., I.Y.-O., S.G., and M.P.; validation, J.E.R.-G., J.M., and M.Y.; formal analysis, J.C.; investigation, J.C., S.G., M.P., J.E.R.-G., M.Y., and J.M.; resources, S.B., J.E.R.-G., J.M., and M.Y.; data curation, J.C.; writing—original draft preparation, J.C.; writing—review and editing, S.B., J.E.R.-G., J.M., and M.Y.; supervision, S.B., J.E.R.-G., J.M., and M.Y.; project administration, S.B., J.E.R.-G., J.M., and M.Y.; funding acquisition, S.B., J.E.R.-G., J.M., and M.Y. All authors have read and agreed to the published version of the manuscript.

Funding: J.C. was funded by the National Agency for Research and Development (ANID), Ministry of Education, Chile (Scheme: Becas Chile Doctorado en el Extranjero, PFCHA; Grant: 2017/72180128). I.Y.-O. was funded by the Secretary of Higher Education, Science, Technology and Innovation (SENESCYT), Ecuador (Scheme: Programa de Becas Internacionales de Posgrado 2019; Grant: CZ02-000507-2019). The authors also acknowledge support from the Ministry of Science and Innovation, Spain (Grants: RYC-2014-15581 and AGL2017-88329-R) and the Regional Government of Catalonia, Spain (2017-SGR-1229).

Institutional Review Board Statement: The study was conducted according to the guidelines of the Declaration of Helsinki, and approved by the Ethics Committee, Autonomous University of Barcelona (Code: CEEAH 1424).

Informed Consent Statement: Not applicable.

Data Availability Statement: The data presented in this study are available on request from the corresponding author.

Acknowledgments: The authors would like to thank Sebastián Bonilla-Correal from Autonomous University of Barcelona, Spain, for his support and Marc Llavanera, Ariadna Delgado-Bermúdez, Sandra Recuero, Yentel Mateo-Otero, Beatriz Fernandez-Fuertes, Estela Garcia, and Isabel Barranco from University of Girona for their technical support.

Conflicts of Interest: J.E.R.-G. and M.Y. are inventors of a patent entitled "Method and apparatus for improving the quality of mammalian sperm" (European Patent Office, No. 16199093.2; EP-3-323-289-A1), which is owned by Instruments Útils de Laboratori Geniul, SL (Barcelona, Spain).

References

1. Kowalczyk, A.; Czerniawska-Piatkowska, E.; Kuczaj, M. Factors influencing the popularity of artificial insemination of mares in Europe. *Animals* **2019**, *9*, 460. [CrossRef] [PubMed]
2. Loomis, P.R. Advanced Methods for Handling and Preparation of Stallion Semen. *Vet. Clin. N. Am. Equine Pract.* **2006**, *22*, 663–676. [CrossRef] [PubMed]
3. Catalán, J.; Papas, M.; Gacem, S.; Noto, F.; Delgado-Bermúdez, A.; Rodríguez-Gil, J.E.; Miró, J.; Yeste, M. Effects of red-light irradiation on the function and survival of fresh and liquid-stored donkey semen. *Theriogenology* **2020**, *149*. [CrossRef]
4. Iaffaldano, N.; Paventi, G.; Pizzuto, R.; Di Iorio, M.; Bailey, J.L.; Manchisi, A.; Passarella, S. Helium-neon laser irradiation of cryopreserved ram sperm enhances cytochrome c oxidase activity and ATP levels improving semen quality. *Theriogenology* **2016**, *86*, 778–784. [CrossRef] [PubMed]
5. Zan-Bar, T.; Bartoov, B.; Segal, R.; Yehuda, R.; Lavi, R.; Lubart, R.; Avtalion, R.R. Influence of Visible Light and Ultraviolet Irradiation on Motility and Fertility of Mammalian and Fish Sperm. *Photomed. Laser Surg.* **2005**, *23*, 549–555. [CrossRef] [PubMed]
6. Iaffaldano, N.; Meluzzi, A.; Manchisi, A.; Passarella, S. Improvement of stored turkey semen quality as a result of He-Ne laser irradiation. *Anim. Reprod. Sci.* **2005**, *85*, 317–325. [CrossRef] [PubMed]
7. Sato, H.; Landthaler, M.; Haina, D.; Schill, W.B. The effects of laser light on sperm motility and velocity in vitro. *Andrologia* **2009**, *16*, 23–25. [CrossRef] [PubMed]
8. Lenzi, A.; Claroni, F.; Gandini, L.; Lombardo, F.; Barbieri, C.; Lino, A.; Dondero, F. Laser radiation and motility patterns of human sperm. *Syst. Biol. Reprod. Med.* **1989**, *23*, 229–234. [CrossRef]
9. Firestone, R.S.; Esfandiari, N.; Moskovtsev, S.I.; Burstein, E.; Videna, G.T.; Librach, C.; Bentov, Y.; Casper, R.F. The Effects of Low-Level Laser Light Exposure on Sperm Motion Characteristics and DNA Damage. *J. Androl.* **2012**, *33*, 469–473. [CrossRef]
10. Salman Yazdi, R.; Bakhshi, S.; Jannat Alipoor, F.; Akhoond, M.R.; Borhani, S.; Farrahi, F.; Lotfi Panah, M.; Sadighi Gilani, M.A. Effect of 830-nm diode laser irradiation on human sperm motility. *Lasers Med. Sci.* **2014**, *29*, 97–104. [CrossRef]
11. Ban Frangez, H.; Frangez, I.; Verdenik, I.; Jansa, V.; Virant Klun, I. Photobiomodulation with light-emitting diodes improves sperm motility in men with asthenozoospermia. *Lasers Med. Sci.* **2015**, *30*, 235–240. [CrossRef] [PubMed]
12. Salama, N.; El-Sawy, M. Light-emitting diode exposure enhances sperm motility in men with and without asthenospermia: Preliminary results. *Arch. Ital. Urol. Androl.* **2015**, *87*, 14–19. [CrossRef] [PubMed]
13. Yeste, M.; Codony, F.; Estrada, E.; Lleonart, M.; Balasch, S.; Peña, A.; Bonet, S.; Rodríguez-Gil, J.E. Specific LED-based red light photo-stimulation procedures improve overall sperm function and reproductive performance of boar ejaculates. *Sci. Rep.* **2016**, *6*, 22569. [CrossRef] [PubMed]
14. Pezo, F.; Zambrano, F.; Uribe, P.; Ramírez-Reveco, A.; Romero, F.; Sanchéz, R. LED-based red light photostimulation improves short-term response of cooled boar semen exposed to thermal stress at 37 °C. *Andrologia* **2019**, *51*, e13237. [CrossRef]
15. Blanco Prieto, O.; Catalán, J.; Lleonart, M.; Bonet, S.; Yeste, M.; Rodríguez-Gil, J.E. Red-light stimulation of boar semen prior to artificial insemination improves field fertility in farms: A worldwide survey. *Reprod. Domest. Anim.* **2019**, *54*, rda.13470. [CrossRef]
16. Iaffaldano, N.; Rosato, M.P.; Paventi, G.; Pizzuto, R.; Gambacorta, M.; Manchisi, A.; Passarella, S. The irradiation of rabbit sperm cells with He–Ne laser prevents their in vitro liquid storage dependent damage. *Anim. Reprod. Sci.* **2010**, *119*, 123–129. [CrossRef]
17. Corral-Baqués, M.I.; Rigau, T.; Rivera, M.; Rodríguez, J.E.; Rigau, J. Effect of 655-nm diode laser on dog sperm motility. *Lasers Med. Sci.* **2005**, *20*, 28–34. [CrossRef]
18. Corral-Baqués, M.I.; Rivera, M.M.; Rigau, T.; Rodríguez-Gil, J.E.; Rigau, J. The effect of low-level laser irradiation on dog spermatozoa motility is dependent on laser output power. *Lasers Med. Sci.* **2009**, *24*, 703–713. [CrossRef]
19. Abdel-Salam, Z.; Dessouki, S.H.M.; Abdel-Salam, S.A.M.; Ibrahim, M.A.M.; Harith, M.A. Green laser irradiation effects on buffalo semen. *Theriogenology* **2011**, *75*, 988–994. [CrossRef]
20. Catalán, J.; Llavanera, M.; Bonilla-Correal, S.; Papas, M.; Gacem, S.; Rodríguez-Gil, J.E.; Yeste, M.; Miró, J. Irradiating frozen-thawed stallion sperm with red-light increases their resilience to withstand post-thaw incubation at 38 °C. *Theriogenology* **2020**, *157*, 85–95. [CrossRef]
21. Catalán, J.; Papas, M.; Gacem, S.; Mateo-Otero, Y.; Rodríguez-Gil, J.E.; Miró, J.; Yeste, M. Red-Light Irradiation of Horse Spermatozoa Increases Mitochondrial Activity and Motility through Changes in the Motile Sperm Subpopulation Structure. *Biology* **2020**, *9*, 254. [CrossRef] [PubMed]
22. Catalán, J.; Papas, M.; Trujillo-Rojas, L.; Blanco-Prieto, O.; Bonilla-Correal, S.; Rodríguez-Gil, J.E.; Miró, J.; Yeste, M. Red LED Light Acts on the Mitochondrial Electron Chain of Donkey Sperm and Its Effects Depend on the Time of Exposure to Light. *Front. Cell Dev. Biol.* **2020**, *8*, 588621. [CrossRef] [PubMed]
23. Yeste, M.; Castillo-Martín, M.; Bonet, S.; Rodríguez-Gil, J.E. Impact of light irradiation on preservation and function of mammalian spermatozoa. *Anim. Reprod. Sci.* **2018**, *194*, 19–32. [CrossRef]
24. De Blas, G.A.; Darszon, A.; Ocampo, A.Y.; Serrano, C.J.; Castellano, L.E.; Hernández-González, E.O.; Chirinos, M.; Larrea, F.; Beltrán, C.; Treviño, C.L. TRPM8, a Versatile Channel in Human Sperm. *PLoS ONE* **2009**, *4*, e6095. [CrossRef] [PubMed]
25. Bahat, A.; Eisenbach, M. Human Sperm Thermotaxis Is Mediated by Phospholipase C and Inositol Trisphosphate Receptor Ca2+ Channel1. *Biol. Reprod.* **2010**, *82*, 606–616. [CrossRef]

26. Gibbs, G.M.; Orta, G.; Reddy, T.; Koppers, A.J.; Martínez-López, P.; De La Vega-Beltràn, J.L.; Lo, J.C.Y.; Veldhuis, N.; Jamsai, D.; McIntyre, P.; et al. Cysteine-rich secretory protein 4 is an inhibitor of transient receptor potential M8 with a role in establishing sperm function. *Proc. Natl. Acad. Sci. USA* **2011**, *108*, 7034–7039. [CrossRef]
27. Wu, L.J.; Sweet, T.B.; Clapham, D.E. International Union of Basic and Clinical Pharmacology. LXXVI. Current progress in the Mammalian TRP ion channel family. *Pharmacol. Rev.* **2010**, *62*, 381–404. [CrossRef]
28. Pérez-Cerezales, S.; Boryshpolets, S.; Afanzar, O.; Brandis, A.; Nevo, R.; Kiss, V.; Eisenbach, M. Involvement of opsins in mammalian sperm thermotaxis. *Sci. Rep.* **2015**, *5*. [CrossRef]
29. Blanco-Prieto, O.; Catalán, J.; Trujillo-Rojas, L.; Peña, A.; Rivera del Álamo, M.M.; Llavanera, M.; Bonet, S.; Fernández-Novell, J.M.; Yeste, M.; Rodríguez-Gil, J.E. Red LED Light Acts on the Mitochondrial Electron Chain of Mammalian Sperm via Light-Time Exposure-Dependent Mechanisms. *Cells* **2020**, *9*, 2546. [CrossRef]
30. Lubart, R.; Friedmann, H.; Levinshal, T.; Lavie, R.; Breitbart, H. Effect of light on calcium transport in bull sperm cells. *J. Photochem. Photobiol. B Biol.* **1992**, *15*, 337–341. [CrossRef]
31. Karu, T. Photobiology of low-power laser effects. *Health Phys.* **1989**, *56*, 691–704. [CrossRef]
32. Gabel, C.P.; Carroll, J.; Harrison, K. Sperm motility is enhanced by Low Level Laser and Light Emitting Diode photobiomodulation with a dose-dependent response and differential effects in fresh and frozen samples. *Laser Ther.* **2018**, *27*, 131–136. [CrossRef]
33. Tuner, J.; Hode, L. *Laser Therapy: Clinical Practice and Scientific*; Prima Books AB: Grängesberg, Sweden, 2002; pp. 12–22.
34. Preece, D.; Chow, K.W.; Gomez-Godinez, V.; Gustafson, K.; Esener, S.; Ravida, N.; Durrant, B.; Berns, M.W. Red light improves spermatozoa motility and does not induce oxidative DNA damage. *Sci. Rep.* **2017**, *7*. [CrossRef]
35. Kenney, M.R. Minimal contamination techniques for breeding mares: Techniques and priliminary findings. *Proc. Am. Assoc. Equine Pract.* **1975**, 327–336.
36. Prieto-Martínez, N.; Vilagran, I.; Morató, R.; Rodríguez-Gil, J.E.; Yeste, M.; Bonet, S. Aquaporins 7 and 11 in boar spermatozoa: Detection, localisation and relationship with sperm quality. *Reprod. Fertil. Dev.* **2016**, *28*, 663–672. [CrossRef] [PubMed]
37. Lee, J.A.; Spidlen, J.; Boyce, K.; Cai, J.; Crosbie, N.; Dalphin, M.; Furlong, J.; Gasparetto, M.; Goldberg, M.; Goralczyk, E.M.; et al. MIFlowCyt: The minimum information about a flow cytometry experiment. *Cytom. Part A* **2008**, *73A*, 926–930. [CrossRef]
38. Petrunkina, A.M.; Waberski, D.; Bollwein, H.; Sieme, H. Identifying non-sperm particles during flow cytometric physiological assessment: A simple approach. *Theriogenology* **2010**, *73*, 995–1000. [CrossRef] [PubMed]
39. Garner, D.L.; Johnson, L.A. Viability assessment of mammalian sperm using SYBR-14 and propidium iodide. *Biol. Reprod.* **1995**, *53*, 276–284. [CrossRef]
40. Rathi, R.; Colenbrander, B.; Bevers, M.M.; Gadella, B.M. Evaluation of in vitro capacitation of stallion spermatozoa. *Biol. Reprod.* **2001**, *65*, 462–470. [CrossRef] [PubMed]
41. Murillo, M.M.; Carmona-Cuenca, I.; Del Castillo, G.; Ortiz, C.; Roncero, C.; Sánchez, A.; Fernández, M.; Fabregat, I. Activation of NADPH oxidase by transforming growth factor-β in hepatocytes mediates up-regulation of epidermal growth factor receptor ligands through a nuclear factor-κB-dependent mechanism. *Biochem. J.* **2007**, *405*, 251–259. [CrossRef]
42. Guthrie, H.D.; Welch, G.R. Determination of intracellular reactive oxygen species and high mitochondrial membrane potential in Percoll-treated viable boar sperm using fluorescence-activated flow cytometry. *J. Anim. Sci.* **2006**, *84*, 2089–2100. [CrossRef] [PubMed]
43. Zhao, H.; Kalivendi, S.; Zhang, H.; Joseph, J.; Nithipatikom, K.; Vásquez-Vivar, J.; Kalyanaraman, B. Superoxide reacts with hydroethidine but forms a fluorescent product that is distinctly different from ethidium: Potential implications in intracellular fluorescence detection of superoxide. *Free Radic. Biol. Med.* **2003**, *34*, 1359–1368. [CrossRef]
44. Yeste, M.; Fernández-Novell, J.M.; Ramió-Lluch, L.; Estrada, E.; Rocha, L.G.; Cebrián-Pérez, J.A.; Muiño-Blanco, T.; Concha, I.I.; Ramírez, A.; Rodríguez-Gil, J.E. Intracellular calcium movements of boar spermatozoa during 'in vitro' capacitation and subsequent acrosome exocytosis follow a multiple-storage place, extracellular calcium-dependent model. *Andrology* **2015**, *3*, 729–747. [CrossRef] [PubMed]
45. Kadirvel, G.; Kumar, S.; Kumaresan, A.; Kathiravan, P. Capacitation status of fresh and frozen-thawed buffalo spermatozoa in relation to cholesterol level, membrane fluidity and intracellular calcium. *Anim. Reprod. Sci.* **2009**, *116*, 244–253. [CrossRef] [PubMed]
46. Luna, C.; Yeste, M.; Rivera Del Alamo, M.M.; Domingo, J.; Casao, A.; Rodriguez-Gil, J.E.; Pérez-Pé, R.; Cebrián-Pérez, J.A.; Muiño-Blanco, T. Effect of seminal plasma proteins on the motile sperm subpopulations in ram ejaculates. *Reprod. Fertil. Dev.* **2017**, *29*, 394–405. [CrossRef]
47. Siqueira, A.F.P.; Maria, F.S.; Mendes, C.M.; Hamilton, T.R.S.; Dalmazzo, A.; Dreyer, T.R.; da Silva, H.M.; Nichi, M.; Milazzotto, M.P.; Visintin, J.A.; et al. Effects of photobiomodulation therapy (PBMT) on bovine sperm function. *Lasers Med. Sci.* **2016**, *31*, 1245–1250. [CrossRef]
48. Quintero-Moreno, A.; Miró, J.; Teresa Rigau, A.; Rodríguez-Gil, J.E. Identification of sperm subpopulations with specific motility characteristics in stallion ejaculates. *Theriogenology* **2003**, *59*, 1973–1990. [CrossRef]
49. Begum, R.; Powner, M.B.; Hudson, N.; Hogg, C.; Jeffery, G. Treatment with 670 nm Light Up Regulates Cytochrome C Oxidase Expression and Reduces Inflammation in an Age-Related Macular Degeneration Model. *PLoS ONE* **2013**, *8*, e57828. [CrossRef]
50. Cohen, N.; Lubart, R.; Rubinstein, S.; Breitbart, H. Light Irradiation of Mouse Spermatozoa: Stimulation of In Vitro Fertilization and Calcium Signals. *Photochem. Photobiol.* **1998**, *68*, 407–413. [CrossRef]

51. De Lamirande, E.; Jiang, H.; Zini, A.; Kodama, H.; Gagnon, C. Reactive oxygen species and sperm physiology. *Rev. Reprod.* **1997**, *2*, 48–54. [CrossRef]
52. Cai, J.; Yang, J.; Jones, D.P. Mitochondrial control of apoptosis: The role of cytochrome c. *Biochim. Biophys. Acta Bioenerg.* **1998**, *1366*, 139–149. [CrossRef]
53. Ortega-Ferrusola, C.; Macías García, B.; Gallardo-Bolaños, J.M.; González-Fernández, L.; Rodríguez-Martinez, H.; Tapia, J.A.; Peña, F.J. Apoptotic markers can be used to forecast the freezeability of stallion spermatozoa. *Anim. Reprod. Sci.* **2009**, *114*, 393–403. [CrossRef]
54. Yeste, M. Boar spermatozoa within the oviductal environment (II): Sperm capacitation. In *Boar Reproduction: Fundamentals and New Biotechnological Trends*; Bonet, S., Casas, I., Holt, W.V., Yeste, M., Eds.; Springer: Berlin, Germany, 2013; pp. 281–342.
55. Correia, J.; Michelangeli, F.; Publicover, S. Regulation and roles of Ca2+ stores in human sperm. *Reproduction* **2015**, *150*, 65–76. [CrossRef]
56. Breitbart, H.; Levinshal, T.; Cohen, N.; Friedmann, H.; Lubart, R. Changes in calcium transport in mammalian sperm mitochondria and plasma membrane irradiated at 633 nm (HeNe laser). *J. Photochem. Photobiol. B Biol.* **1996**, *34*, 117–121. [CrossRef]
57. Blanco-Prieto, O.; Catalán, J.; Rojas, L.; Delgado-Bermúdez, A.; Llavanera, M.; Rigau, T.; Bonet, S.; Yeste, M.; Rivera del Álamo, M.; Rodríguez-Gil, J. Medium-term effects of the diluted pig semen irradiation with red LED light on the integrity of nucleoprotein structure and resilience to withstand thermal stress. *Theriogenology* **2020**. [CrossRef]

Review

The Current Trends in Using Nanoparticles, Liposomes, and Exosomes for Semen Cryopreservation

Islam M. Saadeldin [1,2,*], Wael A. Khalil [3], Mona G. Alharbi [4] and Seok Hee Lee [5,*]

1 Department of Animal Production, College of Food and Agricultural Sciences, King Saud University, Riyadh 11451, Saudi Arabia
2 Department of Comparative Medicine, King Faisal Specialist Hospital & Research Centre, Riyadh 11211, Saudi Arabia
3 Department of Animal Production, Faculty of Agriculture, Mansoura University, Mansoura 35516, Egypt; w-khalil@mans.edu.eg
4 Department of Biochemistry, College of Sciences, King Saud University, Riyadh 11451, Saudi Arabia; mgalharbi@ksu.edu.sa
5 Center for Reproductive Sciences, Department of Obstetrics and Gynecology, University of California San Francisco, San Francisco, CA 94143, USA
* Correspondence: isaadeldin@ksu.edu.sa (I.M.S.); seokhee.lee@ucsf.edu (S.L.)

Received: 19 November 2020; Accepted: 2 December 2020; Published: 3 December 2020

Simple Summary: Long-term preservation of semen is a pivotal step for artificial insemination in most farm animal species, but it is associated with cellular insults at the cell membrane and cytoskeleton level as well as the generation of reactive oxygen species (ROS). We highlight the recent strategies to combat these negative effects through defending against the ROS via antioxidant nanoparticles or through repairing/regenerating the damaged sperm through using liposomes and most recently exosomes derived from the reproductive tract or stem cells.

Abstract: Cryopreservation is an essential tool to preserve sperm cells for zootechnical management and artificial insemination purposes. Cryopreservation is associated with sperm damage via different levels of plasma membrane injury and oxidative stress. Nanoparticles are often used to defend against free radicals and oxidative stress generated through the entire process of cryopreservation. Recently, artificial or natural nanovesicles including liposomes and exosomes, respectively, have shown regenerative capabilities to repair damaged sperm during the freeze–thaw process. Exosomes possess a potential pleiotropic effect because they contain antioxidants, lipids, and other bioactive molecules regulating and repairing spermatozoa. In this review, we highlight the current strategies of using nanoparticles and nanovesicles (liposomes and exosomes) to combat the cryoinjuries associated with semen cryopreservation.

Keywords: nanoparticles; liposomes; exosomes; semen; cryopreservation; livestock production

---

1. Introduction

Semen cryopreservation contributes to genetic improvement through artificial insemination, eliminates geographical barriers in artificial insemination (AI) application, and supports the preservation of endangered breeds, thus the conservation of biodiversity. However, the sperm freezing process induces ultrastructural, biochemical, and functional changes of spermatozoa. Especially, spermatozoa membranes and chromatin can be damaged, sperm membrane permeability is increased, and hyper oxidation and formation of reactive oxygen species takes place, affecting fertilizing ability and subsequent early embryonic development [1].

Cryopreservation of mammalian sperm is a complex process affected by several factors for obtaining good quality semen for AI [2], such as type of cryoprotectants or extenders, rates of cooling and thawing, and method of packaging [3,4]. Cryopreservation is associated with damage on the level of the cell membrane, cytoskeleton, DNA, and mitochondria due to the generation of reactive oxygen species (ROS), which affect the entire cellular functions and genome instability [5]. Post-thawing trauma and cellular injury in gametes have been illustrated to affect the cell membrane, organelles, and biochemical perturbation [6]. Sperm cooling and freezing causes membrane phospholipids to accumulate due to van der Waals forces, and transition occurs from liquid crystal phase to gel phase. During thawing, irregular voids occur in the cell membrane that lead to damage to the membrane structure and irregular ion and water leakage both into and out of the cell [7].

In living organisms, generation of ROS, such as hydrogen peroxide (H_2O_2), superoxide anions (O_2^-), and hydroxyl radicals (OH^-), may be produced as a result of radiation [8], bio-activation of xenobiotics [9], inflammation [10], cell metabolism [11], decompartmentalization of transition metal ions [12], activities of redox enzymes [13], and deficit in the antioxidant defense [14,15]. Physiologically, free radicals level has a positive impact on sperm cells, including capacitation, hyper-activation, and sperm-oocyte fusion [14]. Therefore, ROS with a physiological limit are required for spermatozoa to attain the fertilizing ability [16], acrosome reaction/acrosomal exocytosis, and sperm motility [17]. However, during semen cryopreservation, the cold shock and the atmospheric oxygen [18,19] increase ROS production and cause an imbalance between free radicals and the antioxidant defense in the semen [20]. Increased ROS production can cause toxic effects in the sperm function [21], in terms of inactivating glycolytic enzymes through acrosomal damage [22], lipid peroxidation (LPO), and reducing sperm fertility [23–25]. Notably, LPO is a pathological outcome of several diseases and stress conditions [26]. The LPO process caused by ROS (H_2O_2) is detrimental to sperm survivability. As a result of high contents of polyunsaturated fatty acids in the plasma membrane and lack of antioxidant enzymes, mammalian spermatozoa are susceptible to LPO induced damage and loss of sperm functions [27,28]. Increasing ROS generation under oxidative stress (OS) leads to increased sperm plasma membrane failure, damaged spermatozoa [29], reduced sperm cell cytoplasm [30], and finally a marked reduction in viability, the integrity of the sperm membrane, and fertilizing ability and increased damage to sperm DNA [31].

Moreover, the process of freezing has resulted in a significant reduction in GSH content in frozen semen [32,33]. Baghshahi et al. [34] showed that cryopreservation of ram spermatozoa may cause damage to the function and structure of sperm cells, in terms of reduced semen quality and sperm characteristics. This is due to the reduction of the temperature that is associated with the OS, which has been defined as an imbalance between oxidants and cellular antioxidant mechanisms and is induced by the generation of ROS [35].

In the last few decades, most of the research work was focused on methods/approaches to improve the freezing efficiency of semen, considered to be a significant issue among reproductive biotechnologists. The approaches employed were mostly based on the protection of spermatozoa against the damaging effects of the freezing procedure, including the use of different extenders, cryoprotectant agents, antioxidants, and nutritional components. Moreover, some reports focused on the repair of the damaged spermatozoa during freezing and thawing.

There are many potential applications of nanomaterials in farm animal reproduction such as transgenesis and targeted delivery of substances to a sperm cell, antioxidants, antimicrobial properties, and special surface binding ligand functionalization as well as their application in sperm processing and cryopreservation. The antioxidant properties of some nanoparticles (NPs) are among the most promising characteristics for their application in protecting sperm cell functions during cryopreservation [36]. The use of NPs has markedly increased in various fields of animal reproduction including herd fertility issues [36]. Moreover, recent approaches showed the beneficial effects of using liposomes and extracellular vesicles (EVs) including exosomes of different origins to ameliorate the damaging effects of cryopreservation on spermatozoa. In this review, we highlight the recent strategies to defend against or repair the damage that occurs during cryopreservation of semen such as the use of nanoparticles as a defensive approach and nanovesicles including exosomes and liposomes as a repair and defense mechanism for improving the outcomes of semen cryopreservation in different animal species.

2. Seminal Plasma, Antioxidants, and Their Effect on Sperm Function

Antioxidants are compounds that scavenge or oppose the actions of ROS [37]. Antioxidants work as chelators or binding proteins, and their three main functions are to suppress the generation of ROS and eliminate ROS that are already present [38].

The antioxidant defense system includes an enzymatic mechanism in seminal plasma and sperm cells such as superoxide dismutase, glutathione reductase, glutathione peroxidase, and catalase. However, the nonenzymatic mechanism includes reduced glutathione (GSH), vitamins (A, C, and E), taurine, and hypotaurine. The rate of LPO in sperm cells is determined by the balance between antioxidative and pro-oxidative mechanisms in the semen [32].

Catalase and superoxide dismutase are antioxidant enzymes, which activate scavenging of ROS. Exposure of spermatozoa, primarily to anaerobic conditions during natural mating, may reduce the number of damaged spermatozoa by ROS. Female oviduct fluids contain substantial taurine levels, as it is an important protective factor of spermatozoa from ROS accumulation [39]. For instance, the catalase enzyme exists in ejaculate for the protection of the spermatozoa through the conversion of H_2O_2 into oxygen and water [26]. This prevents the generation of hydroxyl radicals (OH^-), which are powerful oxidants, by the Fenton reaction [40]. However, bull spermatozoa contain little expression of catalase, which makes them prone to OH^- toxicity [41]. Moreover, the concentration of catalase is reduced during semen processing [42]. The addition of antioxidants such as CAT in the buffalo [43], ram [33], boar [44], and bull [45–47] semen protected spermatozoa from the damaging effects of ROS and improved motility and membrane integrity during cooling storage. Elevation of the amount of H_2O_2 can occur as a result of abnormal sperm with residual cytoplasm or abnormal mid-piece [48]. Equine semen is rich in prostate-derived catalase, and therefore dilution or removing the seminal plasma decreases or adversely affects the scavenging capacity of the ROS [49].

Glutathione (GSH) is a tripeptide that comprises cysteine, glutamate, and glycine ubiquitously expressed in the cells. The cysteine subunit plays a pivotal role in scavenging free radicals. GSH acts as an intracellular defense against OS [50].

Exposure of semen to oxygen and visible light radiation during in vitro fertilization or AI resulted in ROS generation and damaged spermatozoa, reduced motility, and reduced membrane integrity in humans and bovines [51–53]. Under these conditions, exogenous addition of catalase, GSH, taurine, superoxide dismutase, and other antioxidants can lead to the maintenance of bovine sperm motility [52]. Supplementation of the whole milk semen extender with hypotaurine or taurine did not improve the motility of bovine spermatozoa in post-thawed semen [54]. In horses, the usage of catalase in extended semen was reported for cooled semen storage [55].

As antioxidants reduce the production of free radicals following the freeze–thaw process [56], the application of ROS scavengers is likely to improve sperm function and protect sperm from the deleterious effects of cryopreservation [57,58]. The detrimental effects of cryopreservation could be ameliorated by adding an exogenous source of antioxidants to the freezing medium to reverse OS [32]. This strategy together with other techniques for the removal of defective spermatozoa and cellular debris from semen could be used for gains in the viability of spermatozoa and reducing the necessary spermatozoa to a minimum number per AI dose [20].

3. Nanoparticles (NPs)

Several factors affect semen quality and fertilizing ability, including genetic, health, nutrition, season, stresses, and semen cryopreservation [59,60]. Multiple factors lead to poor quality semen [61]. The generation of ROS by nonviable sperm cells in the semen samples impairs sperm function [62].

To obtain good male reproduction, removing unviable or degenerated sperm cells and scavenger ROS from semen samples is important. Recent nanotechnologies reflect new prospects for developing novel and noninvasive techniques for sperm manipulation [63–65].

3.1. Definition and Characterization of NPs

NPs are molecules with <100 nm diameter and can be applied for different bioapplications including reproductive biology because they have unique physical and chemical properties [60,66].

Compared to molecules or bulk solids, there are several differences in the structural properties of the NPs [67]. The key factor of NP activity is the characteristics of their surface, such as size, charge density, and hydrophobicity [68,69]. Manipulation into a nanoform can increase the absorption and bioavailability of the functional ingredients [70]. Particle size can affect or change the properties of the original material [71]. The rapid progress in nanotechnology shows great potential for application in both medical and nutritional sciences because NPs possess unusual and advantageous properties that are different from ordinary or microscale materials in terms of their size and high surface reactivity [72]. NPs have been included in pharmaceuticals to increase the bioavailability of drugs and to target particular tissues/organs [73]. Moreover, NPs show increased cellular uptake, binding properties, and reactivity. Furthermore, the antioxidant properties of NPs recently contributed to optimizing the cryopreservation protocols [74].

Small sizes of nanoparticles have shown better integration possibilities in cellular processes and physiological pathways without interfering with the normal biological system. Nanomaterials used in drug delivery have great potential to carry large amounts and different types of biological cargo. The nanosystem abates the drug from rapid degradation and clearance through the reticuloendothelial system. The surface can be modified to react with environmental factors giving responsive drug release [75–77]. Different types of NPs are new forms of materials with promising biological properties and low toxicity and seem to have a high potential for passing through physiological barriers and accessing specific target tissues [78].

3.2. Metal Nanoparticles and Sperm Cryopreservation

Apoptosis, reduced cellular metabolism, and defective acrosome reaction are commonly caused by the increase of ROS levels [79]. Durfey et al. [80] used conjugated magnetic NPs for molecular-based selection of boar spermatozoa, and results showed that the nanoselected spermatozoa had improved motion characteristics with a higher proportion of progressive spermatozoa and straightness. Other reports [81,82] used NPs from FeO conjugated with annexin V to determine the early apoptosis of porcine and bovine sperm cells, respectively.

The use of antioxidants, such as nano-zinc oxide, can be important in reducing ROS generation and increasing sperm survival [75–77,83,84]. Using zinc nanoparticles (50 μg/mL) or selenium nanoparticles (1 μg/mL) in a SHOTOR extender enhanced morphological characteristics and ultrastructure of camel epididymal spermatozoa after cryopreservation via the reduction of apoptosis and lipid peroxidation [60].

In Holstein bulls, supplementing a semen extender with Se-NPs (1.0 μg/mL) improved post-thaw sperm quality and conception rate through reducing apoptosis, LPO, and sperm damage [85]. Moreover, in rams, Hozyen et al. [86] and Nateq et al. [87] used SeNPs (1 μg/mL) and showed improvement in motility, viability index, and membrane integrity, while acrosome defects, DNA fragmentation, and malondialdehyde (MDA) concentrations were reduced.

The addition of green synthesized gold nanoparticles (GSGNPs) (10 ppm) to a Tris-based extender improved buck semen freezing by maintaining the sperm membrane and acrosome integrity post-thawing. In addition, GSGNPs improved antioxidant capacity and consequently scavenged ROS in a buck semen extender [88]. GSGNPs are nontoxic and possess several medical applications [89]. While gold and silver NPs can penetrate the plasma membrane and can be detected inside the human sperm nucleus [90], no evidence regarding their spermatoxicity has been reported (Table 1).

Table 1. Summary of the current reports using nanoparticles (NPs) for semen cryopreservation.

Animal Species	Nanoparticle (NPs)	The Effects	Reference
Goat bucks	Nano-lecithin	Improved motility, viability, and hypo-osmotic swelling test and lower apoptosis.	[91]
Bulls	Nano-lecithin-based extender with glutathione peroxidase	Enhanced plasma membrane integrity and reduced malondialdehyde (MDA) concentration.	[92]
Bulls	Selenium NPs	Improved kinematic sperm quality, antioxidative defense, and decreased apoptotic and necrotic cells.	[85]
	Zinc NPs	Improved plasma membrane integrity and mitochondrial functions.	[93]
Camel	Selenium NPs Zinc NPs	Improved sperm functions (progressive motility, vitality, sperm membrane integrity). Maintained ultrastructural morphology and decreased apoptosis. Increased antioxidative defense.	[60]
Goat	Mint, thyme, and curcumin nanoformulations (NFs)	Improved progressive motility, vitality, and plasma membrane integrity; antioxidative defense; chromatin decondensation. Decreased apoptosis.	[94]
Goat	Green synthesized gold nanoparticles (GSGNPs)	Improved motility, survivability, membrane integrity, acrosome integrity, and antioxidative defense.	[88]
Rabbit	Curcumin NPs	Enhanced sperm motility and antioxidative defense. Reduced apoptotic and necrotic spermatozoa.	[95]

3.3. Herbal Extract Nanoparticles and Sperm Cryopreservation

Recently, several studies examined herbal extracts as natural antioxidants and suppressors of lipid peroxidation in semen preservation of farm animals. For instance, *Moringa oleifera* leaf extract improved the antioxidative defense for cryopreserved ram and buffalo spermatozoa [96,97]. *Arctiumlappa* root extract improved spermatozoa survivability and abnormality with appropriate progressive motility when used as a supplement with cryopreserved ram semen [98]. Curcumin extract exerted antioxidative effects and improved spermatozoa post-thaw quality when used as a supplement with cryopreserved bovine and rabbit semen [95,99,100]. Moreover, *Alnusincana* bark extract [101] and *Albiziaharveyi* leaf extract [102] showed protective antioxidative effects when used as a supplement with cryopreserved ram and bovine semen, respectively. Ginger and echinacea extracts improved the spermatozoa quality and fertilization ability when used as a supplement with cryopreserved ram semen [103]. To this end, Ismail et al. [94] reported that mint, thyme, and curcumin extract nanoformulations enhanced sperm functions and redox status of post-thawed buck semen and decreased sperm apoptosis and chromatin decondensation. Supplementing the extender with curcumin nanoparticles (1.5 μg/mL) also improved the quality of post-thawed rabbit sperm by reducing apoptosis and enhancing antioxidative defense [95] (Table 1).

3.4. Vitamins Nanoparticles and Sperm Cryopreservation

Vitamin E nanoemulsions (NEs) protected red deer epididymal sperm from oxidative damage, maintained mitochondrial activity, protected the acrosome integrity, prevented cell death, and reduced ROS and LPO after OS induction (with 100 μM Fe^{2+}/ 500 μM ascorbate) and hence improved sperm velocity and progressive motility [104].

4. Artificial Exosome-Like Vesicles (Liposomes) for Semen Cryopreservation

A liposome is a spherical nanovesicle with a single lipid bilayer that is produced artificially through disrupting plasma membranes via sonication [105]. Liposomes can be used as a vehicle for delivering nutrients and drugs to target tissues [106,107]. Liposomes can be loaded with antioxidants such as lycopene [108] and quercetin [109] and result in a significant increase in sperm total and progressive motility as well as increased viability, plasma membrane integrity, and mitochondria activity in rooster spermatozoa. Moreover, liposomes can be loaded with lipid-related content (such as lecithin [110]) to improve the plasma membrane regeneration efficacy during the freeze–thaw process of ram spermatozoa. Liposomes were used as a cryoprotectant additives in several animal species including equine [111,112], buffalo [113], ovine [107,114,115], porcine [116], and bovine [117] with reported improvement in fertility after AI [118]. It has been proposed that liposomes with their contents of phospholipids (phosphatidylserine, dioleoylphosphatidylcholine, phosphatidylcholine, dipalmitoylphosphatidylcholine, and dimyristoylphosphocholine) and saturated and unsaturated fatty acids can fuse with the sperm plasma membrane and abate the damage to spermatozoa caused by the freeze–thaw process [119,120] (Figure 1). For instance, in rams, liposomes comprising egg-phosphatidylcholine and dipalmitoylphosphatidylcholine used as a supplement with washed spermatozoa provided immediate protection against cold shock as indicated by motility preservation [121]. Similarly, in stallions, liposomes comprising a mixture of egg phosphatidylcholine and phosphatidylethanolamine (named E80-liposomes) were efficient in preserving post-thaw sperm motility [112]. In contrast, in bovines, liposomes composed of dioleoyl-glycero-phosphocholine and dioleoyl-glycero-phospho-glycerol resulted in higher post-thaw survival, progressive motility, and acrosome reaction when compared to dioleoyl-glycero-phosphocholine alone [117]. The transition of lipid to gel phase during cooling and freezing is highly dependent on the lipid composition of the membranes, and therefore the liposome fusion facilitates lipid and cholesterol transfer, which leads to rearrangement of cell membrane components and modifies the membrane physicochemical properties, thereby improving the cryostability of the spermatozoa [117,118]. OptiXcell® is one such commercial product that uses the liposome-based commercial extender and is currently used for several animal species [122–125].

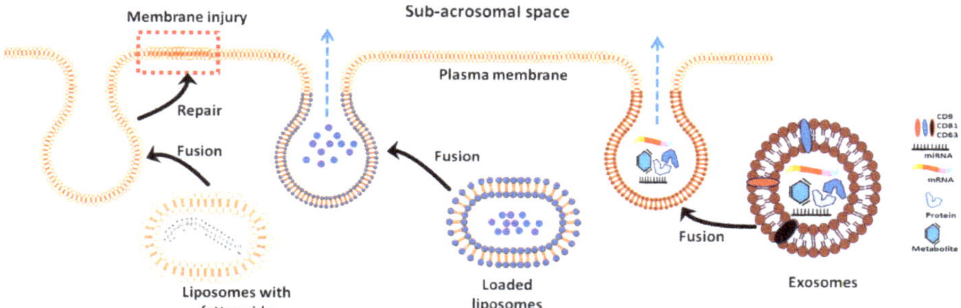

Figure 1. The proposed mechanism of spermatozoa protection through exosomes and liposomes. Liposomes with their contents of fatty acid can replenish the damaged sperm plasma membrane caused by freezing/thawing. Liposomes when artificially loaded with certain chemicals and exosomes with their contents of miRNA, mRNA, proteins, and metabolites can fuse and transfer their cargo into the subacrosomal space and inside the spermatozoa.

5. Potential Uses of Exosomes in Semen Cryopreservation

Extracellular vesicles (EVs) including exosomes are membrane-bounded nanovesicles containing proteins, lipids, and nucleic acids (microRNAs and mRNAs) involved in cellular communication [126,127]. A wide variety of cells release EVs in physiological and pathological circumstances [128]. EVs play major roles in numerous biological communications, including reproduction, serving as potential theranostic candidates for normal and abnormal conditions [129].

Unlike other EVs, exosomes are secreted from cells by the exocytosis pathway. Exosomes are like a snapshot of the originating cells, and the variability of the secreting cell is reflected in the exosomal compositions [126]. Once these exosomes are taken by target cells, they transfer their cargo, which includes proteins [130,131], miRNA [132,133], and mRNA [134–136], to the target cells (Figure 1). This cargo may participate in energy pathways, protein metabolism, and maintenance of recipient cells.

Thus, exosomes confer different epigenetic and phenotypic modifications on recipient cells that affect the viability, tolerance to the external factors, and regenerative capabilities of their target cells [137]. Exosomes have also been found to play important bioactive functions such as sperm maturation, capacitation, acrosome reaction, and fertilization [138]. Recent findings regarding the regenerative potential of exosomes have guided the research towards the exploitation of exosomal potential for improving the outcomes of sperm freezing [137].

Different growth factors associated with exosomes have been reported to play an active role in the repair and accelerated healing of damaged tissue [139]. Additionally, the therapeutic potential of exosomes has also been reported to be effective in arthritis, diabetes [140], immunotherapy, nervous system-related issues [141], cellular aging, and tumors [142]. Similarly, the treatment of spermatozoa with exosomes during the freezing procedure was found effective in improving the post-thaw quality of canine [137], porcine [138], and rat semen [143].

5.1. Effect of Exosomes on Sperm Motility and Viability

Motility and viability of spermatozoa are very important quality-related parameters that have a direct influence on fertility. A strong correlation was found between increasing the concentration of seminal plasma exosomes and the sperm motility and viability of boar spermatozoa [138] when preserved at liquid stage (17 °C for 10 days). Moreover, mesenchymal stem cell (MSC)-derived microvesicles improved the frozen/thawed quality of rat spermatozoa [143]. It has been proposed that MSC-derived microvesicles shuttle surface adhesion molecules, such as CD54 (ICAM-I), CD106 (VCAM-I), CD29 (β1-Integrin), and CD44, and consequently increase the adhesive properties of sperm [143]. This improved motility was demonstrated in liquid storage (17 °C) [138] as well as frozen dog [137] and rat spermatozoa [143]. The amplitude of lateral head displacement also improved in exosome-treated dog spermatozoa [137]. Interestingly, stem-cell-derived conditioned medium and exosomes improved motility, viability, mitochondrial activity, and membrane integrity post-thawing in canine semen cryopreservation [144,145].

5.2. Effect of Exosomes on Sperm Capacitation and Structural Integrity

The structural integrity of spermatozoa is considered imperative for the proper functioning and fertilization of oocytes. The structures including the plasma membrane (physiological barrier [138]), acrosome (sperm penetration), and chromatin (embryo quality [146]) affect gamete interaction and embryonic development. Damage to these structures can lead to fertilization failure. Exosomes could transfer spermadhesins (AWN and porcine seminal protein, PSP-1) to the sperm membrane that could help to maintain sperm function through inhibiting premature capacitation (decapacitation) during long-term liquid storage [138]. Similarly, exosomes derived from mesenchymal stem cells increased the fraction of sperm with an intact acrosome and increased the expression of transcripts related to the repair of the plasma membrane (ANX 1, FN 1, and DYSF) and chromatin material (H3 and HMGB 1) in frozen/thawed dog spermatozoa [137]. In bovines, oviduct-derived EVs significantly stimulated the acrosome reaction by increasing the levels of protein tyrosine phosphorylation (PY) and increasing intracellular calcium levels in frozen/thawed spermatozoa [147]. In wildlife animals (red wolves and cheetahs), oviduct-derived EVs showed improvement in sperm motility and acrosome integrity and prevented the premature acrosome reaction post-thawing [148].

Capacitation is a physiological process that enables the spermatozoa to fertilize the oocytes. Naturally, capacitation occurs during spermatozoa transit through the uterus and oviduct. In vitro storage of spermatozoa requires inhibition of premature capacitation for maintaining sperm survival [138]. The higher concentration of seminal plasma isolated exosomes significantly decreased the percentage of capacitated spermatozoa upon artificially induced capacitation using 3 mg/mL BSA [138].

5.3. Effect of Exosomes on Antioxidant Capacity

Oxidative stress is one of the major causes of low fertility of post-thaw spermatozoa [149]. A. Mokarizadeh et al. [143] reported increased antioxidant activity in frozen/thawed rat spermatozoa treated with exosomes during freezing, i.e., decreased expression of mitochondrial ROS modulator (*ROMO1*) gene in exosome-treated spermatozoa [137]. Moreover, Du et al. [138] showed increased total antioxidant capacity activity and decreased malondialdehyde content when diluents were supplemented with seminal plasma exosomes. It was hypothesized that the enhanced antioxidant capacity of spermatozoa was either due to the horizontally transferred antioxidant and other factors including mRNA and proteins from exosomes or due to the modified hydrophobic character of the membrane. Table 2 describes the available literature that used exosomes for semen preservation either in cooling or in freezing.

Table 2. Main literature reporting the beneficial effects obtained following the supplementation of exosomes for semen preservation.

Species	Sources of EVs/Exosomes	Condition of Storage	The Improved Parameters	References
Pig	Seminal plasma	17 °C for 10 days	Viability, motility, plasma membrane integrity, antioxidant capacity, and MDA reduction	[138]
Pig	Oviduct-derived	Freezing	Survival and motility	[150]
Rat	Bone marrow-derived mesenchymal stem cells	Freezing	Viability, motility, total antioxidant capacity, and increased surface adhesion molecules	[143]
Dog	Amniotic-derived mesenchymal stem cells and conditioned medium	Cooling and freezing	Viability, motility	[144,145]
Dog	Adipose-derived mesenchymal stem cells	Freezing	Viability, motility	[137]
Red wolves and cheetahs	Oviduct-derived	Freezing	Motility and acrosome integrity	[148]
Bovine	Oviduct-derived	Freezing	Viability	[147]

6. Conclusions

Current trends include using nanoparticles and natural or artificial nanovesicles such as exosomes and liposomes to improve the cryopreservation of semen. Nanoparticles mostly work as antioxidants (Figure 2) with significant effects when compared with corresponding metals or herb extracts. The functional molecules present inside the exosomes such as miRNA, mRNA, and proteins (Figure 2) are involved in the proper execution of a wide variety of physiological interactions that can help resolve issues related to the fertility of male gametes. Liposomes, with their contents of phospholipids and lipid chains, can replace the damaged lipid skeletons of the frozen/thawed spermatozoa. The treatment of spermatozoa with exosomes improved the efficiency of freezing procedures; however, further in vivo and fertility studies are essential to investigating the influence of exosome treatment on sperm functions. Since liposomes are currently available as a commercial product for semen cryopreservation, nanoparticles and nanoformulations as well as EVs and exosomes derived from the reproductive tract or stem cells should adhere to the appropriate manufacturing practices, quality control measurements, and safety and efficacy protocols for commercial purposes in AI.

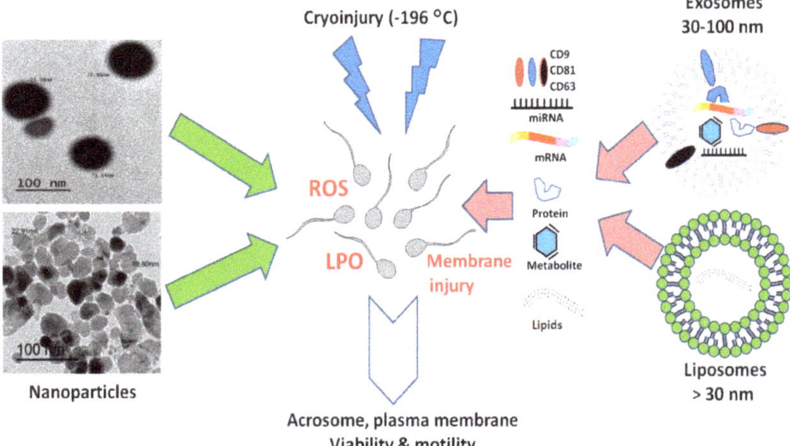

Figure 2. The overall effects of nanoparticles, exosomes, and liposomes in improving semen cryopreservation and reducing cryoinjury. Nanoparticles either from metals or from natural herbs act mainly as antioxidants, while exosomes can deliver bioactive components such as antioxidant enzymes, proteins, lipids, mRNA, and miRNA to protect sperm against cryoinjury such as that caused by reactive oxygen species (ROS) and lipid peroxidation (LPO). Liposomes can fuse with the sperm plasma membrane and replenish the damaged phospholipids caused by freezing/thawing.

Author Contributions: Conceptualization, I.M.S., W.A.K., and S.H.L.; Writing and editing, I.M.S., W.A.K., M.G.A., and S.H.L. Project administration, I.M.S.; funding acquisition, I.M.S. All authors have read and agreed to the published version of the manuscript.

Funding: The work is supported by the Deputyship for Research & Innovation, Ministry of Education, Saudi Arabia through the project number IFKSURP-13.

Acknowledgments: The authors extend their appreciation to the Deputyship for Research & Innovation, Ministry of Education, in Saudi Arabia for funding this research work through the project number IFKSURP-13.

Conflicts of Interest: The authors declare no conflict of interest.

References

1. Ntemka, A.; Tsakmakidis, I.A.; Kiossis, E.; Milovanović, A.; Boscos, C.M. Current status and advances in ram semen cryopreservation. *J. Hell. Vet. Med. Soc.* **2018**, *69*, 911. [CrossRef]
2. Ferdinand, N.; Ngwa, T.D.; Augustave, K.; Dieudonné, B.P.H.; Willington, B.O.; D'Alex, T.C.; Pierre, K.; Joseph, T. Effect of egg yolk concentration in semen extender, pH adjustment of extender and semen cooling methods on bovine semen characteristics. *Glob. Vet.* **2014**, *12*, 292–298.
3. Clulow, J.; Mansfield, L.; Morris, L.; Evans, G.; Maxwell, W. A comparison between freezing methods for the cryopreservation of stallion spermatozoa. *Anim. Reprod. Sci.* **2008**, *108*, 298–308. [CrossRef]
4. Yousif, A.I.A. Studying the Negative Effects of Traditional Sperm Cryopreservation Methods in Ossimi and Finnish Rams Using Advanced Techniques. Ph.D. Thesis, Animal Production Department, Faculty of Agriculture, Masoura University, Mansoura, Egypt, 2018.
5. Len, J.S.; Koh, W.S.D.; Tan, S.-X. The roles of reactive oxygen species and antioxidants in cryopreservation. *Biosci. Rep.* **2019**, *39*. [CrossRef]
6. Tatone, C.; Di Emidio, G.; Vento, M.; Ciriminna, R.; Artini, P.G. Cryopreservation and oxidative stress in reproductive cells. *Gynecol. Endocrinol.* **2010**, *26*, 563–567. [CrossRef] [PubMed]
7. Patist, A.; Zoerb, H. Preservation mechanisms of trehalose in food and biosystems. *Colloids Surf. B Biointerfaces* **2005**, *40*, 107–113. [CrossRef]

8. Zhang, S.; Hu, J.; Li, Q.; Jiang, Z.; Zhang, X. The cryoprotective effects of soybean lecithin on boar spermatozoa quality. *Afr. J. Biotechnol.* **2009**, *8*, 6476–6480.
9. Akiyama, M. In vivo scavenging effect of ethylcysteine on reactive oxygen species in human semen. *Jpn. J. Urol.* **1999**, *90*, 421–428. [CrossRef] [PubMed]
10. Villegas, J.; Kehr, K.; Soto, L.; Henkel, R.; Miska, W.; Sanchez, R. Reactive oxygen species induce reversible capacitation in human spermatozoa. *Andrologia* **2003**, *35*, 227–232. [CrossRef] [PubMed]
11. Hollan, S. Membrane fluidity of blood cells. *Haematologia* **1996**, *27*, 109–127.
12. Huang, Y.-L.; Tseng, W.-C.; Lin, T.-H. In vitro effects of metal ions (Fe^{2+}, Mn^{2+}, Pb^{2+}) on sperm motility and lipid peroxidation in human semen. *J. Toxicol. Environ. Health A* **2001**, *62*, 259–267. [CrossRef]
13. Davydov, D.R. Microsomal monooxygenase in apoptosis: Another target for cytochrome c signaling? *Trends Biochem. Sci.* **2001**, *26*, 155–160. [CrossRef]
14. Aitken, R.J.; Sawyer, D. The human spermatozoon—Not waving but drowning. In *Advances in Male Mediated Developmental Toxicity*; Springer: Boston, MA, USA; Berlin/Heidelberg, Germany, 2003; pp. 85–98.
15. Khalifa, T.; El-Saidy, B. Pellet-freezing of Damascus goat semen in a chemically defined extender. *Anim. Reprod. Sci.* **2006**, *93*, 303–315. [CrossRef] [PubMed]
16. Aitken, R.J. The Amoroso Lecture The human spermatozoon–a cell in crisis? *Reprod. Fertil. Dev.* **1999**, *115*, 1–7. [CrossRef] [PubMed]
17. Agarwal, A.; Nallella, K.P.; Allamaneni, S.S.; Said, T.M. Role of antioxidants in treatment of male infertility: An overview of the literature. *Reprod. Biomed. Online* **2004**, *8*, 616–627. [CrossRef]
18. Arav, A.; Yavin, S.; Zeron, Y.; Natan, D.; Dekel, I.; Gacitua, H. New trends in gamete's cryopreservation. *Mol. Cell. Endocrinol.* **2002**, *187*, 77–81. [CrossRef]
19. Medeiros, C.; Forell, F.; Oliveira, A.; Rodrigues, J. Current status of sperm cryopreservation: Why isn't it better? *Theriogenology* **2002**, *57*, 327–344. [CrossRef]
20. Petruska, P.; Capcarova, M.; Sutovsky, P. Antioxidant supplementation and purification of semen for improved artificial insemination in livestock species. *Turk. J. Vet. Anim. Sci.* **2014**, *38*, 643–652. [CrossRef]
21. Sikka, S.C. Oxidative stress and role of antioxidants in normal and abnormal sperm function. *Front. Biosci.* **1996**, *1*, e78–e86. [CrossRef]
22. Alvarez, J.G.; Storey, B.T. Assessment of cell damage caused by spontaneous lipid peroxidation in rabbit spermatozoa. *Biol. Reprod.* **1984**, *30*, 323–331. [CrossRef]
23. Gil-Guzman, E.; Ollero, M.; Lopez, M.; Sharma, R.; Alvarez, J.; Thomas, A., Jr.; Agarwal, A. Differential production of reactive oxygen species by subsets of human spermatozoa at different stages of maturation. *Hum. Reprod.* **2001**, *16*, 1922–1930. [CrossRef] [PubMed]
24. Bilodeau, J.F.; Chatterjee, S.; Sirard, M.A.; Gagnon, C. Levels of antioxidant defenses are decreased in bovine spermatozoa after a cycle of freezing and thawing. *Mol. Reprod. Dev.* **2000**, *55*, 282–288. [CrossRef]
25. Ball, B.; Medina, V.; Gravance, C.; Baumber, J. Effect of antioxidants on preservation of motility, viability and acrosomal integrity of equine spermatozoa during storage at 5 C. *Theriogenology* **2001**, *56*, 577–589. [CrossRef]
26. Aitken, R.J.; Clarkson, J.S.; Fishel, S. Generation of reactive oxygen species, lipid peroxidation, and human sperm function. *Biol. Reprod.* **1989**, *41*, 183–197. [CrossRef] [PubMed]
27. Ziaullah, M.; Ijaz, A.; Aleem, M.; Mahmood, A.; Rehman, H.; Bhatti, S.; Farooq, U.; Sohail, M. Optimal inclusion level of butylated hydroxytoluene in semen extender improves the quality of post-thawed canine sperm. *Czech J. Anim. Sci.* **2012**, *57*, 377–381. [CrossRef]
28. Nair, S.J.; Brar, A.; Ahuja, C.; Sangha, S.; Chaudhary, K. A comparative study on lipid peroxidation, activities of antioxidant enzymes and viability of cattle and buffalo bull spermatozoa during storage at refrigeration temperature. *Anim. Reprod. Sci.* **2006**, *96*, 21–29. [CrossRef]
29. Pagl, R.; Aurich, J.E.; Müller-Schlösser, F.; Kankofer, M.; Aurich, C. Comparison of an extender containing defined milk protein fractions with a skim milk-based extender for storage of equine semen at 5 °C. *Theriogenology* **2006**, *66*, 1115–1122. [CrossRef]
30. Ortega, A.M.; Izquierdo, A.C.; Gómez, J.J.H.; Corichi, I.M.O.; Torres, V.M.M.; Méndez, J.d.J.V. Peroxidación lipídica y antioxidantes en la preservación de semen. Una revisión. *Interciencia* **2003**, *28*, 699–704.
31. Bucak, M.N.; Tuncer, P.B.; Sarıözkan, S.; Başpınar, N.; Taşpınar, M.; Çoyan, K.; Bilgili, A.; Akalın, P.P.; Büyükleblebici, S.; Aydos, S. Effects of antioxidants on post-thawed bovine sperm and oxidative stress parameters: Antioxidants protect DNA integrity against cryodamage. *Cryobiology* **2010**, *61*, 248–253. [CrossRef]

32. Gadea, J.; Sellés, E.; Marco, M.A.; Coy, P.; Matás, C.; Romar, R.; Ruiz, S. Decrease in glutathione content in boar sperm after cryopreservation: Effect of the addition of reduced glutathione to the freezing and thawing extenders. *Theriogenology* **2004**, *62*, 690–701. [CrossRef]
33. Maxwell, W.; Stojanov, T. Liquid storage of ram semen in the absence or presence of some antioxidants. *Reprod. Fertil. Dev.* **1996**, *8*, 1013–1020. [CrossRef] [PubMed]
34. Baghshahi, H.; Riasi, A.; Mahdavi, A.; Shirazi, A. Antioxidant effects of clove bud (*Syzygium aromaticum*) extract used with different extenders on ram spermatozoa during cryopreservation. *Cryobiology* **2014**, *69*, 482–487. [CrossRef]
35. Stradaioli, G.; Noro, T.; Sylla, L.; Monaci, M. Decrease in glutathione (GSH) content in bovine sperm after cryopreservation: Comparison between two extenders. *Theriogenclogy* **2007**, *67*, 1249–1255. [CrossRef]
36. Hashem, N.M.; Gonzalez-Bulnes, A. State-of-the-Art and Prospective of Nanotechnologies for Smart Reproductive Management of Farm Animals. *Animals* **2020**, *10*, 840. [CrossRef]
37. Bansal, A.K.; Bilaspuri, G. Impacts of oxidative stress and antioxidants on semen functions. *Vet. Med. Int.* **2011**, *2011*, 1–7. [CrossRef]
38. Agarwal, A.; Virk, G.; Ong, C.; du Plessis, S.S. Effect of oxidative stress on male reproduction. *World J. Men Health* **2014**, *32*, 1–17. [CrossRef]
39. Alvarez, J.G.; Storey, B.T. Evidence for increased lipid peroxidative damage and loss of superoxide dismutase activity as a mode of sublethal cryodamage to human sperm during cryopreservation. *J. Androl.* **1992**, *13*, 232–241. [PubMed]
40. Partyka, J.; Sitarz, M.; Leśniak, M.; Gasek, K.; Jeleń, P. The effect of SiO_2/Al_2O_3 ratio on the structure and microstructure of the glazes from SiO_2–Al_2O_3–CaO–MgO–Na_2O–K_2O system. *Spectrochim. Acta A Mol. Biomol. Spectrosc.* **2015**, *134*, 621–630. [CrossRef]
41. Holland, M.K.; Alvarez, J.G.; Storey, B.T. Production of superoxide and activity of superoxide dismutase in rabbit epididymal spermatozoa. *Biol. Reprod.* **1982**, *27*, 1109–1118. [CrossRef] [PubMed]
42. Perumal, P.; Vupru, K.; Khate, K. Effect of addition of melatonin on the liquid storage (5 °C) of Mithun (*Bos frontalis*) Semen. *Int. J. Zool.* **2013**, *2013*, 1–10. [CrossRef]
43. El-Sisy, G.; El-Nattat, W.; El-Sheshtawy, R.J. Effect of superoxide dismutase and catalase on viability of cryopreserved buffalo spermatozoa. *Glob. Vet.* **2008**, *2*, 61–65.
44. Roca, J.; Rodríguez, M.J.; Gil, M.A.; Carvajal, G.; Garcia, E.M.; Cuello, C.; Vazquez, J.M.; Martinez, E.A. Survival and in vitro fertility of boar spermatozoa frozen in the presence of superoxide dismutase and/or catalase. *J. Androl.* **2005**, *26*, 15–24. [PubMed]
45. Abdulkareem, T.A.; Alzaidi, O.H. Effect of adding aqueous extract of *Melissa officinalis* leaves and some other antioxidants to milk–based extender on post-cooling and post-cryopreservative sperm's individual motility and live sperm percentage of Holstein bulls. *Al-Anbar J. Vet. Sci.* **2018**, *11*, 37–53. [CrossRef]
46. Abdulkareem, T.A.; Khalil, R.I.; Salman, A.H. Effect of adding *Ferula hermonis Boiss* roots and some antioxidants to Tris extender on post-cryopreserved sperm abnormalities percentage of Holstein bulls. *Al-Anbar J. Vet. Sci.* **2018**, *11*, 70–81.
47. Eidan, S.M.; Abdulkareem, T.A.; Sultan, O.A.A. Influence of adding manganese to Tris extender on some post-cryopreservation semen attributes of Holstein bulls. *Int. J. Appl. Agric. Sci.* **2015**, *1*, 26–30.
48. Ball, B.A.; Vo, A.T.; Baumber, J. Generation of reactive oxygen species by equine spermatozoa. *Am. J. Vet. Res.* **2001**, *62*, 508–515. [CrossRef]
49. Ball, B.A.; Gravance, C.G.; Medina, V.; Baumber, J.; Liu, I.K. Catalase activity in equine semen. *Am. J. Vet. Res.* **2000**, *61*, 1026–1030. [CrossRef]
50. Irvine, D.S. Glutathione as a treatment for male infertility. *Rev. Reprod.* **1996**, *1*, 6–12. [CrossRef]
51. Aitken, R.J.; Gordon, E.; Harkiss, D.; Twigg, J.P.; Milne, P.; Jennings, Z.; Irvine, D.S. Relative impact of oxidative stress on the functional competence and genomic integrity of human spermatozoa. *Biol. Reprod.* **1998**, *59*, 1037–1046. [CrossRef]
52. Bilodeau, J.-F.; Blanchette, S.; Gagnon, C.; Sirard, M.-A. Thiols prevent H_2O_2-mediated loss of sperm motility in cryopreserved bull semen. *Theriogenology* **2001**, *56*, 275–286. [CrossRef]
53. Storey, B.T. Biochemistry of the induction and prevention of lipoperoxidative damage in human spermatozoa. *Mol. Hum. Reprod.* **1997**, *3*, 203–213. [CrossRef] [PubMed]
54. Chen, Y.; Foote, R.; Brockett, C. Effect of sucrose, trehalose, hypotaurine, taurine, and blood serum on survival of frozen bull sperm. *Cryobiology* **1993**, *30*, 423–431. [CrossRef] [PubMed]

55. Aurich, J.; Schönherr, U.; Hoppe, H.; Aurich, C. Effects of antioxidants on motility and membrane integrity of chilled-stored stallion semen. *Theriogenology* **1997**, *48*, 185–192. [CrossRef]
56. Ashrafi, I.; Kohram, H.; Ardabili, F.F. Antioxidative effects of melatonin on kinetics, microscopic and oxidative parameters of cryopreserved bull spermatozoa. *Anim. Reprod. Sci.* **2013**, *139*, 25–30. [CrossRef]
57. Watson, P. The causes of reduced fertility with cryopreserved semen. *Anim. Reprod. Sci.* **2000**, *60*, 481–492. [CrossRef]
58. Varnet, P.; Aitken, R.; Drevet, J. Antioxidant strategies in the epidydimis. *Mol. Cell. Endocrinol.* **2004**, *216*, 31–39. [CrossRef] [PubMed]
59. Bungum, M.; Humaidan, P.; Axmon, A.; Spano, M.; Bungum, L.; Erenpreiss, J.; Giwercman, A. Sperm DNA integrity assessment in prediction of assisted reproduction technology outcome. *Hum. Reprod.* **2007**, *22*, 174–179. [CrossRef] [PubMed]
60. Shahin, M.A.; Khalil, W.A.; Saadeldin, I.M.; Swelum, A.A.-A.; El-Harairy, M.A. Comparison between the Effects of Adding Vitamins, Trace Elements, and Nanoparticles to SHOTOR Extender on the Cryopreservation of Dromedary Camel Epididymal Spermatozoa. *Animals* **2020**, *10*, 78. [CrossRef]
61. Broekhuijse, M.; Gaustad, A.; Bolarin Guillén, A.; Knol, E. Efficient Boar Semen Production and Genetic Contribution: The Impact of Low Dose Artificial Insemination on Fertility. *Reprod. Domest. Anim.* **2015**, *50*, 103–109. [CrossRef] [PubMed]
62. Sharma, R.K.; Agarwal, A. Role of reactive oxygen species in male infertility. *Urology* **1996**, *48*, 835–850. [CrossRef]
63. Feugang, J.M. Novel agents for sperm purification, sorting, and imaging. *Mol. Reprod. Dev.* **2017**, *84*, 832–841. [CrossRef] [PubMed]
64. Said, T.M.; Land, J.A. Effects of advanced selection methods on sperm quality and ART outcome: A systematic review. *Hum. Reprod. Update* **2011**, *17*, 719–733. [CrossRef] [PubMed]
65. Hozaien, M.; Elqusi, K.; Hassanen, E.; Hussin, A.; Alkhader, H.; El Tanbouly, S.; El-qassaby, S.; Zaki, H. A comparison of reproductive outcome using different sperm selection techniques; density gradient, testicular sperm, PICSI, and MACS for ICSI patients with abnormal DNA fragmentation index. *Fertil. Steril.* **2018**, *110*, 19–20. [CrossRef]
66. Jain, S.; Park, S.B.; Pillai, S.R.; Ryan, P.L.; Willard, S.T.; Feugang, J.M. Applications of fluorescent quantum dots for reproductive medicine and disease detection. In *Unraveling the Safety Profile of Nanoscale Particles and Materials—From Biomedical to Environmental Applications*; IntechOpen: London, UK, 2018. [CrossRef]
67. Huber, D.L.C. Synthesis, properties, and applications of iron nanoparticles. *Small* **2005**, *1*, 482–501. [CrossRef]
68. Gessner, A.; Lieske, A.; Paulke, B.R.; Müller, R.H.J. Influence of surface charge density on protein adsorption on polymeric nanoparticles: Analysis by two-dimensional electrophoresis. *Eur. J. Pharm. Biopharm.* **2002**, *54*, 165–170. [CrossRef]
69. Thorek, D.L.; Tsourkas, A. Size, charge and concentration dependent uptake of iron oxide particles by non-phagocytic cells. *Biomaterials* **2008**, *29*, 3583–3590. [CrossRef]
70. Zhang, N.; Ping, Q.; Huang, G.; Xu, W.; Cheng, Y.; Han, X. Lectin-modified solid lipid nanoparticles as carriers for oral administration of insulin. *Int. J. Pharm.* **2006**, *327*, 153–159. [CrossRef]
71. Chen, H.; Seiber, J.N.; Hotze, M. ACS Select on Nanotechnology in Food and Agriculture: A Perspective on Implications and Applications. *J. Agric. Food Chem.* **2014**, *62*, 1209–1212. [CrossRef]
72. Albrecht, M.A.; Evans, C.W.; Raston, C.L. Green chemistry and the health implications of nanoparticles. *Green Chem.* **2006**, *8*, 417–432. [CrossRef]
73. Davda, J.; Labhasetwar, V. Characterization of nanoparticle uptake by endothelial cells. *Int. J. Pharm.* **2002**, *233*, 51–59. [CrossRef]
74. Nel, A.; Xia, T.; Mädler, L.; Li, N. Toxic potential of materials at the nano level. *Science* **2006**, *311*, 622–627. [CrossRef] [PubMed]
75. Barkalina, N.; Charalambous, C.; Jones, C.; Coward, K. Nanotechnology in reproductive medicine: Emerging applications of nanomaterials. *Nanomed. Nanotechnol. Biol. Med.* **2014**, *10*, 921–938. [CrossRef] [PubMed]
76. Naahidi, S.; Jafari, M.; Edalat, F.; Raymond, K.; Khademhosseini, A.; Chen, P. Biocompatibility of engineered nanoparticles for drug delivery. *J. Control. Release* **2013**, *166*, 182–194. [CrossRef] [PubMed]
77. Wilczewska, A.Z.; Niemirowicz, K.; Markiewicz, K.H.; Car, H. Nanoparticles as drug delivery systems. *Pharmacol. Rep.* **2012**, *64*, 1020–1037. [CrossRef]

78. Suri, S.S.; Fenniri, H.; Singh, B. Nanotechnology-based drug delivery systems. *J. Occupat. Med. Toxicol.* **2007**, *2*, 1–6. [CrossRef]
79. Niżański, W.; Partyka, A.; Prochowska, S. Evaluation of spermatozoal function—Useful tools or just science. *Reprod. Domest. Anim.* **2016**, *51*, 37–45. [CrossRef]
80. Durfey, C.L.; Burnett, D.D.; Liao, S.F.; Steadman, C.S.; Crenshaw, M.A.; Clemente, H.J.; Willard, S.T.; Ryan, P.L.; Feugang, J.M. Nanotechnology-based selection of boar spermatozoa: Growth development and health assessments of produced offspring. *Livest. Sci.* **2017**, *205*, 137–142. [CrossRef]
81. Feugang, J.M.; Youngblood, R.C.; Greene, J.M.; Willard, S.T.; Ryan, P.L. Self-illuminating quantum dots for non-invasive bioluminescence imaging of mammalian gametes. *J. Nanobiotechnol.* **2015**, *13*, 38. [CrossRef]
82. Odhiambo, J.F.; DeJarnette, J.; Geary, T.W.; Kennedy, C.E.; Suarez, S.S.; Sutovsky, M.; Sutovsky, P. Increased conception rates in beef cattle inseminated with nanopurified bull semen. *Biol. Reprod.* **2014**, *97*, 1–10.
83. Ozsoy, N.; Can, A.; Mutlu, O.; Akev, N.; Yanardag, R. Oral zinc supplementation protects rat kidney tissue from oxidative stress in diabetic rats. *Kafkas Univ. Vet. Fak. Derg.* **2012**, *18*, 545–550.
84. Heidari, J.; Seifdavati, J.; Mohebodini, H.; Sharifi, R.S.; Benemar, H.A. Effect of nano zinc oxide on post-thaw variables and oxidative status of Moghani ram semen. *Kafkas Univ. Vet. Fak. Derg.* **2018**, *25*, 71–76.
85. Khalil, W.A.; El-Harairy, M.A.; Zeidan, A.E.; Hassan, M.A. Impact of selenium nano-particles in semen extender on bull sperm quality after cryopreservation. *Theriogenology* **2019**, *126*, 121–127. [CrossRef] [PubMed]
86. Hozyen, H.; El-Shamy, A.; Farghali, A. In vitro supplementation of nano selenium minimizes freeze-thaw induced damage to ram spermatozoa. *Int. J. Vet. Sci.* **2019**, *8*, 249–254.
87. Nateq, S.; Moghaddam, G.; Alijani, S.; Behnam, M. The effects of different levels of Nano selenium on the quality of frozen-thawed sperm in ram. *J. Appl. Anim. Res.* **2020**, *48*, 434–439. [CrossRef]
88. Ismail, A.A.; Abdel-Khalek, A.; Khalil, W.; El-Harairy, M. Influence of adding green synthesized gold nanoparticles to tris-extender on sperm characteristics of cryopreserved goat semen. *J. Anim. Poult. Prod. Mansoura Univ.* **2020**, *11*, 39–45. [CrossRef]
89. Tripathi, R.; Shrivastav, A.; Shrivastav, B. Biogenic gold nanoparticles: As a potential candidate for brain tumor directed drug delivery. *Artif. Cells Nanomed. Biotechnol.* **2015**, *43*, 311–317. [CrossRef]
90. Moretti, E.; Terzuoli, G.; Renieri, T.; Iacoponi, F.; Castellini, C.; Giordano, C.; Collodel, G. In vitro effect of gold and silver nanoparticles on human spermatozoa. *Andrologia* **2013**, *45*, 392–396. [CrossRef]
91. Nadri, T.; Towhidi, A.; Zeinoaldini, S.; Martínez-Pastor, F.; Mousavi, M.; Noei, R.; Tar, M.; Mohammadi Sangcheshmeh, A. Lecithin nanoparticles enhance the cryosurvival of caprine sperm. *Theriogenology* **2019**, *133*, 38–44. [CrossRef]
92. Mousavi, S.M.; Towhidi, A.; Zhandi, M.; Amoabediny, G.; Mohammadi-Sangcheshmeh, A.; Sharafi, M.; Hussaini, S.M.H. Comparison of two different antioxidants in a nano lecithin-based extender for bull sperm cryopreservation. *Anim. Reprod. Sci.* **2019**, *209*, 106171. [CrossRef]
93. Jahanbin, R.; Yazdanshenas, P.; Amin, A.M.; Mohammadi, S.A.; Varnaseri, H.; Chamani, M.; Nazaran, M.H.; Bakhtiyarizadeh, M.R. Effect of zinc nano-complex on bull semen quality after freeze-thawing process. *J. Anim. Prod.* **2016**, *17*, 371–380.
94. Ismail, A.A.; Abdel-Khalek, A.-K.E.; Khalil, W.A.; Yousif, A.I.; Saadeldin, I.M.; Abomughaid, M.M.; El-Harairy, M.A. Effects of mint, thyme, and curcumin extract nanoformulations on the sperm quality, apoptosis, chromatin decondensation, enzyme activity, and oxidative status of cryopreserved goat semen. *Cryobiology* **2020**. [CrossRef]
95. Abdelnour, S.A.; Hassan, M.A.; Mohammed, A.K.; Alhimaidi, A.R.; Al-Gabri, N.; Al-Khaldi, K.O.; Swelum, A.A. The Effect of Adding Different Levels of Curcumin and Its Nanoparticles to Extender on Post-Thaw Quality of Cryopreserved Rabbit Sperm. *Animals* **2020**, *10*, 1508. [CrossRef] [PubMed]
96. Shokry, D.M.; Badr, M.R.; Orabi, S.H.; Khalifa, H.K.; El-Seedi, H.R.; Abd Eldaim, M.A. Moringa oleifera leaves extract enhances fresh and cryopreserved semen characters of Barki rams. *Theriogenology* **2020**, *153*, 133–142. [CrossRef] [PubMed]
97. Elsheshtawy, R.; Elnattat, W.S. Assessment of buffalo semen preservability using tris extender enriched with *Moringa oleifera* extract. *Egypt. J. Vet. Sci.* **2020**, *51*, 235–239. [CrossRef]
98. El-Harairy, M.; Abdel-Khalek, A.; Khalil, W.; Khalifa, E.; El-Khateeb, A.; Abdulrhmn, A. Effect of aqueous extracts of Moringa oleifera leaves or Arctium lappa roots on lipid peroxidation and membrane integrity of ram sperm preserved at cool temperature. *J. Anim. Poult. Prod. Mansoura Univ.* **2016**, *7*, 467–473. [CrossRef]

99. Tvrdá, E.; Greifová, H.; Mackovich, A.; Hashim, F.; Lukáč, N. Curcumin offers antioxidant protection to cryopreserved bovine semen. *Czech J. Anim. Sci.* **2018**, *63*, 247–255. [CrossRef]
100. Tvrdá, E.; Halenár, M.; Greifová, H.; Mackovich, A.; Hashim, F.; Lukáč, N. The effect of curcumin on cryopreserved bovine semen. *Int. J. Anim. Vet. Sci.* **2016**, *10*, 707–711.
101. Abadjieva, D.; Yotov, S.; Mladenova, V.; Lauberte, L.; Kalvanov, I.; Krasilnikova, J.; Telesheva, G.; Kistanova, E. Positive effect of natural antioxidant oregonin from *Alnus incana* bark on ram semen quality stored at 5 °C for 48 h. *Res. Vet. Sci.* **2020**, *131*, 153–158. [CrossRef]
102. Sobeh, M.; Hassan, S.; El Raey, M.; Khalil, W.; Hassan, M.; Wink, M. Polyphenolics from *Albizia harveyi* exhibit antioxidant activities and counteract oxidative damage and ultra-structural changes of cryopreserved bull semen. *Molecules* **2017**, *22*, 1993. [CrossRef]
103. Merati, Z.; Farshad, A. Ginger and echinacea extracts improve the quality and fertility potential of frozen-thawed ram epididymal spermatozoa. *Cryobiology* **2020**, *92*, 138–145. [CrossRef]
104. Sanchez-Rubio, F.; Soria-Meneses, P.J.; Jurado-Campos, A.; Bartolome-Garcia, J.; Gomez-Rubio, V.; Soler, A.J.; Arroyo-Jimenez, M.M.; Santander-Ortega, M.J.; Plaza-Oliver, M.; Lozano, M.V.; et al. Nanotechnology in reproduction: Vitamin E nanoemulsions for reducing oxidative stress in sperm cells. *Free Radic. Biol. Med.* **2020**, *160*, 47–56. [CrossRef] [PubMed]
105. Stremersch, S.; Vandenbroucke, R.E.; Van Wonterghem, E.; Hendrix, A.; De Smedt, S.C.; Raemdonck, K. Comparing exosome-like vesicles with liposomes for the functional cellular delivery of small RNAs. *J. Control. Release* **2016**, *232*, 51–61. [CrossRef] [PubMed]
106. Antimisiaris, S.; Mourtas, S.; Marazioti, A. Exosomes and Exosome-Inspired Vesicles for Targeted Drug Delivery. *Pharmaceutics* **2018**, *10*, 218. [CrossRef] [PubMed]
107. Mortazavi, S.-H.; Eslami, M.; Farrokhi-Ardabili, F. Comparison of different carrier-compounds and varying concentrations of oleic acid on freezing tolerance of ram spermatozoa in tris-citric acid-egg yolk plasma semen diluent. *Anim. Reprod. Sci.* **2020**, *219*, 106533. [CrossRef]
108. Najafi, A.; Taheri, R.A.; Mehdipour, M.; Farnoosh, G.; Martínez-Pastor, F. Lycopene-loaded nanoliposomes improve the performance of a modified Beltsville extender broiler breeder roosters. *Anim. Reprod. Sci.* **2018**, *195*, 168–175. [CrossRef] [PubMed]
109. Najafi, A.; Kia, H.D.; Mehdipour, M.; Hamishehkar, H.; Alvarez-Rodriguez, M. Effect of quercetin loaded liposomes or nanostructured lipid carrier (NLC) on post-thawed sperm quality and fertility of rooster sperm. *Theriogenology* **2020**, *152*, 122–128. [CrossRef]
110. Mehdipour, M.; Daghigh Kia, H.; Nazari, M.; Najafi, A. Effect of lecithin nanoliposome or soybean lecithin supplemented by pomegranate extract on post-thaw flow cytometric, microscopic and oxidative parameters in ram semen. *Cryobiology* **2017**, *78*, 34–40. [CrossRef]
111. Medina-León, A.Z.; Domínguez-Mancera, B.; Cazalez-Penino, N.; Cervantes-Acosta, P.; Jácome-Sosa, E.; Romero-Salas, D.; Barrientos-Morales, M. Cryopreservation of horse semen with a liposome and trehalose added extender. *Aust. J. Vet. Sci.* **2019**, *51*, 119–123. [CrossRef]
112. Pillet, E.; Labbe, C.; Batellier, F.; Duchamp, G.; Beaumal, V.; Anton, M.; Desherces, S.; Schmitt, E.; Magistrini, M. Liposomes as an alternative to egg yolk in stallion freezing extender. *Theriogenology* **2012**, *77*, 268–279. [CrossRef]
113. Kumar, P.; Saini, M.; Kumar, D.; Balhara, A.K.; Yadav, S.P.; Singh, P.; Yadav, P.S. Liposome-based semen extender is suitable alternative to egg yolk-based extender for cryopreservation of buffalo (Bubalus bubalis) semen. *Anim. Reprod. Sci.* **2015**, *159*, 38–45. [CrossRef]
114. Mafolo, K.S.; Pilane, C.M.; Chitura, T.; Nedambale, T.L. Use of phosphatidylcholine in Tris-based extender with or without egg yolk to freeze Bapedi ram semen. *S. Afr. J. Anim. Sci.* **2020**, *50*, 389–396. [CrossRef]
115. Luna-Orozco, J.R.; Gonzalez-Ramos, M.A.; Calderon-Leyva, G.; Gaytan-Aleman, L.R.; Arellano-Rodriguez, F.; Angel-Garcia, O.; Veliz-Deras, F.G. Comparison of different diluents based on liposomes and egg yolk for ram semen cooling and cryopreservation. *Iran. J. Vet. Res.* **2019**, *20*, 126–130. [PubMed]
116. He, L.; Bailey, J.L.; Buhr, M.M. Incorporating Lipids into Boar Sperm Decreases Chilling Sensitivity but Not Capacitation Potential1. *Biol. Reprod.* **2001**, *64*, 69–79. [CrossRef] [PubMed]
117. Röpke, T.; Oldenhof, H.; Leiding, C.; Sieme, H.; Bollwein, H.; Wolkers, W.F. Liposomes for cryopreservation of bovine sperm. *Theriogenology* **2011**, *76*, 1465–1472. [CrossRef] [PubMed]
118. Sullivan, R.; Saez, F. Epididymosomes, prostasomes, and liposomes: Their roles in mammalian male reproductive physiology. *Reproduction* **2013**, *146*, R21–R35. [CrossRef] [PubMed]

119. Köse, G.T.; Arica, M.Y.; Hasirci, V. Low-Molecular-Weight Heparin-Conjugated Liposomes with Improved Stability and Hemocompatibility. *Drug Deliv.* **1998**, *5*, 257–264. [CrossRef] [PubMed]
120. Purdy, P.H.; Graham, J.K. Membrane Modification Strategies for Cryopreservation. *Cryopreserv. Freeze. Dry. Protoc.* **2015**, *1257*, 337–342. [CrossRef]
121. Quinn, P.J.; Chow, P.Y.W.; White, I.G. Evidence that phospholipid protects ram spermatozoa from cold shock at a plasma membrane site. *Reproduction* **1980**, *60*, 403–407. [CrossRef]
122. Ansari, M.A.; Rakha, B.A.; Akhter, S. Cryopreservation of bull semen in OptiXcell®and conventional extenders: Comparison of semen quality and fertility. *Anim. Sci. Pap. Rep.* **2017**, *35*, 317–328.
123. Ansari, M.S.; Rakha, B.A.; Akhter, S.; Ashiq, M. OPTIXcell improves the postthaw quality and fertility of buffalo bull sperm. *Theriogenology* **2016**, *85*, 528–532. [CrossRef]
124. Abdel-Aziz Swelum, A.; Saadeldin, I.M.; Ba-Awadh, H.; Al-Mutary, G.M.; Moumen, A.F.; Alowaimer, A.N.; Abdalla, H. Efficiency of Commercial Egg Yolk-Free and Egg Yolk-Supplemented Tris-Based Extenders for Dromedary Camel Semen Cryopreservation. *Animals* **2019**, *9*, 999. [CrossRef]
125. Al-Bulushi, S.; Manjunatha, B.M.; Bathgate, R.; Rickard, J.P.; de Graaf, S.P. Liquid storage of dromedary camel semen in different extenders. *Anim. Reprod. Sci.* **2019**, *207*, 95–106. [CrossRef]
126. Raposo, G.; Stoorvogel, W. Extracellular vesicles: Exosomes, microvesicles, and friends. *J. Cell Biol.* **2013**, *200*, 373–383. [CrossRef]
127. D'Souza-Schorey, C.; Clancy, J.W. Tumor-derived microvesicles: Shedding light on novel microenvironment modulators and prospective cancer biomarkers. *Genes Dev.* **2012**, *26*, 1287–1299. [CrossRef] [PubMed]
128. Machtinger, R.; Laurent, L.C.; Baccarelli, A.A. Extracellular vesicles: Roles in gamete maturation, fertilization and embryo implantation. *Hum. Reprod. Update* **2015**, *22*, 182–193. [CrossRef]
129. Gould, S.J.; Raposo, G. As we wait: Coping with an imperfect nomenclature for extracellular vesicles. *J. Extracell. Vesicles* **2013**, *2*, 20389. [CrossRef] [PubMed]
130. Iglesias, D.M.; El-Kares, R.; Taranta, A.; Bellomo, F.; Emma, F.; Besouw, M.; Levtchenko, E.; Toelen, J.; van den Heuvel, L.; Chu, L.; et al. Stem Cell Microvesicles Transfer Cystinosin to Human Cystinotic Cells and Reduce Cystine Accumulation In Vitro. *PLoS ONE* **2012**, *7*, e42840. [CrossRef]
131. Charrier, A.; Chen, R.; Chen, L.; Kemper, S.; Hattori, T.; Takigawa, M.; Brigstock, D.R. Exosomes mediate intercellular transfer of pro-fibrogenic connective tissue growth factor (CCN2) between hepatic stellate cells, the principal fibrotic cells in the liver. *Surgery* **2014**, *156*, 548–555. [CrossRef] [PubMed]
132. Shtam, T.A.; Kovalev, R.A.; Varfolomeeva, E.; Makarov, E.M.; Kil, Y.V.; Filatov, M.V. Exosomes are natural carriers of exogenous siRNA to human cells in vitro. *Cell Commun. Signal.* **2013**, *11*, 88. [CrossRef] [PubMed]
133. Ong, S.G.; Lee, W.H.; Huang, M.; Dey, D.; Kodo, K.; Sanchez-Freire, V.; Gold, J.D.; Wu, J.C. Cross Talk of Combined Gene and Cell Therapy in Ischemic Heart Disease: Role of Exosomal MicroRNA Transfer. *Circulation* **2014**, *130*, S60–S69. [CrossRef]
134. Tomasoni, S.; Longaretti, L.; Rota, C.; Morigi, M.; Conti, S.; Gotti, E.; Capelli, C.; Introna, M.; Remuzzi, G.; Benigni, A. Transfer of Growth Factor Receptor mRNA Via Exosomes Unravels the Regenerative Effect of Mesenchymal Stem Cells. *Stem Cell Dev.* **2013**, *22*, 772–780. [CrossRef] [PubMed]
135. Saadeldin, I.M.; Oh, H.J.; Lee, B.C. Embryonic-maternal cross-talk via exosomes: Potential implications. *Stem Cell Cloning Adv. Appl.* **2015**, 103–107. [CrossRef]
136. Saadeldin, I.M.; Kim, S.J.; Choi, Y.B.; Lee, B.C. Improvement of Cloned Embryos Development by Co-Culturing with Parthenotes: A Possible Role of Exosomes/Microvesicles for Embryos Paracrine Communication. *Cell. Reprogram.* **2014**, *16*, 223–234. [CrossRef] [PubMed]
137. Qamar, A.Y.; Fang, X.; Kim, M.J.; Cho, J. Improved post-thaw quality of canine semen after treatment with exosomes from conditioned medium of adipose-derived mesenchymal stem cells. *Animals* **2019**, *9*, 865. [CrossRef] [PubMed]
138. Du, J.; Shen, J.; Wang, Y.; Pan, C.; Pang, W.; Diao, H.; Dong, W. Boar seminal plasma exosomes maintain sperm function by infiltrating into the sperm membrane. *Oncotarget* **2016**, *7*, 58832–58847. [CrossRef] [PubMed]
139. Kim, W.-S.; Park, B.-S.; Sung, J.-H.; Yang, J.-M.; Park, S.-B.; Kwak, S.-J.; Park, J.-S. Wound healing effect of adipose-derived stem cells: A critical role of secretory factors on human dermal fibroblasts. *J. Dermatol. Sci.* **2007**, *48*, 15–24. [CrossRef]
140. Ahmad, M.; Ahmad, R.; Murtaza, G. Comparative bioavailability and pharmacokinetics of flurbiprofen 200 mg SR pellets from India and France. *Adv. Clin. Exp. Med.* **2011**, *20*, 599–604.

141. Escudier, B.; Dorval, T.; Chaput, N.; André, F.; Caby, M.-P.; Novault, S.; Flament, C.; Leboulaire, C.; Borg, C.; Amigorena, S. Vaccination of metastatic melanoma patients with autologous dendritic cell (DC) derived-exosomes: Results of the first phase I clinical trial. *J. Transl. Med.* **2005**, *3*, 10. [CrossRef]
142. Shedden, K.; Xie, X.T.; Chandaroy, P.; Chang, Y.T.; Rosania, G.R. Expulsion of small molecules in vesicles shed by cancer cells: Association with gene expression and chemosensitivity profiles. *Cancer Res.* **2003**, *63*, 4331–4337.
143. Mokarizadeh, A.; Rezvanfar, M.-A.; Dorostkar, K.; Abdollahi, M. Mesenchymal stem cell derived microvesicles: Trophic shuttles for enhancement of sperm quality parameters. *Reprod. Toxicol.* **2013**, *42*, 78–84. [CrossRef]
144. Mahiddine, F.Y.; Kim, J.W.; Qamar, A.Y.; Ra, J.C.; Kim, S.H.; Jung, E.J.; Kim, M.J. Conditioned Medium from Canine Amniotic Membrane-Derived Mesenchymal Stem Cells Improved Dog Sperm Post-Thaw Quality-Related Parameters. *Animals* **2020**, *10*, 1899. [CrossRef]
145. Mahiddine, F.Y.; Qamar, A.Y.; Kim, M.J. Canine amniotic membrane derived mesenchymal stem cells exosomes addition in canine sperm freezing medium. *J. Anim. Reprod. Biotechnol.* **2020**, *35*, 268–272. [CrossRef]
146. Simon, L.; Murphy, K.; Shamsi, M.; Liu, L.; Emery, B.; Aston, K.; Hotaling, J.; Carrell, D. Paternal influence of sperm DNA integrity on early embryonic development. *Hum. Reprod.* **2014**, *29*, 2402–2412. [CrossRef] [PubMed]
147. Franchi, A.; Moreno-Irusta, A.; Domínguez, E.M.; Adre, A.J.; Giojalas, L.C. Extracellular vesicles from oviductal isthmus and ampulla stimulate the induced acrosome reaction and signaling events associated with capacitation in bovine spermatozoa. *J. Cell. Biochem.* **2019**, *121*, 2877–2888. [CrossRef] [PubMed]
148. De Almeida Monteiro Melo Ferraz, M.; Nagashima, J.B.; Noonan, M.J.; Crosier, A.E.; Songsasen, N. Oviductal Extracellular Vesicles Improve Post-Thaw Sperm Function in Red Wolves and Cheetahs. *Int. J. Mol. Sci.* **2020**, *21*, 3733. [CrossRef] [PubMed]
149. Naresh, S.; Atreja, S.K. The protein tyrosine phosphorylation during in vitro capacitation and cryopreservation of mammalian spermatozoa. *Cryobiology* **2015**, *70*, 211–216. [CrossRef]
150. Alcântara-Neto, A.S.; Schmaltz, L.; Caldas, E.; Blache, M.-C.; Mermillod, P.; Almiñana, C. Porcine oviductal extracellular vesicles interact with gametes and regulate sperm motility and survival. *Theriogenology* **2020**, *155*, 240–255. [CrossRef]

Review

Extracellular Vesicles, the Road toward the Improvement of ART Outcomes

Maria G. Gervasi [1], Ana J. Soler [2], Lauro González-Fernández [3,4], Marco G. Alves [5], Pedro F. Oliveira [6] and David Martín-Hidalgo [3,7,*]

1 Department of Veterinary & Animal Sciences, University of Massachusetts, Amherst, MA 01003-9301, USA; mariagracia@vasci.umass.edu
2 IREC (CSIC-UCLM-JCCM), ETSIAM, Campus Universitario, s/n, 02071 Albacete, Spain; anajosefa.soler@uclm.es
3 Research Group of Intracellular Signalling and Technology of Reproduction (Research Institute INBIO G+C), University of Extremadura, 10003 Cáceres, Spain; lgonfer@unex.es
4 Department of Biochemistry and Molecular Biology and Genetics, Faculty of Veterinary Sciences, University of Extremadura, 10003 Cáceres, Spain
5 Unit for Multidisciplinary Research in Biomedicine (UMIB), Laboratory of Cell Biology, Department of Microscopy, Institute of Biomedical Sciences Abel Salazar (ICBAS), University of Porto, 4050-313 Porto, Portugal; alvesmarc@gmail.com
6 Department of Chemistry, University of Aveiro, 3810-193 Aveiro, Portugal; p.foliveira@ua.pt
7 Department of Physiology, Faculty of Veterinary Sciences, University of Extremadura, 10003 Cáceres, Spain
* Correspondence: davidmh@unex.es

Received: 24 October 2020; Accepted: 19 November 2020; Published: 21 November 2020

Simple Summary: Nowadays, the farm and pet industries cannot be sustained without assisted reproductive technologies (ART). Nevertheless, ART outcomes still are far from ideal. Recently, the emerging role of bioactive molecules—known as "extracellular vesicles" (EVs)—in the reproductive processes has been reported. EVs originate in different sections of the reproductive tract, and they can be isolated from reproductive fluids. Here, we update recent advances in the use of EVs as additive to media used in ART to enhance their reproductive outcome, mainly in domestic mammal animals.

Abstract: Nowadays, farm animal industries use assisted reproductive technologies (ART) as a tool to manage herds' reproductive outcomes, for a fast dissemination of genetic improvement as well as to bypass subfertility issues. ART comprise at least one of the following procedures: collection and handling of oocytes, sperm, and embryos in in vitro conditions. Therefore, in these conditions, the interaction with the oviductal environment of gametes and early embryos during fertilization and the first stages of embryo development is lost. As a result, embryos obtained in in vitro fertilization (IVF) have less quality in comparison with those obtained in vivo, and have lower chances to implant and develop into viable offspring. In addition, media currently used for IVF are very similar to those empirically developed more than five decades ago. Recently, the importance of extracellular vesicles (EVs) in the fertility process has flourished. EVs are recognized as effective intercellular vehicles for communication as they deliver their cargo of proteins, lipids, and genetic material. Thus, during their transit through the female reproductive tract both gametes, oocyte and spermatozoa (that previously encountered EVs produced by male reproductive tract) interact with EVs produced by the female reproductive tract, passing them important information that contributes to a successful fertilization and embryo development. This fact highlights that the reproductive tract EVs cargo has an important role in reproductive events, which is missing in current ART media. This review aims to recapitulate recent advances in EVs functions on the fertilization process, highlighting the latest proposals with an applied approach to enhance ART outcome through EV utilization as an additive to the media of current ART procedures.

Keywords: spermatozoa; oocyte; in vitro fertilization; extracellular vesicles; assisted reproductive technologies; embryo

1. Assisted Reproductive Technologies and Their Handicaps

Countless advantages can be quoted for the use of assisted reproductive technologies (ART). ART have been used to preserve valuable genetic material (cryobiology), to perform offspring sex selection (by sperm sorting), to reduce the incidence of venereal diseases (by artificial insemination (AI)), to bypass sub-fertility issues (by in vitro fertilization (IVF) or intracytoplasmic sperm injection (ICSI)), to increase reproductive outcomes and maximize the number of offspring that can be obtained by a single female (by inducing superovulation, performing IVF and eventually transferring embryos to female recipients by embryo transference (ET)), and to enhance reproduction of males (by increasing the performance of a single ejaculate that can be cryopreserved and then used by AI). Consequently, ART have a critical role on the management of the herd; for example, a high proportion of pigs and bovines are produced by AI [1,2].

One of the most popular ART used is IVF. Current IVF protocols are based on the basic knowledge provided in 1951 by two investigators that independently discovered that ejaculated mammalian sperm require a period of incubation in the female reproductive tract to acquire the ability to fertilize [3,4]. This phenomenon is described as sperm capacitation [5], and it settled the cornerstone for the development of IVF [6]. Twenty years after the description of sperm capacitation, the first successful IVF in mouse using a defined medium in absence of female fluids was performed [7]. Before that, successful IVF were performed by including female fluids from the reproductive tract in the incubation media [8,9]. These results highlighted the importance of factors present in female reproductive fluids to accomplish fertilization. Nevertheless, current IVF media do not differ than those developed empirically more than 50 years ago [7].

It is well documented that the quantity and quality of embryos obtained by ART are lower in comparison to those obtained in vivo by mating [10,11]. In addition, ART-derived embryos present lower chances to fully develop and derive in live offspring [10,11], underlining the need to enhance the current media used during ART protocols. It is clear that current ART procedures lack the interaction of gametes with several components present in the reproductive tract during fertilization and the first stages of development. Recent advances point out that extracellular vesicles (EVs) present in the reproductive environment help to achieve in vitro-derived embryos with development levels similar to in vivo-derived embryos. In the following sections, novel manuscripts that emphasize the roles of EVs from male and female reproductive fluids on reproductive processes are summarized. These insights on gametes and embryo production and culture must be considered to enhance the poor success rates of some of the ART procedures. For instance, an effective method for horse IVF [12] or a method to improve the less than 45 % of IVF/ICSI blastocyst rate in the ovine model [13] have not been established yet, despite decades of research.

2. Extracellular Vesicles (EVs) and Their Role on ART Outcome Improvement

Nowadays, neither the farm animal industry nor pet reproductive management can be conceived without the use of ART. However, improvement of IVF and embryo culture media are required to overcome differences in quality and developmental potential between in vivo- and in vitro-derived embryos. EVs, present in both the female and male reproductive tracts, play important roles in gamete maturation and ultimately in the fertilization process. The omission of EVs in current media used for ART is one of the causes that might explain the lower embryo quality obtained in these in vitro conditions.

EVs are defined as spherical bilayers containing proteins, genetic material and lipids that transport their cellular content (cargo) to other cells acting as intercellular communicators [14]. EVs can be classified according to the organ where are originated, in particular, for the reproductive tracts EVs are called: epididymosomes if derived from the epididymis; prostasomes if derived from the prostate; vaginosomes if derived from the vagina; uterosomes if derived from the uterus; and oviductosomes if derived from the oviduct. In addition, EVs can be classified according to their diameter size in microvesicles or ectosomes (100–1000 nm) or exosomes (30–100 nm) [15]. To avoid confusion, in this manuscript we refer to EVs based on their origin without considering the diameter size classification (see Table 1 for more classification details). In this section we focus on the role of EVs on the gain of gametes function as well as embryo development competence with special emphasis in those manuscripts whose findings are applied to the improvement of ART outcome.

Table 1. Classification of extracellular vesicles (EVs) according to their localization in the reproductive tract or their diameter size.

Subsection	Name
Epididymis	Epididymosome
Prostate	Prostasome
Vagina	Vaginosome
Uterus	Uterosome
Oviduct	Oviductosome
Size (nm)	**Name**
100–1000	Microvesicles or ectosomes
30–100	Exosomes

2.1. Relationship between Spermatozoa and EVs as a Tool to Enhance ART Results

The spermatozoon is an extraordinary cell that is designed to survive in different microenvironments including a body where it was not created with the ultimate goal of fertilization. Due to the large amount of protamines that replace histones, sperm chromatin is highly compacted [16]. Thus, sperm are not able to transcribe gene information, do not synthesize proteins, and therefore, regulate their function by post-translational modifications [17]. Nevertheless, sperm are surrounded by a plasma membrane; and during their maturation in the epididymis, the ejaculation process, and along their journey through the female reproductive tract, sperm can acquire information from the surrounding milieu by exchanging information with EVs found in these fluctuating environments [18]. It has been shown that EVs cargos are up taken by sperm by a fusogenic mechanism of their membranes [19,20] or by lipid-rafts domain mediated-endocytosis [21].

Sperm are produced in the testis and transported to the epididymis. Sperm maturation occurs while the sperm transit from the caput region of the epididymis towards the cauda, where they are ultimately stored until ejaculation takes place [22]. Several works described the role of epididymosomes on the sperm maturation process through the transfer of cargo proteins and small RNAs to sperm (for review see [23]) while others focused on their applied roles on ART. Here we focus on the latter (Summarized in Table 2).

Table 2. Effects of co-incubation of extracellular vesicles and spermatozoa on ART outcome.

Specie	EVs Origen	ART used	Output	Reference
Human	Prostate	In vitro incubation in acidic media	↑ % of motile spermatozoa	[24]
Human	Prostate	In vitro capacitation	Inhibit sperm capacitation Inhibit spontaneous acrosome reaction	[25]
Human	EECs	In vitro capacitation	Enhance sperm capacitation status	[26]
Human	Prostate	In vitro capacitation	Enhance acrosome reaction response to calcium ionophore	[27]
Human/ mouse	Prostate	In vitro incubation	↑ Hypermotility ↑ IVF fertility	[28]
Mouse	Vagina from superovulated females	In vitro capacitation	Enhance sperm responsiveness to progesterone Incorporation of several sperm proteins with roles on calcium homeostasis (SPAM1, PMCA1/4, PMCA4) and capacitation process (protein tyrosine phosphorylation)	[29]
Pig	Prostate	In vitro incubation	Enhance sperm acrosome reaction	[30]
Pig	Prostate	Preservation at low temperature	Prolonged sperm motility ↑ Sperm antioxidative capacity ↓ Lipid peroxidation Protect plasma membrane Protect against premature capacitation	[31]
Stallion	Prostate	In vitro capacitation	Inhibit sperm capacitation events as protein tyrosine phosphorylation	[32]
Feline	Epididymis	In vitro incubation	↑ % of motile spermatozoa for a short period of time (up to 1 h) ↑ Forward motility (1.5 to 3 h of co-incubation)	[33]
Feline	Oviduct (different follicular phases)	IVF	↑ % Motile spermatozoa Protect again premature acrosome reaction Enhanced IVF outcome	[34]
Dog	ASCs	Cryopreservation	↑ Sperm motility and viability ↑ Mucus penetration ability or ↓ Acrosome and chromatin damaged	[35]
Bovine	Oviduct (different sections)	Cryopreservation	↑Protein tyrosine phosphorylation ↑ Responsiveness to progesterone Maintain sperm survival	[36]

List of abbreviations: EECs: endometrial epithelial cells; ASCs: adipose-derived mesenchymal stem cells; SPAM1: sperm adhesion molecule 1; PMCA: plasma membrane calcium-transporting ATPase; ↑: increase; ↓: decrease.

It has been shown in cats that epididymosomes affect sperm motility in vitro. Co-incubation of immature sperm obtained from the caput epididymis with epididymosomes for a short period of time (up to 1 h), showed a modest enhancement of total motility. Interestingly, longer exposure to epididymosomes (1.5 to 3 h) increased the percentage of sperm displaying progressive motility [33]. This could help maximize the chances of obtaining sperm with fertilization potential from epididymis of animals with high genetic value that present ejaculation issues or sudden deaths of endangered species.

At the moment of ejaculation, sperm get exposed to EVs-containing seminal fluid originated in three accessory glands: the seminiferous vesicle, the bulbourethral glands, and the prostate. Nevertheless, prostasomes are the most widely studied EVs in the seminal fluid. Once ejaculated, sperm initiate their journey through the female reproductive tract. The acquisition of hyperactivated motility, the ability to undergo the acrosome reaction, and, at the molecular level, the increase of protein tyrosine phosphorylation have been historically used as hallmarks of capacitation status [37,38].

Contradictory results have been found on the role of prostasomes on sperm capacitation. On one hand, a protective function against premature capacitation and acrosome reaction has been described on human and stallion sperm [24,25,39,40]. These findings were associated to the membrane composition of EVs with high content of cholesterol and sphingomyelin that decrease fluidity of sperm plasma membrane once the EVs and sperm fusion occurs [40,41]. The inhibition of premature capacitation might have a functional application on ART. For instance, the prevention of premature sperm capacitation is desirable when seminal doses are stored before performing AI or to counteract capacitation-like events during the sperm cryopreservation procedure [42]. In addition, an exhaustive work described that prostasomes added to boar seminal doses preserved at 17 °C for a long-term period was able to: (1) prolong sperm motility; (2) increase the total sperm antioxidant capacity, and; (3) protect plasma membrane integrity [31]. In that work, they also showed that prostasomes protection of sperm against premature capacitation was associated to their seminal plasma protein 1 (PSP-1) and carbohydrate-binding protein AWN (AWN) cargo. Nevertheless, prostasomes did not affect sperm ability to undergo capacitation when they were stimulated [31]. On the other hand, other authors described that in vitro incubation of boar sperm with isolated prostasomes enhanced the acrosome reaction [30]. Interestingly, qualitative but not quantitative differences were found on prostasomes of normozoospermic and severe asthenozoospermic men [43]. Prostasomes from normozoospermic men transferred cysteine-rich secretory protein 1 (CRISP1) to sperm [43]. CRISP1 is a protein with the ability to regulate murine CatSper channel enhancing the acrosome reaction induced by Ca^{2+} ionophore [27] and also to participate in the sperm-zona pellucida binding through the interaction with ZP3 [44]. Similarly, the plasma membrane calcium ATPase pump 4 (PMCA4), a vital machinery to regulate sperm calcium homeostasis was delivered in vitro into spermatozoa either by epididymosomes [45] or by prostasomes [46]. Future research on prostasomes must explore their applied function on ART considering that variation between species might be found that could be related to differences in their evolutive reproductive strategies.

In most species, sperm are deposited in the vagina during ejaculation and spend there a short period of time before continuing to travel through the female reproductive tract. This short time in the vagina is sufficient for sperm to be exposed to vaginosomes (VGS), which could have an effect on sperm functionality. Mice sperm incubated in non-capacitating conditions exposed for 30 min to VGS presented an enhanced progesterone-induced acrosome reaction [29]. In addition, co-incubation of sperm with vaginal luminal fluid (containing VGS) resulted in sperm incorporation of SPAM1, PMCA1/4, PMCA4, all proteins with roles on calcium homeostasis and the capacitation process as well as an overall increase in sperm protein tyrosine phosphorylation [29]. The transfer of tyrosine phosphorylated proteins by VGS could explain why sperm lacking the tyrosine kinase FER do not display tyrosine phosphorylation and do not fertilize in vitro although they are able to fertilize in vivo [47].

After leaving the vagina sperm enter the uterus. In vitro studies have shown that a short exposure (15 min) of sperm with uterosomes secreted by endometrial epithelial cells simulating the time that they spend in the uterus is enough to enhance sperm capacitation status in human spermatozoa [26]. Others authors have described the same results but with longer exposure times [21].

Finally, sperm pass through the uterotubal junction and reach the oviduct where the encounter with the oocyte and fertilization take place. Here, sperm interact with oviductosomes (OVS) produced by the oviductal epithelium. The impact of OVS on sperm function has been studied in several species. In mice, the OVS cargo is incorporated into sperm and is responsible for the rise of the protein PMCA1 and an increase in tyrosine-phosphorylated proteins levels in sperm [48]. Interestingly, it was observed that capacitated spermatozoa uptake higher quantity of OVS cargo that their non-capacitated counterparts [48]. These results were associated to a higher plasma membrane permeability found in capacitated spermatozoa since they lost sterol during the capacitation process [48].

In the bovine model, OVS have been used as a supplement of non-capacitating media for thawing cryopreserved sperm. The results obtained were dependent on the OVS origin: ampulla or isthmus. After incubation with isthmus-originated OVS, sperm displayed characteristic features associated to control capacitated sperm such as high levels of protein tyrosine phosphorylation, increased acrosome reaction and intracellular calcium responsiveness to progesterone [36]. The co-incubation of sperm with ampulla-originated OVS displayed an augmented capacitation response even when compared to capacitated control [36]. In summary, incubation of non-capacitated spermatozoa with OVS induced sperm capacitation and enhanced sperm survival with similar levels to those described in capacitated spermatozoa [36].

The role of OVS on sperm function was also studied in cats. In vitro experiments showed that cat's OVS bind to the acrosomal region of the sperm head and to the mid-piece of the sperm tail. In addition, OVS used as additive to regular capacitating media enhanced the percentage of motile sperm as well as increased the rates of cleavage and blastocyst formation (23% and 8%, respectively) in comparison with control (no OVS co-incubated) [34]. The authors then investigated the cargo proteins in the OVS by mass spectrometry. The analysis of protein content of OVS identified a total of 4879 proteins, and between other functions, proteins involved on the sperm-oocyte interaction and fertilization process as cluster differentiation 9 (CD9), CCTs (cytosolic chaperonin containing TCP-1;) and TCP1 were found [34].

It has been shown in mice that EVs derived from either the epididymis or the oviduct can transfer miRNAs into the sperm [49,50]. For instance, OVS miR-34c-5p is delivered to the sperm [50] and has an important role on the fertilization process by initiating the first cleavage division. Bypassing the interaction between sperm and OVS as occurs in in vitro conditions could lead to the failure of first cleavage division. Consequently, it might negatively impact ART results where for example a single semen donor can be used to inseminate hundreds of females.

Besides the studies focusing on EVs obtained from the reproductive tract, other study evaluated the effect of EVs obtained from adipose-derived mesenchymal stem cells (ASCs) cultivated in vitro and used as additive to dog sperm cryopreservation media [35]. Surprisingly, ASCs-EV lead to a significant improvement of sperm motility and mucus penetration ability [35]. In addition, ASCs-EV protected spermatozoa against damage of the plasma membrane, the acrosome membrane, and the chromatin. The authors associated EVs beneficial effects during the freezing/thawing process to their protein and mRNA cargo associated to plasma membrane and chromatin repair process [35].

In summary, sperm receive pivotal elements through their interaction with EVs produced in the different sections of the male and female reproductive tracts (Figure 1), and these evidences should not be underestimated for the development of future enhanced sperm culture media that mimic physiological environmental conditions.

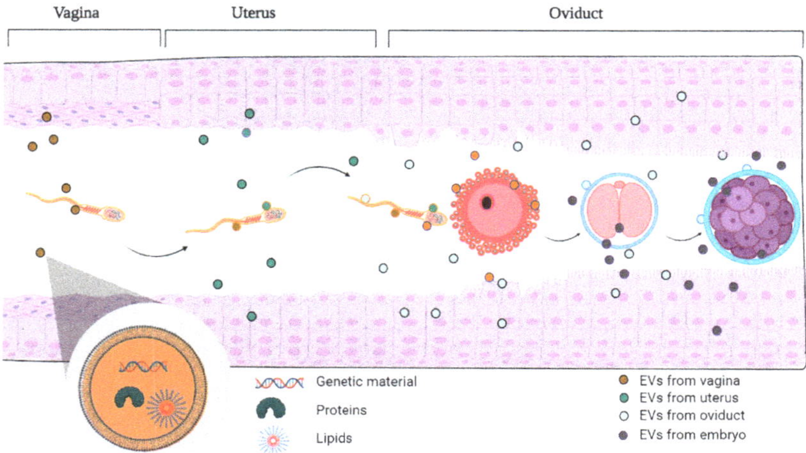

Figure 1. Schematic representation of the sperm travel through the female reproductive tract. Sperm enter in contact with the different extracellular vesicles produced in the vagina, the uterus and the oviduct. Once fertilization takes place, the embryo will come into contact with the EVs produced by the oviduct and the uterus where the embryo and the future fetus will remain for the rest of the pregnancy until delivery. Note that the embryo also produces EVs that allow bidirectional communication with the mother tissue (oviduct).

2.2. Relationship between Oocyte Maturation and EVs Used as a Tool to Enhance ART Results

In most mammalian species, before ovulation, oocytes are in an immature stage (germinal vesicle (GV)), in order to be competent for fertilization, they need to undergo meiotic resumption and arrive to the meiotic competence stage (metaphase II (MII)). Collection of immature GV oocytes from the ovaries followed by incubation in specific conditions that allow for the resumption of meiosis is a common practice in ART such as IVF [51]. This process is named in vitro maturation (IVM) and current IVM media described are deficient to obtain a proper oocyte maturation in some species [52,53]. For instance, the canine industry also has to address the issue of low efficiency of oocytes IVM. Thus, OVS used as additive along the canine oocyte IVM renders better results in comparison with control 21.82% and 8.66%, respectively [52]. Recently, it has been elegantly demonstrated that the percentage of mature canine oocytes after IVM in presence of OVS is enhanced through OVS cargo [54,55]. Leet et al., demonstrated that EGFR/MAPK signaling is the responsible for this improvement by the use of an inhibitor of this pathway (gefitinib) and an inhibitor of exosomes generation (GW4869) [54]. In addition, canine OVS enhance antioxidant capacity, viability and proliferation of canine cumulus cells [54].

In the ovary, oocytes are contained in follicles that will gradually mature from primordial into pre-ovulatory. The latter are follicles larger in size, filled with follicular fluid, and composed by theca and granulosa cells that surround an oocyte. It has been shown in several species that the follicular fluid contains EVs and that these EVs may have a role in cellular communication within the follicle [56,57]. In bovines, EVs from follicular fluid induce granulose cells proliferation through Src, PI3K/Akt and MAPK signaling pathways [58]. In addition, granulose cells preferentially uptake EVs from small over EVs from larger follicles [58]. The follicular fluid of mares contains EVs, and their proteins and miRNAs cargo were analyzed and described as a pathway of communication between oocytes and ovaries [56]. It was elegantly shown—in both in vivo and in vitro conditions—that EVs from follicular fluid are uptaken by the granulosa cells that surround the oocyte [56]. Interestingly, the authors described that miRNAs' EVs cargo from follicular fluid changes along with the age of the mare and this fact might explain age-related decline of oocyte quality in this species [56]. Interestingly, EVs isolated from ovarian follicular fluid used as an additive during bovine oocyte maturation and embryo development in in vitro conditions enhanced blastocyst rate and decreased global DNA methylation and hydroxymethylation levels [59]. Nevertheless, caution needs to be taken when using follicular fluid EVs as additive to enhance oocyte competence, as it was shown that the EVs cargo vary along the estrus cycle [60–62], the size of the follicle [63], and with the age of the female [56].

Another clear example of EVs used to enhance current ART was shown when EVs isolated from follicular fluid were added during the process of vitrification/thawing of immature cat oocytes, a procedure that compromises oocytes ability to undergo meiotic resumption [64]. In this case, the addition of EVs did not protect against the loss of oocyte viability during vitrification/thawing but enhanced the oocyte IVM rate after vitrification as higher numbers of oocytes arrived to MII stage (28.3%) in comparison with controls (8.6%) [64].

2.3. Relationship between EVs Used as a Tool to Enhance Embryos and Conceptus Development Obtained by ART

After ovulation, oocytes transit through the oviduct from the ampulla towards the isthmus. Fertilization occurs in the ampulla and is followed by the initiation of embryo development. Early embryo development and the transit of the embryo towards the uterus occur simultaneously. Valuable information was acquired in the 90s when it was described that embryos cultivated in groups displayed better cleavage and blastocyst formation rates than those incubated individually [65]. This fact has been associated to embryo secretion of autocrine/paracrine growth factors (secretome) that lead to a better embryo development [66,67]. In addition to the secretome, it was confirmed that EVs produced by the embryos contribute to the better embryo competence when they are cultured together [68]. Hence, information carried by embryo-derived EVs is not only important to communicate with the female tract, but also to communicate between them and achieve better embryo competence.

The oviduct has an important role in fertilization and embryo development. It is not surprising to find that the oviductal fluid of several species contains EVs [19,34,48,69,70], and as mentioned above these EVs were specifically named oviductosomes (OVS). The OVS cargo varies along the estrus cycle [48,71]. For example, OVS cargo of plasma membrane Ca^{2+}-ATPase (PMCA), with Ca^{2+} clearance-homeostasis role [72], changes along the estrus cycle where the PMCA levels in proestrus/estrus are higher than in metestrus/diestrus [48]. Interestingly, it was shown that the concentration and size of OVS is stable along the bovine estrus cycle; however the OVS content varies [71]. For example, higher mRNA composition was found in OVS recovered in the post-ovulatory stages when compared to OVS recovered during the rest of the cycle [71]. Differences were also found at the protein level, and the major differences were found between OVS recovered at the post-ovulatory and pre-ovulatory stages [71]. Similarly, differences were found along the estrus cycle when the protein cargo of porcine OVS was analyzed [62]. Due to the variations found during the estrous cycle, the authors hypothesized that OVS cargo changes are regulated by hormonal changes during the estrous cycle [62,71].

Primary cultures of bovine oviductal epithelial cells (BOECs) in monolayers are commonly used as an in vitro model for the study of gametes/embryo interaction with the oviduct in the bovine model [73,74]. It has been shown that BOECs secrete EVs [75]. The supplementation of the embryo culture media with BOEC-derived EVs did not affect embryo development outcome. However, these BOEC-derived EVs improved cryotolerance of embryos vitrified as they increased survival rate and number of cells, and upregulated genes related to implantation (PAG1) and metabolism (GADPH) [75]. In that work, the authors also described a negative effect of EVs present in fetal calf serum (FCS), a common component used on embryo culture media, over the bovine embryo vitrification/thawing process [75]. Oviductal region specificity was evidenced as isthmus-derived EVs were more effective than ampulla-derived EVs [76]. Interestingly, similar results were found when OVS were used to supplement in vitro bovine embryo cultures: the OVS did not have any effect on embryo development rate but improve embryo cryotolerance [76]. Almiñana and collaborators showed that the addition of oviduct-derived EVs to the culture media did not improve IVF fertilization rates; although, it enhanced the blastocyst embryo quality and the embryo hatching rate [70]. They also showed that frozen EVs had better reproductive outcome (as hatching rate) when used as additive to the embryo culture media in comparison to fresh EVs [70]. These results highlight that EVs can be frozen without any detriment in their cargo capacities simplifying the logistics of their application to different ART.

Despite similar results found by addition of BOECs-derived or oviduct-derived EVs to enhance ART outcome, it was revealed that there are qualitative and quantitative differences in the protein cargo between them [70]. A total of 319 proteins (47 only expressed in BOEC derived and 97 only expressed from oviduct derived) were identified by mass spectrometry, where only 175 of them were common to both populations [70]. Oviduct-specific glycoprotein 1 (OVGP1), a protein that enhances embryo development [77] and quality [78] was only present in in vivo EVs. One explanation for this discrepancy between in vitro and in vivo could be that static BOEC cultures induce cell dedifferentiation and therefore these cultured cells might lose some functions. Then, invaluable information could be obtained when using the novel dynamic culture system oviduct-in-a-chip developed by Ferraz an collaborators (2018). This culture system of oviductal epithelial cells allows the investigators to simulate physiological conditions by applying hormonal waves, and has shown to maintain epithelial cell differentiation and presence of cilia [79]. Interestingly, OVGP1 levels also increased when 3D-Chip were used to grow BOECs [79]. More cell culture studies using the chip strategy could help extrapolate results to in vivo conditions (see Figure 2).

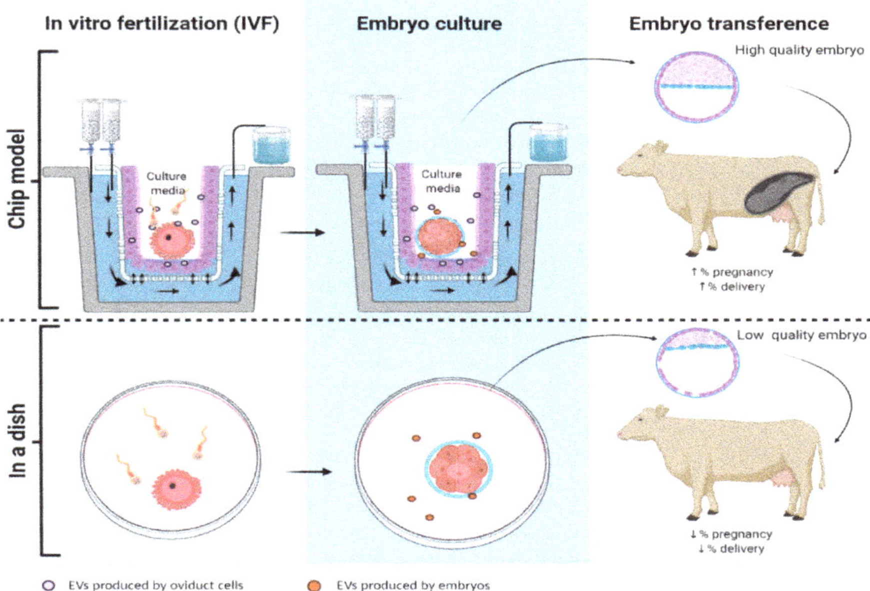

Figure 2. Schematic representation of embryo production using a Chip-model or an in vitro model (in a dish). Hypothetically in a Chip-model, the system mimics estrus cycle hormones concentrations that allows oviductal cells to develop cilia and produce EVs that will vary their cargo along the estrus cycle, increasing the pregnancy and delivery rate. In contrast, in embryo production by classic ART in a dish there is no interaction with EVs oviduct, consequently the embryos produced have lower quality and less chance of implantation and ending in delivery.

The use of OVS in ART was proven beneficial also in other species. In rabbits, embryos co-incubated with OVS decreased reactive oxygen species (ROS) and DNA methylation levels that lead to an increase of the blastocyst development rate [80]. The authors showed that the antioxidant properties were associated to the melatonin OVS cargo by the use of luzindole, a selective melatonin receptor antagonist, that attenuated the positive effect of OVS on embryos [80]. In pigs, OVS have been used successfully to face the problem of polyspermy in porcine IVF, doubling the percentage of oocyte penetrated by a single spermatozoon [69].

Another study using a murine model described that OVS obtained from pregnant females used as additive for the IVF procedure enhanced embryo transference efficiency in comparison with the supplementation with OVS obtained from pseudo-pregnant females [81]. Pregnant females OVS increased the percentage of blastocysts, and the embryo quality determined by an increase of gene expression related to successful embryo development (Bcl-2 and Oct-4), an increased inner cell mass: trophectoderm ratio, and a decreased embryo apoptosis. In addition, birth rates after embryo transference were also enhanced [81].

An appealing application of EVs for human ART was found by the use of EVs obtained from human endometrial-derived mesenchymal stem cells (EV-endMSCs) isolated from menstrual blood [82–84]. IVF-embryos obtained from aged murine females and supplemented with human EV-endMSCs in the embryo culture media presented enhanced embryo development. Furthermore, blastocyst rate was doubled by the addition of 20 μg/mL of EV-endMSC in comparison to controls [83,84]. Positive effects of EV-endMSC in the embryo competence and quality of those embryos obtained in aged oocyte were associated to different levels of mRNA expression of genes associated to cellular response to oxidative stress (Sod1), metabolism (Gadph), placentation (Vegfa) and trophectoderm/ICM formation (Sox2) [83]. Moreover, human EV-endMSC used as additive to the culture media of mouse zygotes obtained in vivo increased the number of total cells by blastocyst and the hatching rate [82]. Recently, EVs derived from human umbilical cord mesenchymal stem cells (HUCMSCs) were successfully used to restore fertility on female mice with premature ovarian insufficiency (POI) [85]. Isolated EVs-HUCMSCs administered by only one intravenous injection rescued ovarian function, hormones levels (FSH and E2), as well as enhanced the mating behavior and the numbers of pups born after four weeks of treatment [85]. In addition, the POI group treated with EVs-HUCMSCs increased parameters associated to IVF procedures such as number of oocytes retrieved, percentage of fertilized oocytes, cleaved embryos, and blastocyst rates although did not reach values of the control group [85]. The effects of EVs on ART and embryo development are summarized in Table 3.

Once fertilization has occured, the embryo starts a series of developmental changes that orchestrate the transformation from a zygote into a blastocyst. During this period of early embryonic development, the embryo is still enclosed in the zona pellucida (ZP, a glycoproteic layer that surrounds the plasma membrane of the oocyte). Embryos in blastocyst stage will hatch from the ZP and implant by attaching and invading the uterus of the mother. Thus, embryo-uterus interaction plays vital roles in the recognition of pregnancy, maintenance of the corpus luteum, and consequently the allowing for the progression of pregnancy [86]. It has been shown that exist differences between non and pregnates mares' miRNA EVs cargo obtained from serum [87]. In adddition this EVs cargo might contribute to the pathway of maternal recognition of the pregnancy [87]. In other work it was shown that murine EVs produced by endometrium that contained miRNAs involved in the implantation process were internalized by embryo through the trophectoderm and increased embryo adhesion levels [88].

On the other hand, it was shown that conceptus-derived EVs also promote communication between the conceptus and the maternal tissue during the establishment of pregnancy in ovines [89]. In that study, the conceptus (14 days) originated in vivo were cultivated for a day in a dish and then the EVs were collected from the medium. Later on, the EVs originated by the conceptus in in vitro conditions were labelled (PKH67) and transferred to pregnant ewes. The authors showed that the embryos and the uterine epithelium of the pregnant female receptors internalized the conceptus-derived EVs [89]. In the same work, it was also shown that EVs obtained from the uterine luminal fluid are internalized by the conceptus trophectoderm and uterine epithelia [89]. This work ilustrates that embryo-derived EVs might have an impact on the receptivity of the uterus and that maternal-fetal communication could be key for a successful implantation and progression of pregnancy. Considering this, implantation failure of certain ART such as embryo transference might be related to the lack of embryo-mother communication during the first stages of development.

Table 3. Effects of co-incubation of extracellular vesicles and embryo on ART outcome.

Specie	EVs Origen	ART Used	Output	Reference
Bovine	BOEC	IVC Vitrification	No differences on embryo development Enhance vitrification outcome: ↑ Embryo quality ↑ Cryo-survival rate ↑ Number of cells	[75]
Bovine	BOEC	IVP IVC	Enhanced the embryo quality ↑ Number of cells ↑ Hatching rate = Fertilization rate	[70]
Bovine	Oviduct (ampulla and isthmus)	IVC Vitrification	No differences on embryo development ↑ Cryo-survival rate	[76]
Mouse	endMSCs	Embryo culture obtained in vivo	↑ Number of total cells by blastocyst ↑ Hatching rate	[82]
Mouse (ageing)	endMSCs	IVF IVC	Enhance embryo competence and quality ×2 blastocyst rate ↑ mRNA expression of Sod1, Gadph, Vegfa and Sox2	[83,84]
Mouse	Oviduct from pregnant females	IVF ET	↑ Embryo quality (↑ Bcl-2; Oct-4↓Bax) ↑ ICM ↑ Blastocyst and birth rates	[81]
Mouse (POI)	HUCMSCs	IVF	Rescue ovary function, hormones levels (FSH and E2), natural fertility. ↑ oocytes retrieved, fertilized zygotes, cleaved embryos and blastocysts	[85]
Porcine	Oviduct	IVF	↓ Polyspermia	[69]
Feline	Ovarian fluid	Vitrification IVM	= Vitrification survival rate ↑ Oocyte IVM from 8.6% in control to 28.3% in supplemented with EVs	[64]
Canine	Oviduct	IVM	↑ Oocyte IVM 21.82% vs. control 8.66%	[52]
Canine	Oviduct	IVM	↑ Cumulus cell viability, and proliferation rate ↓ ROS and apoptotic rate	[54]
Canine	Oviduct	IVM	↑ Maturation rate of oocytes	[55]
Rabbit	Oviduct	IVF IVC	↓ ROS and DNA methylation levels ↑ Blastocyst rate	[80]

List of abbreviation: Bax: BCL2-associated X protein; BOEC: bovine oviduct epithelial cells; E2: estradiol; ET: embryo transference; endMSCs: human endometrial-derived Mesenchymal Stem Cell; FSH: follicle stimulating hormone; Gadph: glyceraldehyde-3-phosphate dehydrogenase; HUCMSCs: human umbilical cord mesenchymal stem cells; ICM: inner cell mass; IVC: in vitro culture; IVM: in vitro maturation; Oct-4: transcription factor-1; POI: premature ovarian insufficiency; ROS: reactive oxygen species; Sod1: superoxide dismutase 1; Sox2: SRY-related HMG-box gene 2; Vegfa: vascular endothelial growth factor A. ↑: increase; ↓: decrease.

3. Challenges for the New Era of ART Development

Based on the broad-spectrum possibilities for EVs application, EVs have been postulated as potential non-invasive biomarkers. For example, EVs can be used as a biomarker for embryo quality [90–92] to select good embryos before transferring them to female recipients or to know the receptivity/health of a uterus before performing an embryo transference [93]. In addition, the increasing knowledge about the role of EVs on reproductive events and ART efficiency will lead to the next challenge for the biology of reproduction field: the large-scale production of EVs to use them as additives to gametes and embryo culture media. This might be addressed easily in livestock animals by the use of animal's remains from abattoirs; where moreover this strategy is in accordance with the three rules of animal welfare. Similarly, EVs can be obtained from the biological material originated in the daily practice of vet clinics, as it is the castration of domestic animals (i.e., cats and dogs). Nevertheless, both strategies are time consuming and it might present contaminations with other biological fluids. The future of EVs development as a real alternative tool to boost ART results involves the optimization of the oviduct-in-a-chip model prototype to culture in vitro embryos before transferring them to female recipients (See Figure 2).

Another possible therapeutic application of EVs for ART is to use it as cargo of information that the gametes or embryos are lacking in current ART protocols to succeed in fertilization and embryo development. It has been shown that EVs can be loaded with the desired information by electroporation [94]. Then, EVs could be preloaded with the biological deficient information found for instance in proteins of sperm of asthenozoospermic men or with those miRNAs transported by sperm that have shown to enhance embryo development.

The study of EVs for ART should also contemplate possible adverse effects of EVs on specific ART outcome. EVs overdose presented detrimental effects over embryo development due to elevated levels of ammonium that might be correlated to the EVs protein cargo in mice [81]. Negative effects of OVS overdose were also found when these EVs were used for canine oocyte IVM (less oocytes arrived to MII phase), and this effect was associated to higher levels of miR-375 [52], a microRNA that has been linked to poor oocyte quality in aged mares [56]. Consequently, EVs titration obtained from every subsection of the reproductive tract must be performed, and their effects on gametes and embryos must be determined before the universalization of the use of EVs protocols. In addition, these protocols must be adapted to the specific species under study.

4. Concluding Remarks

Although this review focused on EVs used to enhance ART outcome in mammals, it should be noted that EVs have also drawn the attention of the avian reproduction sector as it has been recently postulated that uterosomes might have a role on avian sperm survival [95]. All the information described above should be considered for the development of improved media such as gamete collection, IVF, and embryo culture media used in vitro for ART. Besides the supplementation of media with specific EVs, the possibility of re-use EVs should be explored. For instance, once an in vitro embryo culture is finalized, the EVs contained in the media could be recovered, isolated and used as additive for a new culture media in those females with a record of low oocytes and embryo quality.

We expect that researchers working in the reproductive field will increase their interest in the study and use of EVs. This is a fundamental step for the development of new tools that could improve ART outcomes. From our point of view, the broad applications of EVs will have a lasting impact on the field of reproductive biotechnology.

Author Contributions: Conceptualization, M.G.G., A.J.S. and D.M.-H.; writing—original draft preparation, M.G.G. and D.M.-H.; writing—review and editing, A.J.S., L.G.-F., P.F.O. and M.G.A. All authors have read and agreed to the published version of the manuscript.

Funding: David Martín-Hidalgo was recipient of a post-doctoral grant from the Government of Extremadura (Spain) and "Fondo Social Europeo"; Reference: PO17020. L. González-Fernández was supported by the regional grant "Atracción y retorno de talento investigador a Centros de I+D+i pertenecientes al Sistema Extremeño de Ciencia, Tecnología e Innovación" from "Junta de Extremadura" (Spain); Reference: TA18008.

Acknowledgments: Figures were made using www.biorender.com software.

Conflicts of Interest: The authors declare no conflict of interest.

References

1. Dahlen, C.; Larson, J.; Lamb, G.C. Impacts of reproductive technologies on beef production in the United States. *Adv. Exp. Med. Biol.* **2014**, *752*, 97–114. [CrossRef] [PubMed]
2. Bortolozzo, F.P.; Menegat, M.B.; Mellagi, A.P.; Bernardi, M.L.; Wentz, I. New Artificial Insemination Technologies for Swine. *Reprod. Domest. Anim.* **2015**, *50*, 80–84. [CrossRef] [PubMed]
3. Austin, C.R. Observations on the penetration of the sperm in the mammalian egg. *Aust. J. Sci. Res. B* **1951**, *4*, 581–596. [CrossRef] [PubMed]
4. Chang, M.C. Fertilizing capacity of spermatozoa deposited into the fallopian tubes. *Nature* **1951**, *168*, 697–698. [CrossRef]
5. Austin, C.R. The capacitation of the mammalian sperm. *Nature* **1952**, *170*, 326. [CrossRef]
6. Gervasi, M.G.; Visconti, P.E. Chang's meaning of capacitation: A molecular perspective. *Mol. Reprod. Dev.* **2016**, *83*, 860–874. [CrossRef]
7. Toyoda, Y.; Yokoyama, M.; Hosi, T. Studies on the fertilization of mouse eggs in vitro. *J. STAGE* **1971**, *16*, 147–157. [CrossRef]

8. Chang, M.C. Fertilization of rabbit ova in vitro. *Nature* **1959**, *184*, 466–467. [CrossRef]
9. Yanagimachi, R.; Chang, M.C. Fertilization of hamster eggs in vitro. *Nature* **1963**, *200*, 281–282. [CrossRef]
10. Rizos, D.; Clemente, M.; Bermejo-Alvarez, P.; de La Fuente, J.; Lonergan, P.; Gutiérrez-Adán, A. Consequences of in vitro culture conditions on embryo development and quality. *Reprod. Domest. Anim.* **2008**, *43*, 44–50. [CrossRef]
11. Rizos, D.; Ward, F.; Duffy, P.; Boland, M.P.; Lonergan, P. Consequences of bovine oocyte maturation, fertilization or early embryo development in vitro versus in vivo: Implications for blastocyst yield and blastocyst quality. *Mol. Reprod. Dev.* **2002**, *61*, 234–248. [CrossRef] [PubMed]
12. Leemans, B.; Gadella, B.M.; Stout, T.A.; De Schauwer, C.; Nelis, H.; Hoogewijs, M.; Van Soom, A. Why doesn't conventional IVF work in the horse? The equine oviduct as a microenvironment for capacitation/fertilization. *Reproduction* **2016**, *152*, R233–r245. [CrossRef] [PubMed]
13. Zhu, J.; Moawad, A.R.; Wang, C.Y.; Li, H.F.; Ren, J.Y.; Dai, Y.F. Advances in in vitro production of sheep embryos. *Int. J. Vet. Sci. Med.* **2018**, *6*, S15–s26. [CrossRef] [PubMed]
14. Théry, C.; Zitvogel, L.; Amigorena, S. Exosomes: Composition, biogenesis and function. *Nat. Rev. Immunol.* **2002**, *2*, 569–579. [CrossRef]
15. Colombo, M.; Raposo, G.; Thery, C. Biogenesis, secretion, and intercellular interactions of exosomes and other extracellular vesicles. *Annu. Rev. Cell. Dev. Biol.* **2014**, *30*, 255–289. [CrossRef]
16. Braun, R. Packaging paternal chromosomes with protamine. *Nat. Genet.* **2001**, *28*, 10–12. [CrossRef]
17. Samanta, L.; Swain, N.; Ayaz, A.; Venugopal, V.; Agarwal, A. Post-Translational Modifications in sperm Proteome: The Chemistry of Proteome diversifications in the Pathophysiology of male factor infertility. *Biochim. Biophys. Acta* **2016**, *1860*, 1450–1465. [CrossRef]
18. Caballero, J.; Frenette, G.; Sullivan, R. Post testicular sperm maturational changes in the bull: Important role of the epididymosomes and prostasomes. *Vet. Med. Int.* **2010**, *2011*, 757194. [CrossRef]
19. Al-Dossary, A.A.; Bathala, P.; Caplan, J.L.; Martin-DeLeon, P.A. Oviductosome-Sperm Membrane Interaction in Cargo Delivery: Detection of fusion and underlying molecular players using three-dimensional super-resolution structure illumination microscopy (SR-SIM). *J. Biol. Chem.* **2015**, *290*, 17710–17723. [CrossRef]
20. Arienti, G.; Carlini, E.; Palmerini, C.A. Fusion of human sperm to prostasomes at acidic pH. *J. Membr. Biol.* **1997**, *155*, 89–94. [CrossRef]
21. Murdica, V.; Giacomini, E.; Makieva, S.; Zarovni, N.; Candiani, M.; Salonia, A.; Vago, R.; Viganò, P. In vitro cultured human endometrial cells release extracellular vesicles that can be uptaken by spermatozoa. *Sci. Rep.* **2020**, *10*, 8856. [CrossRef] [PubMed]
22. Gervasi, M.G.; Visconti, P.E. Molecular changes and signaling events occurring in spermatozoa during epididymal maturation. *Andrology* **2017**, *5*, 204–218. [CrossRef] [PubMed]
23. Trigg, N.A.; Eamens, A.L.; Nixon, B. The contribution of epididymosomes to the sperm small RNA profile. *Reproduction* **2019**, *157*, R209–r223. [CrossRef] [PubMed]
24. Arienti, G.; Carlini, E.; Nicolucci, A.; Cosmi, E.V.; Santi, F.; Palmerini, C.A. The motility of human spermatozoa as influenced by prostasomes at various pH levels. *Biol. Cell.* **1999**, *91*, 51–54. [CrossRef] [PubMed]
25. Pons-Rejraji, H.; Artonne, C.; Sion, B.; Brugnon, F.; Canis, M.; Janny, L.; Grizard, G. Prostasomes: Inhibitors of capacitation and modulators of cellular signalling in human sperm. *Int. J. Androl.* **2011**, *34*, 568–580. [CrossRef] [PubMed]
26. Franchi, A.; Cubilla, M.; Guidobaldi, H.A.; Bravo, A.A.; Giojalas, L.C. Uterosome-like vesicles prompt human sperm fertilizing capability. *Mol. Hum. Reprod.* **2016**, *22*, 833–841. [CrossRef]
27. Weigel Muñoz, M.; Battistone, M.A.; Carvajal, G.; Maldera, J.A.; Curci, L.; Torres, P.; Lombardo, D.; Pignataro, O.P.; Da Ros, V.G.; Cuasnicú, P.S. Influence of the genetic background on the reproductive phenotype of mice lacking Cysteine-Rich Secretory Protein 1 (CRISP1). *Biol. Reprod.* **2018**, *99*, 373–383. [CrossRef]
28. Park, K.H.; Kim, B.J.; Kang, J.; Nam, T.S.; Lim, J.M.; Kim, H.T.; Park, J.K.; Kim, Y.G.; Chae, S.W.; Kim, U.H. Ca^{2+} signaling tools acquired from prostasomes are required for progesterone-induced sperm motility. *Sci. Signal.* **2011**, *4*, ra31. [CrossRef]
29. Fereshteh, Z.; Bathala, P.; Galileo, D.S.; Martin-DeLeon, P.A. Detection of extracellular vesicles in the mouse vaginal fluid: Their delivery of sperm proteins that stimulate capacitation and modulate fertility. *J. Cell. Physiol.* **2019**, *234*, 12745–12756. [CrossRef]

30. Siciliano, L.; Marcianò, V.; Carpino, A. Prostasome-like vesicles stimulate acrosome reaction of pig spermatozoa. *Reprod. Biol. Endocrinol.* **2008**, *6*, 5. [CrossRef]
31. Du, J.; Shen, J.; Wang, Y.; Pan, C.; Pang, W.; Diao, H.; Dong, W. Boar seminal plasma exosomes maintain sperm function by infiltrating into the sperm membrane. *Oncotarget* **2016**, *7*, 58832–58847. [CrossRef] [PubMed]
32. Aalberts, M.; Sostaric, E.; Wubbolts, R.; Wauben, M.W.; Nolte-'t Hoen, E.N.; Gadella, B.M.; Stout, T.A.; Stoorvogel, W. Spermatozoa recruit prostasomes in response to capacitation induction. *Biochim. Biophys. Acta* **2013**, *1834*, 2326–2335. [CrossRef] [PubMed]
33. Rowlison, T.; Ottinger, M.A.; Comizzoli, P. Key factors enhancing sperm fertilizing ability are transferred from the epididymis to the spermatozoa via epididymosomes in the domestic cat model. *J. Assist. Reprod. Genet.* **2018**, *35*, 221–228. [CrossRef] [PubMed]
34. Ferraz, M.d.A.M.M.; Carothers, A.; Dahal, R.; Noonan, M.J.; Songsasen, N. Oviductal extracellular vesicles interact with the spermatozoon's head and mid-piece and improves its motility and fertilizing ability in the domestic cat. *Sci. Rep.* **2019**, *9*, 1–12. [CrossRef] [PubMed]
35. Qamar, A.Y.; Fang, X.; Kim, M.J.; Cho, J. Improved Post-Thaw Quality of Canine Semen after Treatment with Exosomes from Conditioned Medium of Adipose-Derived Mesenchymal Stem Cells. *Animals* **2019**, *9*, 865. [CrossRef] [PubMed]
36. Franchi, A.; Moreno-Irusta, A.; Dominguez, E.M.; Adre, A.J.; Giojalas, L.C. Extracellular vesicles from oviductal isthmus and ampulla stimulate the induced acrosome reaction and signaling events associated with capacitation in bovine spermatozoa. *J. Cell. Biochem.* **2020**, *121*, 2877–2888. [CrossRef] [PubMed]
37. Visconti, P.E.; Bailey, J.L.; Moore, G.D.; Pan, D.; Olds-Clarke, P.; Kopf, G.S. Capacitation of mouse spermatozoa. I. Correlation between the capacitation state and protein tyrosine phosphorylation. *Development* **1995**, *121*, 1129–1137.
38. Yanagimachi, R. The movement of golden hamster spermatozoa before and after capacitation. *J. Reprod. Fertil.* **1970**, *23*, 193–196. [CrossRef]
39. Arienti, G.; Carlini, E.; De Cosmo, A.M.; Di Profio, P.; Palmerini, C.A. Prostasome-like particles in stallion semen. *Biol. Reprod.* **1998**, *59*, 309–313. [CrossRef]
40. Arienti, G.; Carlini, E.; Polci, A.; Cosmi, E.V.; Palmerini, C.A. Fatty acid pattern of human prostasome lipid. *Arch. Biochem. Biophys.* **1998**, *358*, 391–395. [CrossRef]
41. Carlini, E.; Palmerini, C.A.; Cosmi, E.V.; Arienti, G. Fusion of sperm with prostasomes: Effects on membrane fluidity. *Arch. Biochem. Biophys.* **1997**, *343*, 6–12. [CrossRef] [PubMed]
42. Kravets, F.G.; Lee, J.; Singh, B.; Trocchia, A.; Pentyala, S.N.; Khan, S.A. Prostasomes: Current concepts. *Prostate* **2000**, *43*, 169–174. [CrossRef]
43. Murdica, V.; Giacomini, E.; Alteri, A.; Bartolacci, A.; Cermisoni, G.C.; Zarovni, N.; Papaleo, E.; Montorsi, F.; Salonia, A.; Viganò, P.; et al. Seminal plasma of men with severe asthenozoospermia contain exosomes that affect spermatozoa motility and capacitation. *Fertil. Steril.* **2019**, *111*, 897–908. [CrossRef] [PubMed]
44. Maldera, J.A.; Weigel Muñoz, M.; Chirinos, M.; Busso, D.; Raffo, F.G.E.; Battistone, M.A.; Blaquier, J.A.; Larrea, F.; Cuasnicu, P.S. Human fertilization: Epididymal hCRISP1 mediates sperm-zona pellucida binding through its interaction with ZP3. *Mol. Hum. Reprod.* **2014**, *20*, 341–349. [CrossRef] [PubMed]
45. Patel, R.; Al-Dossary, A.A.; Stabley, D.L.; Barone, C.; Galileo, D.S.; Strehler, E.E.; Martin-DeLeon, P.A. Plasma membrane Ca^{2+}-ATPase 4 in murine epididymis: Secretion of splice variants in the luminal fluid and a role in sperm maturation. *Biol. Reprod.* **2013**, *89*, 6. [CrossRef] [PubMed]
46. Andrews, R.E.; Galileo, D.S.; Martin-DeLeon, P.A. Plasma membrane Ca^{2+}-ATPase 4: Interaction with constitutive nitric oxide synthases in human sperm and prostasomes which carry Ca^{2+}/CaM-dependent serine kinase. *Mol. Hum. Reprod.* **2015**, *21*, 832–843. [CrossRef] [PubMed]
47. Alvau, A.; Battistone, M.A.; Gervasi, M.G.; Navarrete, F.A.; Xu, X.; Sánchez-Cárdenas, C.; De la Vega-Beltran, J.L.; Da Ros, V.G.; Greer, P.A.; Darszon, A.; et al. The tyrosine kinase FER is responsible for the capacitation-associated increase in tyrosine phosphorylation in murine sperm. *Development* **2016**, *143*, 2325–2333. [CrossRef]
48. Bathala, P.; Fereshteh, Z.; Li, K.; Al-Dossary, A.A.; Galileo, D.S.; Martin-DeLeon, P.A. Oviductal extracellular vesicles (oviductosomes, OVS) are conserved in humans: Murine OVS play a pivotal role in sperm capacitation and fertility. *Mol. Hum. Reprod.* **2018**, *24*, 143–157. [CrossRef]

49. Reilly, J.N.; McLaughlin, E.A.; Stanger, S.J.; Anderson, A.L.; Hutcheon, K.; Church, K.; Mihalas, B.P.; Tyagi, S.; Holt, J.E.; Eamens, A.L.; et al. Characterisation of mouse epididymosomes reveals a complex profile of microRNAs and a potential mechanism for modification of the sperm epigenome. *Sci. Rep.* **2016**, *6*, 31794. [CrossRef]
50. Fereshteh, Z.; Schmidt, S.A.; Al-Dossary, A.A.; Accerbi, M.; Arighi, C.; Cowart, J.; Song, J.L.; Green, P.J.; Choi, K.; Yoo, S.; et al. Murine Oviductosomes (OVS) microRNA profiling during the estrous cycle: Delivery of OVS-borne microRNAs to sperm where miR-34c-5p localizes at the centrosome. *Sci. Rep.* **2018**, *8*, 16094. [CrossRef]
51. Aguila, L.; Felmer, R.; Arias, M.E.; Navarrete, F.; Martin-Hidalgo, D.; Lee, H.C.; Visconti, P.; Fissore, R. Defective sperm head decondensation undermines the success of ICSI in the bovine. *Reproduction* **2017**, *154*, 307–318. [CrossRef] [PubMed]
52. Lange-Consiglio, A.; Perrini, C.; Albini, G.; Modina, S.; Lodde, V.; Orsini, E.; Esposti, P.; Cremonesi, F. Oviductal microvesicles and their effect on in vitro maturation of canine oocytes. *Reproduction* **2017**, *154*, 167–180. [CrossRef] [PubMed]
53. Luvoni, G.C.; Chigioni, S.; Allievi, E.; Macis, D. Factors involved in vivo and in vitro maturation of canine oocytes. *Theriogenology* **2005**, *63*, 41–59. [CrossRef] [PubMed]
54. Lee, S.H.; Oh, H.J.; Kim, M.J.; Lee, B.C. Exosomes derived from oviduct cells mediate the EGFR/MAPK signaling pathway in cumulus cells. *J. Cell. Physiol.* **2020**, *235*, 1386–1404. [CrossRef] [PubMed]
55. Lee, S.H.; Oh, H.J.; Kim, M.J.; Lee, B.C. Canine oviductal exosomes improve oocyte development via EGFR/MAPK signaling pathway. *Reproduction* **2020**, *160*, 613–625. [CrossRef] [PubMed]
56. da Silveira, J.C.; Veeramachaneni, D.N.; Winger, Q.A.; Carnevale, E.M.; Bouma, G.J. Cell-secreted vesicles in equine ovarian follicular fluid contain miRNAs and proteins: A possible new form of cell communication within the ovarian follicle. *Biol. Reprod.* **2012**, *86*, 71. [CrossRef] [PubMed]
57. Santonocito, M.; Vento, M.; Guglielmino, M.R.; Battaglia, R.; Wahlgren, J.; Ragusa, M.; Barbagallo, D.; Borzì, P.; Rizzari, S.; Maugeri, M.; et al. Molecular characterization of exosomes and their microRNA cargo in human follicular fluid: Bioinformatic analysis reveals that exosomal microRNAs control pathways involved in follicular maturation. *Fertil. Steril.* **2014**, *102*, 1751–1761. [CrossRef]
58. Hung, W.T.; Navakanitworakul, R.; Khan, T.; Zhang, P.; Davis, J.S.; McGinnis, L.K.; Christenson, L.K. Stage-specific follicular extracellular vesicle uptake and regulation of bovine granulosa cell proliferation. *Biol. Reprod.* **2017**, *97*, 644–655. [CrossRef]
59. da Silveira, J.C.; Andrade, G.M.; Del Collado, M.; Sampaio, R.V.; Sangalli, J.R.; Silva, L.A.; Pinaffi, F.V.L.; Jardim, I.B.; Cesar, M.C.; Nogueira, M.F.G.; et al. Supplementation with small-extracellular vesicles from ovarian follicular fluid during in vitro production modulates bovine embryo development. *PLoS ONE* **2017**, *12*, e0179451. [CrossRef]
60. da Silveira, J.C.; Winger, Q.A.; Bouma, G.J.; Carnevale, E.M. Effects of age on follicular fluid exosomal microRNAs and granulosa cell transforming growth factor-β signalling during follicle development in the mare. *Reprod. Fertil. Dev.* **2015**, *27*, 897–905. [CrossRef]
61. de Ávila, A.; Bridi, A.; Andrade, G.M.; Del Collado, M.; Sangalli, J.R.; Nociti, R.P.; da Silva Junior, W.A.; Bastien, A.; Robert, C.; Meirelles, F.V.; et al. Estrous cycle impacts microRNA content in extracellular vesicles that modulate bovine cumulus cell transcripts during in vitro maturation†. *Biol. Reprod.* **2020**, *102*, 362–375. [CrossRef] [PubMed]
62. Laezer, I.; Palma-Vera, S.E.; Liu, F.; Frank, M.; Trakooljul, N.; Vernunft, A.; Schoen, J.; Chen, S. Dynamic profile of EVs in porcine oviductal fluid during the periovulatory period. *Reproduction* **2020**, *159*, 371–382. [CrossRef] [PubMed]
63. Navakanitworakul, R.; Hung, W.T.; Gunewardena, S.; Davis, J.S.; Chotigeat, W.; Christenson, L.K. Characterization and Small RNA Content of Extracellular Vesicles in Follicular Fluid of Developing Bovine Antral Follicles. *Sci. Rep.* **2016**, *6*, 25486. [CrossRef] [PubMed]
64. Ferraz, M.d.A.M.M.; Fujihara, M.; Nagashima, J.B.; Noonan, M.J.; Inoue-Murayama, M.; Songsasen, N. Follicular extracellular vesicles enhance meiotic resumption of domestic cat vitrified oocytes. *Sci. Rep.* **2020**, *10*, 8619. [CrossRef]
65. O'Doherty, E.M.; Wade, M.G.; Hill, J.L.; Boland, M.P. Effects of culturing bovine oocytes either singly or in groups on development to blastocysts. *Theriogenology* **1997**, *48*, 161–169. [CrossRef]

66. Cortezzi, S.S.; Garcia, J.S.; Ferreira, C.R.; Braga, D.P.; Figueira, R.C.; Iaconelli, A., Jr.; Souza, G.H.; Borges, E., Jr.; Eberlin, M.N. Secretome of the preimplantation human embryo by bottom-up label-free proteomics. *Anal. Bioanal. Chem.* **2011**, *401*, 1331–1339. [CrossRef]
67. Katz-Jaffe, M.G.; Schoolcraft, W.B.; Gardner, D.K. Analysis of protein expression (secretome) by human and mouse preimplantation embryos. *Fertil. Steril.* **2006**, *86*, 678–685. [CrossRef]
68. Saadeldin, I.M.; Kim, S.J.; Choi, Y.B.; Lee, B.C. Improvement of cloned embryos development by co-culturing with parthenotes: A possible role of exosomes/microvesicles for embryos paracrine communication. *Cell. Reprogram.* **2014**, *16*, 223–234. [CrossRef]
69. Alcântara-Neto, A.S.; Fernandez-Rufete, M.; Corbin, E.; Tsikis, G.; Uzbekov, R.; Garanina, A.S.; Coy, P.; Almiñana, C.; Mermillod, P. Oviduct fluid extracellular vesicles regulate polyspermy during porcine in vitro fertilisation. *Reprod. Fertil. Dev.* **2020**, *32*, 409–418. [CrossRef]
70. Alminana, C.; Corbin, E.; Tsikis, G.; Alcantara-Neto, A.S.; Labas, V.; Reynaud, K.; Galio, L.; Uzbekov, R.; Garanina, A.S.; Druart, X.; et al. Oviduct extracellular vesicles protein content and their role during oviduct-embryo cross-talk. *Reproduction* **2017**, *154*, 153–168. [CrossRef]
71. Alminana, C.; Tsikis, G.; Labas, V.; Uzbekov, R.; da Silveira, J.C.; Bauersachs, S.; Mermillod, P. Deciphering the oviductal extracellular vesicles content across the estrous cycle: Implications for the gametes-oviduct interactions and the environment of the potential embryo. *BMC Genom.* **2018**, *19*, 622. [CrossRef] [PubMed]
72. Wennemuth, G.; Babcock, D.F.; Hille, B. Calcium clearance mechanisms of mouse sperm. *J. Gen. Physiol.* **2003**, *122*, 115–128. [CrossRef] [PubMed]
73. Gervasi, M.G.; Rapanelli, M.; Ribeiro, M.L.; Farina, M.; Billi, S.; Franchi, A.M.; Perez Martinez, S. The endocannabinoid system in bull sperm and bovine oviductal epithelium: Role of anandamide in sperm-oviduct interaction. *Reproduction* **2009**, *137*, 403–414. [CrossRef] [PubMed]
74. Walter, I. Culture of bovine oviduct epithelial cells (BOEC). *Anat. Rec.* **1995**, *243*, 347–356. [CrossRef] [PubMed]
75. Lopera-Vásquez, R.; Hamdi, M.; Fernandez-Fuertes, B.; Maillo, V.; Beltrán-Breña, P.; Calle, A.; Redruello, A.; López-Martín, S.; Gutierrez-Adán, A.; Yañez-Mó, M.; et al. Extracellular Vesicles from BOEC in In Vitro Embryo Development and Quality. *PLoS ONE* **2016**, *11*, e0148083. [CrossRef] [PubMed]
76. Lopera-Vasquez, R.; Hamdi, M.; Maillo, V.; Gutierrez-Adan, A.; Bermejo-Alvarez, P.; Ramírez, M.; Yáñez-Mó, M.; Rizos, D. Effect of bovine oviductal extracellular vesicles on embryo development and quality in vitro. *Reproduction* **2017**, *153*, 461–470. [CrossRef]
77. Kouba, A.J.; Abeydeera, L.R.; Alvarez, I.M.; Day, B.N.; Buhi, W.C. Effects of the porcine oviduct-specific glycoprotein on fertilization, polyspermy, and embryonic development in vitro. *Biol. Reprod.* **2000**, *63*, 242–250. [CrossRef]
78. Algarra, B.; Maillo, V.; Aviles, M.; Gutierrez-Adan, A.; Rizos, D.; Jimenez-Movilla, M. Effects of recombinant OVGP1 protein on in vitro bovine embryo development. *J. Reprod. Dev.* **2018**, *64*, 433–443. [CrossRef]
79. Ferraz, M.d.A.M.M.; Rho, H.S.; Hemerich, D.; Henning, H.H.W.; van Tol, H.T.A.; Hölker, M.; Besenfelder, U.; Mokry, M.; Vos, P.; Stout, T.A.E.; et al. An oviduct-on-a-chip provides an enhanced in vitro environment for zygote genome reprogramming. *Nat. Commun.* **2018**, *9*, 4934. [CrossRef]
80. Qu, P.; Luo, S.; Du, Y.; Zhang, Y.; Song, X.; Yuan, X.; Lin, Z.; Li, Y.; Liu, E. Extracellular vesicles and melatonin benefit embryonic develop by regulating reactive oxygen species and 5-methylcytosine. *J. Pineal Res.* **2020**, *68*, e12635. [CrossRef]
81. Qu, P.; Zhao, Y.; Wang, R.; Zhang, Y.; Li, L.; Fan, J.; Liu, E. Extracellular vesicles derived from donor oviduct fluid improved birth rates after embryo transfer in mice. *Reprod. Fertil. Dev.* **2019**, *31*, 324–332. [CrossRef] [PubMed]
82. Blázquez, R.; Sánchez-Margallo, F.M.; Álvarez, V.; Matilla, E.; Hernández, N.; Marinaro, F.; Gómez-Serrano, M.; Jorge, I.; Casado, J.G.; Macías-García, B. Murine embryos exposed to human endometrial MSCs-derived extracellular vesicles exhibit higher VEGF/PDGF AA release, increased blastomere count and hatching rates. *PLoS ONE* **2018**, *13*, e0196080. [CrossRef] [PubMed]
83. Marinaro, F.; Macias-Garcia, B.; Sanchez-Margallo, F.M.; Blazquez, R.; Alvarez, V.; Matilla, E.; Hernandez, N.; Gomez-Serrano, M.; Jorge, I.; Vazquez, J.; et al. Extracellular vesicles derived from endometrial human mesenchymal stem cells enhance embryo yield and quality in an aged murine model. *Biol. Reprod.* **2019**, *100*, 1180–1192. [CrossRef] [PubMed]

84. Marinaro, F.; Pericuesta, E.; Sanchez-Margallo, F.M.; Casado, J.G.; Alvarez, V.; Matilla, E.; Hernandez, N.; Blazquez, R.; Gonzalez-Fernandez, L.; Gutierrez-Adan, A.; et al. Extracellular vesicles derived from endometrial human mesenchymal stem cells improve IVF outcome in an aged murine model. *Reprod. Domest. Anim.* **2018**, *53*, 46–49. [CrossRef] [PubMed]
85. Liu, C.; Yin, H.; Jiang, H.; Du, X.; Wang, C.; Liu, Y.; Li, Y.; Yang, Z. Extracellular Vesicles Derived from Mesenchymal Stem Cells Recover Fertility of Premature Ovarian Insufficiency Mice and the Effects on their Offspring. *Cell Transplant.* **2020**, *29*. [CrossRef]
86. Wang, H.; Dey, S.K. Roadmap to embryo implantation: Clues from mouse models. *Nat. Rev. Genet.* **2006**, *7*, 185–199. [CrossRef]
87. Klohonatz, K.M.; Cameron, A.D.; Hergenreder, J.R.; da Silveira, J.C.; Belk, A.D.; Veeramachaneni, D.N.; Bouma, G.J.; Bruemmer, J.E. Circulating miRNAs as Potential Alternative Cell Signaling Associated with Maternal Recognition of Pregnancy in the Mare. *Biol. Reprod.* **2016**, *95*, 124. [CrossRef]
88. Vilella, F.; Moreno-Moya, J.M.; Balaguer, N.; Grasso, A.; Herrero, M.; Martínez, S.; Marcilla, A.; Simón, C. Hsa-miR-30d, secreted by the human endometrium, is taken up by the pre-implantation embryo and might modify its transcriptome. *Development* **2015**, *142*, 3210–3221. [CrossRef]
89. Burns, G.W.; Brooks, K.E.; Spencer, T.E. Extracellular Vesicles Originate from the Conceptus and Uterus During Early Pregnancy in Sheep. *Biol. Reprod.* **2016**, *94*, 56. [CrossRef]
90. Pallinger, E.; Bognar, Z.; Bodis, J.; Csabai, T.; Farkas, N.; Godony, K.; Varnagy, A.; Buzas, E.; Szekeres-Bartho, J. A simple and rapid flow cytometry-based assay to identify a competent embryo prior to embryo transfer. *Sci. Rep.* **2017**, *7*, 39927. [CrossRef]
91. Mellisho, E.A.; Briones, M.A.; Velásquez, A.E.; Cabezas, J.; Castro, F.O.; Rodríguez-Álvarez, L. Extracellular vesicles secreted during blastulation show viability of bovine embryos. *Reproduction* **2019**, *158*, 477–492. [CrossRef] [PubMed]
92. Dissanayake, K.; Nõmm, M.; Lättekivi, F.; Ressaissi, Y.; Godakumara, K.; Lavrits, A.; Midekessa, G.; Viil, J.; Bæk, R.; Jørgensen, M.M.; et al. Individually cultured bovine embryos produce extracellular vesicles that have the potential to be used as non-invasive embryo quality markers. *Theriogenology* **2020**, *149*, 104–116. [CrossRef] [PubMed]
93. Giacomini, E.; Makieva, S.; Murdica, V.; Vago, R.; Viganó, P. Extracellular vesicles as a potential diagnostic tool in assisted reproduction. *Curr. Opin. Obstet. Gynecol.* **2020**, *32*, 179–184. [CrossRef] [PubMed]
94. Pomatto, M.A.C.; Bussolati, B.; D'Antico, S.; Ghiotto, S.; Tetta, C.; Brizzi, M.F.; Camussi, G. Improved Loading of Plasma-Derived Extracellular Vesicles to Encapsulate Antitumor miRNAs. *Mol. Ther. Methods Clin. Dev.* **2019**, *13*, 133–144. [CrossRef]
95. Riou, C.; Brionne, A.; Cordeiro, L.; Harichaux, G.; Gargaros, A.; Labas, V.; Gautron, J.; Gérard, N. Avian uterine fluid proteome: Exosomes and biological processes potentially involved in sperm survival. *Mol. Reprod. Dev.* **2020**, *87*, 454–470. [CrossRef]

Publisher's Note: MDPI stays neutral with regard to jurisdictional claims in published maps and institutional affiliations.

MDPI

Article

Selection of Boar Sperm by Reproductive Biofluids as Chemoattractants

Luis Alberto Vieira [1,2], Alessia Diana [1,2], Cristina Soriano-Úbeda [1,2,3] and Carmen Matás [1,2,*]

1 Department of Physiology, Faculty of Veterinary Science, International Excellence Campus for Higher Education and Research Campus Mare Nostrum, University of Murcia, 30100 Murcia, Spain; tuncan24@yahoo.es (L.A.V.); alessia.diana@um.es (A.D.); cmsu1@um.es (C.S.-Ú.)
2 Institute for Biomedical Research of Murcia (IMIB-Arrixaca), 30120 Murcia, Spain
3 Department of Veterinary and Animal Sciences, University of Massachusetts, Amherst, MA 01002, USA
* Correspondence: cmatas@um.es

Simple Summary: Both in natural breeding and some assisted reproduction technologies, spermatozoa are deposited into the uterus. The journey the spermatozoa must take from the place of semen deposition to the fertilization site is long, hostile, and selective of the best spermatozoa. For the fertilization to succeed, spermatozoa are guided by chemical stimuli (chemoattractants) to the fertilization site, mainly secreted by the oocyte, cumulus cells, and other substances poured into the oviduct in the periovulatory period. This work studied some sources of chemotactic factors and their action on spermatozoa functionality in vitro, including the fertility. A special chemotactic chamber for spermatozoa selection was designed which consists of two wells communicated by a tube. The spermatozoa are deposited in well A, and the chemoattractants in well B. This study focuses on the use of follicular fluid (FF), periovulatory oviductal fluid (pOF), conditioned medium from the in vitro maturation of oocytes (CM), and progesterone (P4) as chemoattractants to sperm. The chemotactic potential of these substances is also investigated as related to their action on CatSper which is a calcium channel in the spermatozoa known to be sensitive to chemoattractants and essential for motility.

Abstract: Chemotaxis is a spermatozoa guidance mechanism demonstrated in vitro in several mammalian species including porcine. This work focused on follicular fluid (FF), periovulatory oviductal fluid (pOF), the medium surrounding oocytes during in vitro maturation (conditioned medium; CM), progesterone (P4), and the combination of those biofluids (Σ) as chemotactic agents and modulators of spermatozoa fertility in vitro. A chemotaxis chamber was designed consisting of two independent wells, A and B, connected by a tube. The spermatozoa are deposited in well A, and the chemoattractants in well B. The concentrations of biofluids that attracted a higher proportion of spermatozoa to well B were 0.25% FF, 0.25% OF, 0.06% CM, 10 pM P4 and 0.25% of a combination of biofluids (Σ2), which attracted between 3.3 and 12.3% of spermatozoa ($p < 0.05$). The motility of spermatozoa recovered in well B was determined and the chemotactic potential when the sperm calcium channel CatSper was inhibited, which significantly reduced the % of spermatozoa attracted ($p < 0.05$). Regarding the in vitro fertility, the spermatozoa attracted by FF produced higher rates of penetration of oocytes and development of expanded blastocysts. In conclusion, porcine reproductive biofluids show an in vitro chemotactic effect on spermatozoa and modulate their fertilizing potential.

Keywords: sperm selection; chemotaxis; reproductive fluids; IVF; IVC; porcine

Citation: Vieira, L.A.; Diana, A.; Soriano-Úbeda, C.; Matás, C. Selection of Boar Sperm by Reproductive Biofluids as Chemoattractants. *Animals* **2021**, *11*, 53. https://doi.org/10.3390/ani11010053

Received: 5 December 2020
Accepted: 21 December 2020
Published: 30 December 2020

1. Introduction

Assisted reproductive technologies (ART) have been widely used in humans and farm animals in the last decades. Even though the female plays a critical role in the success of an ART program, we cannot ignore that the spermatozoa preparation has been essential to selecting the best and most capable spermatozoa population in an ejaculated sample [1], thus contributing to the increasing success of ART. However, the efficiency of many ART is still low, in part due to spermatozoa quality which directly influences fertilization and embryo development.

In standard methodologies of in vitro fertilization (IVF) in mammals, both male and female gametes are co-cultured without barriers regulating their interaction which causes a high number of spermatozoa to be simultaneously at the place of fertilization. Consequently, increased rates of polyspermy (more than one spermatozoon penetrating the oocyte) are achieved, compromising the embryo development. This is of particular interest to the porcine species, representing one of the major detrimental factors for an efficient and successful porcine IVF (reviewed by Romar et al. [2]).

There are several mechanisms within the female reproductive tract that regulate the number and quality of spermatozoa reaching the fertilization site to reduce the risk of polyspermy [3]. The in vitro simulation of the physico-chemical conditions in the oviduct at the time of fertilization can reduce the high incidence of polyspermy in porcine animals [4].

Under in vivo conditions, only progressively motile spermatozoa will move toward the oocytes. Current sperm selection assays based on sperm chemotaxis towards progesterone (P4) provide a sperm subpopulation enriched with spermatozoa that are capacitated, with intact DNA and low levels of oxidative stress [1]. Since the selected subpopulation of spermatozoa are in an optimum physiological state, it would be reasonable to suggest that the application of spermatozoa selection methods may improve the efficiency of the current ART. Despite the fact that P4 has been one of the most studied chemoattractant agents, other substances have also been shown to have this effect such as atrial natriuretic peptide (ANP), heparin, adrenaline, oxytocin, calcitonin, acetylcholine, and nitric oxide [5,6]. On the other hand, it has been suggested that estradiol (E2), cyclic AMP (cAMP) and cyclic GMP (cGMP) could be essential for chemotaxis because they increase the levels of intracellular calcium (Ca^{2+}) in the sperm [7–9].

Villanueva-Diaz et al. [10], showed for the first time that crude human follicular fluid (FF) produced a chemical attraction for spermatozoa. FF increases the in vitro motility of spermatozoa especially since the original follicle is mature and stimulates capacitation and the acrosomal reaction of human sperm [11]. The presence of chemotactic agents has been shown in the FF of several species [5]. Therefore, FF facilitates spermatozoa reaching the fertilization site. Likewise, oviductal fluid (OF) and secretions of the cumulus cells (conditioned medium, CM) have been demonstrated to have chemotactic components [12–15].

In an in vitro chemotactic system, it is important to consider the optimum concentration at which a chemoattractant has an effect. A typical chemotaxis response is represented by a bell-shaped curve [12–14,16,17]. The present work analyzed the chemotactic effect of different biofluids (FF, OF, CM) and P4 individually or in combination on spermatozoa as well as their potential as sperm fertility modulators.

2. Material and Methods

2.1. Ethics Statement

The study was carried out in accordance with the Spanish Policy for Animal Protection RD 53/2013, which meets the European Union Directive 2010/63/UE on animal protection. All experimental protocols were approved by the Ethical Committee of Animal Experimentation of the University of Murcia and by the Animal Production Service of the Agriculture Department of the Region of Murcia (Spain) (ref. no. A13160609).

2.2. Reagents, Culture Media, and Solutions

All the chemicals used in this study were purchased from Sigma-Aldrich Química S.A. (Madrid, Spain) unless otherwise indicated. Tyrode's albumin lactate pyruvate medium (TALP; [18]) was supplemented with 1.10 mmol/L sodium pyruvate and 3 mg/mL bovine serum albumin (BSA; A9647). Dulbecco's PBS (DPBS) was supplemented with 1 mg/mL PVA and 0.005 mg/mL red phenol. North Carolina State University 37 medium (NCSU37; [19]) was supplemented with 0.57 mmol/L cysteine, 1 mmol/L dibutyryl cAMP, 5 μg/mL insulin, 50 μmol/L β-mercaptoethanol, 1 mmol/L glutamine, 10 IU/mL equine chorionic gonadotropin (eCG; Foligon; Intervet International BV, Boxmeer, Holland), 10 IU/mL human chorionic gonadotropin (hCG; Veterin Corion; Divasa Farmavic, Barcelona, Spain), and 10% (*v*/*v*) porcine FF [20]. A fixation solution of glutaraldehyde was prepared at 0.5% in phosphate-buffered saline (PBS). A staining solution of bisbenzimide (Hoechst 33342; Invitrogen, Thermo Fisher Scientific, Waltham, MA, USA, H1399) was prepared at 1% (*w*/*v*) in PBS. The P4 (P8783) solution was prepared in DMSO (D2650) at 1mg/mL and frozen at −20 °C until use. The solution of the CatSper channel inhibitor NNC 55-0396 (NNC; Cayman 17216) was prepared in DMSO at 8.8 mM and frozen at −20 °C until use.

2.3. Spermatozoa Collection and Preparation

Fresh ejaculated spermatozoa were obtained by the manual method from mature and tested fertility boars. Once in the laboratory, the spermatozoa were separated from the seminal plasma (SP) by centrifugation on a discontinuous density gradient (Percoll®, Pharmacia, Uppsala, Sweden) 45/90% (*v*/*v*) at 700× *g* for 30 min [21]. The pellet was resuspended in TALP medium previously equilibrated in a humified atmosphere of 5% CO_2 in air at 38.5°C and centrifuged at 700× *g* for 5 min. Finally, the pellet of spermatozoa was resuspended in 5 mL of fresh TALP medium and the concentration of spermatozoa/mL was determined.

2.4. In Vitro Maturation (IVM) of Oocytes

Ovaries were obtained from prepuberal gilts at the local slaughterhouse and transported to the laboratory in saline solution at 38.5 °C within 1 h after the death of the animals. Cumulus oocyte complexes (COCs) were collected from antral follicles (3–6 mm diameter) and in vitro matured as described previously [4].

2.5. Follicular Fluid (FF), Periovulatory Oviductal Fluid (pOF) and Conditioned Medium (CM) Collection and Preparation

The FF was obtained making a pool by aspirating the liquid content of antral follicles (3–6 mm diameter) of prepuberal gilts ovaries, as described previously [22]. Periovulatory oviductal fluid (pOF) was obtained from a pool of porcine oviducts with ovaries close to ovulation according to the classification of Carrasco et al. [23]. The conditioned medium (CM), consisting of the pooled secretions of the cumulus cells, was obtained by pippeting NCSU37 medium from wells where groups of 50 COCs had completed the second phase of IVM (without dbAMPc, eCG and hCG) [4]. After collection, FF, pOF and CM were centrifugated at 7000× *g* and 4 °C for 10 min. After centrifugation, the supernatants were collected discarding the cellular debris and/or mucus at the bottom of the centrifuge tubes. All fluids, FF, pOF, and CM were aliquoted and frozen at −20 °C until use, avoiding repeated freezing-thawing cycles.

2.6. HPLC-MS Analysis

The separation and analysis of samples were performed with a HPLC-MS system consisting of an Agilent 1290 Infinity II Series HPLC (Agilent Technologies, Santa Clara, CA, USA) equipped with an Automated Multisampler module and a High-Speed Binary Pump and connected to an Agilent 6550 Q-TOF Mass Spectrometer (Agilent Technologies, Santa Clara, CA, USA) using an Agilent Jet Stream Dual electrospray (AJS-Dual ESI) interface.

Standards or samples (20 uL) were injected onto an Agilent Zorbax Eclipse Plus C18 (2.1 × 100 mm, 1.8 um) HPLC column, at a flow rate of 0.4 mL/min. The column was equilibrated at 40 °C. Solvents A (MilliQ water + 0.1% formic acid) and B (acetonitrile + 0.1% formic acid) were used for the compound separation with the following elution program: 2 min at 3% B, linear gradient from 3 to 100% B in 9 min, and 1 min at 100% B.

The mass spectrometer was operated in the positive mode. Extracted ion chromatograms of the following compounds were analyzed: 273.1849 > 255.1749 m/z for β-estradiol (E2), 315.2319 > 109.0660 for P4, 330.0566 > 136.0623 for AMPc and 346.0547 > 152.05686 for GMPc.

2.7. Sperm Chemotaxis System

For this study, a new chemotaxis chamber for spermatozoa was designed (Figure S1). It consisted of two 500 μL wells with stoppers connected by a tube of 250 μm in diameter and 850 μm in length. Spermatozoa suspension in TALP were added to well A in a final concentration of 20×10^6 spermatozoa/mL. The chemoattractants under study were added to well B in TALP. This system was kept for 20 min in a humified atmosphere of 5% CO_2 in air at 38.5 °C. After that incubation, the percentage of spermatozoa recovered in well B were determined.

2.8. Spermatozoa Motility

Spermatozoa motility was evaluated by computer assisted semen analysis (CASA) using the ISAS® system (PROISER R + D S.L., Valencia, Spain), as protocolized in our laboratory [24]. The motion parameters determined into 3 different fields per sample were: the percentage of total motile spermatozoa (Mot, %), motile progressive spermatozoa (MotPro, %), curvilinear velocity (VCL, μm/s), straight line velocity (VSL, μm/s), average path velocity (VAP, μm/s), linearity of the curvilinear trajectory (LIN, ratio of VSL/VCL, %), straightness (STR, ratio of VSL/VAP, %), wobble of the curvilinear trajectory (WOB, ratio of VAP/VCL, %), amplitude of lateral head displacement (ALH, μm), and beat cross-frequency (BCF, Hz).

2.9. Spermatozoa Plasma Membrane Integrity

The plasma membrane integrity was analyzed by eosin-nigrosin staining as a reflection of spermatozoa viability, as described by Soriano-Úbeda et al. [24]. The percentage of membrane-intact spermatozoa (non-stained spermatozoa) were determined and considered viable spermatozoa.

2.10. In Vitro Fertilization (IVF) and Embryo Development (IVC)

The matured oocytes from IVM were mechanically denuded, washed in fresh TALP, and transferred in groups of 50 oocytes to 4-well plates (Nunc, Roskilde, Denmark) with 500 μL/well TALP. The insemination of oocytes was carried out giving a final concentration of 25×10^3 spermatozoa/mL. The spermatozoa for insemination were those recovered in well B of the chemotactic system. Eighteen hours after insemination, putative zygotes were fixed in a 0.5% glutaraldehyde solution, stained in a bisbenzimide solution (Hoechst 33342) and examined under an epifluorescence microscopy (Leica® DMR, USA), as it has been previously described [25]. The IVF parameters analyzed were the percentage of penetrated oocytes (Pen, %), percentage of monospermy of penetrated oocytes (Mon, %), mean number of spermatozoa per penetrated oocyte (Spz/O) and mean number of spermatozoa bound to zona pellucida (Spz/ZP). For the embryo development assessment, 18 h after insemination, putative zygotes were transferred to a culture dish for 7 days (168 h), as previously described [20]. For the first 48 h, the media used for IVC was NCSU23 [19]. At 48 h after the addition of spermatozoa to the chemotactic system, the cleavage was assessed under the stereomicroscope. On day 7 (168 h after the addition of spermatozoa to the chemotactic system), the blastocyst stage of development was assessed under the stereomicroscope and classified as in Bó and Mapletoft [26]. Blastocysts and expanded blastocysts were fixed and stained as described for putative zygotes, and the mean number of blastomeres was determined under an epifluorescence microscope (Leica® DMR, Richmond, IL, USA).

2.11. Experimental Design

The present study investigated the chemotactic ability of the female reproductive fluids (FF, pOF, CM) and P4 on the selection of porcine spermatozoa and the fertility of those selected spermatozoa (Figure 1). Firstly, the characterization in content of E2, P4, cAMP, and cGMP in each pool of biofluids used in the present work was carried out. The concentration of those components was measured twice by HPLC-MS in the pools of FF, pOF and CM, and the results are shown in Table 1. Once characterized, the same pools of biofluids were used in the two experiments of this work.

Table 1. Concentration of β-estradiol (E2), progesterone (P4), 3′,5′-cyclic adenosine monophosphate (cAMP), and guanosine 3′-5′-cyclic monophosphate (cGMP) in reproductive biofluids used as chemoattractants. HPLC-MS was performed twice in each pool of biofluids.

	FF (ng/mL)	pOF (ng/mL)	CM (ng/mL)
E2	0.71 ± 0.60	<0.01 ± 0.00	0.10 ± 0.10
P4	0.53 ± 0.01	0.05 ± 0.04	0.08 ± 0.05
cAMP	0.11 ± 0.04	0.28 ± 0.10	0.05 ± 0.04
cGMP	2.10 ± 0.80	0.39 ± 0.20	0.10 ± 0.08

FF: Follicular fluid; pOF: periovulatory oviductal fluid; CM: secretions of the cumulus cells. Results are expressed as mean ± standard deviation (SD).

Experiment 1. Spermatozoa selection by FF, pOF, CM, and P4

(1.1) Effective concentration of chemoattractants. Increasing concentrations of the chemoattractants (FF at 0.13, 0.25, 0.50, 1.00 and 1.50%; pOF at 0.13, 0.25, 0.50, 1.00 and 1.50%; CM at 0.03, 0.06, 0.13, 0.25 and 0.50%; P4 at 1.00, 2.50, 5.00, 7.50 and 10.00 pM) were added to well B of the chemotactic system. One experimental group in which there was no supplementation with any chemoattractant was included as control (C). Spermatozoa were added to the well A and incubated for 20 min. The percentage of spermatozoa recovered in well B at the end of the incubation period was determined, and the most effective concentration of each chemoattractant was established as the lowest concentration that attracted the highest proportion of spermatozoa. Four replicates were performed.

(1.2) Chemotactic potential of chemoattractants. Since all the chemoattractants coexist in combination in the female reproductive tract during the periovulatory stage, this work compared the effective concentration of the chemoattractants (FF, pOF, CM, and P4) and their possible synergy. For this purpose, a combination of the most effective concentrations of FF, pOF and CM were studied as the summation of the most effective concentrations of chemoattractants obtained in experiment 1.1 (Σ1 = 0.25% FF + 0.25% pOF + 0.06% CM). Due to the possible saturation of receptors [27], a second Σ group (Σ2) was included in which the total proportion of chemoattractants was limited to 0.25%. The proportion of each chemoattractant was maintained at 1/3 of the total 0.25% of biofluids in the chemotactic system (Σ2 = 1/3 of 0.25% FF + 1/3 of 0.25% OF + 1/3 of 0.06% CM). The chemotactic potential of chemoattractants and their combination was compared. A group in which no chemoattractants were added to well B was included as a control (C). Four replicates were performed.

(1.3) Motility of spermatozoa selected by chemoattractants. The possible influence of chemoattractants in motility of spermatozoa was studied in those spermatozoa selected in the chemotactic system by the chemoattractants FF, pOF, CM, P4 and Σ2. A group in which no chemoattractants were added to well B was included as a control (C). Four replicates were performed.

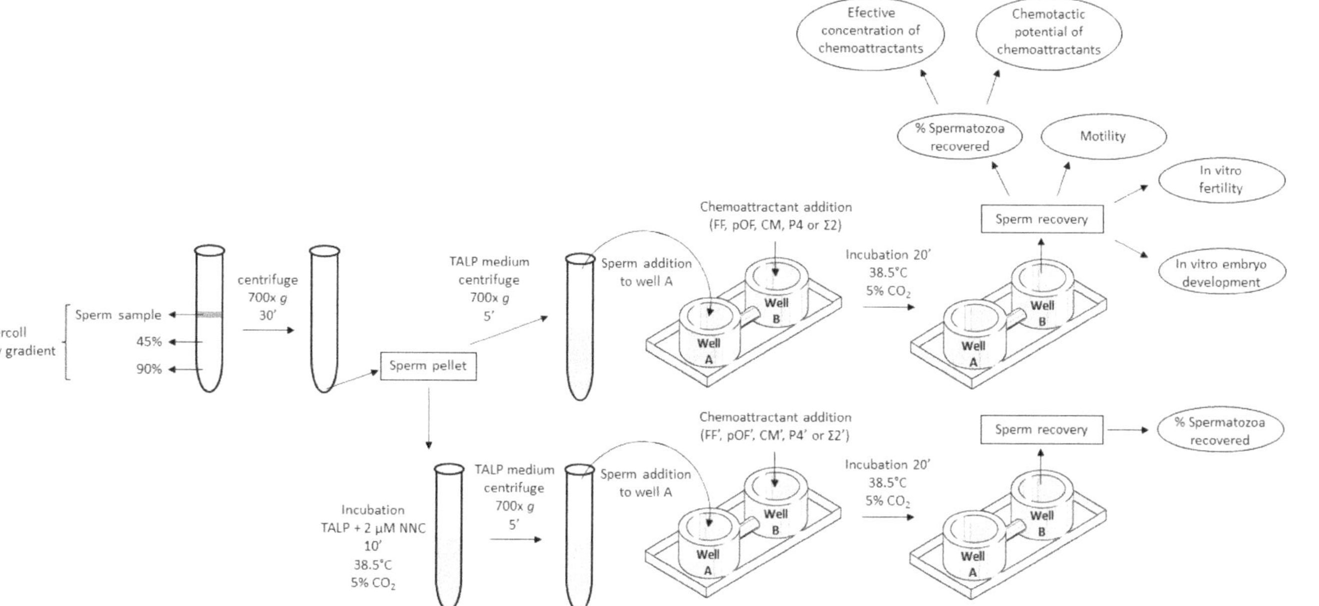

Figure 1. Diagram of the experimental design. Boar ejaculated spermatozoa were separated from seminal plasma by centrifugation on a discontinuous density gradient (Percoll). The pellet was resuspended in TALP medium supplemented or not with 2 µM of the CatSper inhibitor NNC 55-0396 (NNC). Spermatozoa resuspended in TALP + NNC were incubated for 10 min (38.5 °C, 5% CO_2). Chemotactic chamber was loaded with 500 µL TALP medium per well. Chemoattractants were added to well B of the chemotactic chamber (in different concentrations depending on the experiment). Sperm samples in fresh TALP were centrifuged 700× *g*, 5 min and added to the well A of the chamber. The chemotactic system was incubated for 20 min (38.5 °C, 5% CO_2). Spermatozoa recovered in well B were analyzed: % of spermatozoa recovered, motility, in vitro fertility, and in vitro development of blastocyst [just for spermatozoa selected by follicular fluid (FF)]. With the % of spermatozoa recovered in well B, it was also determined the effective concentration of chemoattractants (lowest concentration of chemoattractant that attracted the highest proportion of spermatozoa), and the chemotactic potential of chemoattractants [comparison of the most effective concentration of chemoattractants through the % of spermatozoa recovered in well B attracted by each chemoattractant and a combination of them (Σ)].

(1.4) Mechanism of action of chemoattractants. The P4 triggering of Ca^{2+} uptake into the spermatozoa takes place through CatSper [28] and seems to play a key role in chemotaxis [14,29–32]. To elucidate P4 dependence on the chemotactic potential, spermatozoa selected by the most effective concentration of each chemoattractant (FF, pOF, CM, P4) were compared to the selected spermatozoa preincubated for 10 min with 2 μM of the CatSper inhibitor NNC 55-0396 [28] (FF′, pOF′, CM′, P4′). The most effective combination of chemoattractants (Σ2 and Σ2′) were included in this study, and groups without chemoattractant supplementation in well B were included as controls (C and C′). Four replicates were performed. Additionally, the possible collateral effect of NNC on spermatozoa motility by CASA and the membrane integrity was evaluated in three replicates. For the membrane integrity study, 200 spermatozoa per experimental group and replicate were analyzed.

Experiment 2. In vitro fertility of spermatozoa selected by chemotaxis and in vitro development of embryos.

To study the functionality of the selected spermatozoa, IVF was performed with those spermatozoa recovered in well B of the chemotactic chamber. The experimental groups were established according to the chemoattractant used for sperm selection: 0.25% FF, 0.25% pOF, 0.06% CM, 10.00 pM P4, and Σ2. One group in which there was no supplementation with any chemoattractant in well B was included as control (C). A total of 702 oocytes in four replicates were inseminated, and they were fixed at 18 h post-insemination for IVF parameter evaluation. The experimental group with higher penetration in IVF, the FF, was tested also in IVC. For that purpose, a total of 1,520 oocytes in six replicates were inseminated and 18 h post-insemination they were transferred to NCSU23 medium for up to 7 days of culture.

2.12. Statistical Analysis

The statistical analyses were performed using SPSS v.20 (SPSS Inc. Chicago, IL, USA). The variables were analyzed by analysis of variance (ANOVA). When ANOVA revealed a significant effect, values were compared by the least significant difference pairwise multiple comparison Tukey post hoc test. The results were expressed as the mean ± standard error of the mean (SEM) and $p < 0.05$ was established to indicate statistical significance.

3. Results

3.1. Concentration of E2, P4, cAMP and cGMP in FF, pOF and CM

Despite chemotaxis being mainly attributed to P4, other components have been suggested as potential chemoattractants in the porcine species. This work characterized the pools of biofluids used in all experiments through the concentration of some molecules responsible for chemotaxis in some reproductive biofluids (Table 1). FF showed the highest amount of E2, P4, and cGMP, followed by pOF which showed the highest amount of cAMP. CM showed the lowest cAMP values but higher than pOF for E2 and P4. The concentration of cGMP was higher in FF than in E2, P4, and cAMP. cGMP represented the main component in pOF, also, followed by cAMP, P4, and E2. On the other hand, CM was richer in E2 than in the rest of the components analyzed. Interestingly, P4 was the third highest component in concentration of each biofluid analyzed.

3.2. Effective Concentration of Chemoattractants Selecting Spermatozoa

High concentrations of chemoattractants can saturate the cognate receptors and consequently the chemotactic response decreases. The results of the percentage of spermatozoa recovered in well B from those initially added to well A (called most effective concentration) for each chemoattractant are shown in Figure 2. The most effective concentration ($p < 0.05$) of each chemoattractant was: 0.25% FF (7.6 ± 1.6%), 0.25% pOF (8.4 ± 1.0%), 0.06% CM (3.3 ± 0.7%), 10.00 pM P4 (9.6 ± 1.6%).

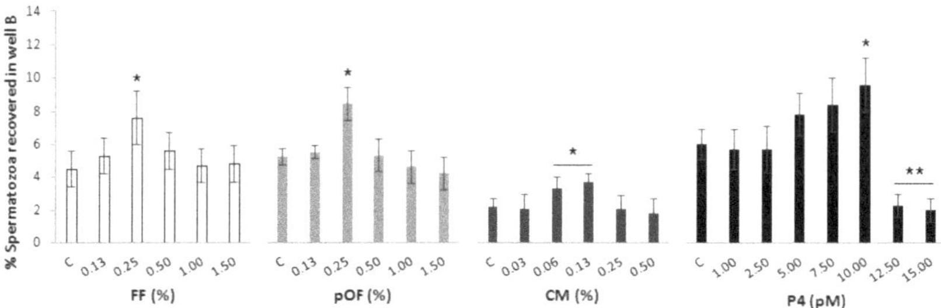

Figure 2. Effective concentration of chemoattractants. Increasing concentrations of the chemoattractants were added to well B of the chemotactic system. A group in which there was not a supplementation of any chemoattractant in well B was included as control (C) for each chemoattractant. Spermatozoa were added to well A in a concentration of 20×10^6 spermatozoa/mL. The system was incubated for 20 min, 38 °C, 5% CO_2 and 95% humidity. The results are expressed as the percentage of spermatozoa recovered in well B. The most effective concentration of each chemoattractant was established as the lowest concentration that attracted the highest proportion of spermatozoa. Four replicates were performed. Asterisks (*, **) indicate statistical differences between groups within the same chemoattractant ($p < 0.05$).

3.3. Chemotactic Potential of Chemoattractants

The possible synergies between chemoattractants on spermatozoa are shown in Figure S2. The combination of a lower concentration of each chemoattractant (FF, pOF, CM) to a final concentration of 0.25% (Σ2) attracted a higher percentage of spermatozoa (Σ2 = 12.3 ± 1.3%; $p < 0.05$) than Σ1 (Σ1 = 8.0 ± 2.1%) and C (C = 8.9 ± 1.8%).

The results of the chemotactic potential of chemoattractants are shown in Figure 3. The % of spermatozoa recovered when a chemoattractant (FF: 6.5 ± 1.0%; pOF: 6.0 ± 0.9%; CM: 6.7 ± 0.8%; P4: 6.6 ± 0.6%; Σ2: 6.1 ± 0.5%) was added to well B was higher ($p < 0.05$) than the control (C: 4.4 ± 0.4%).

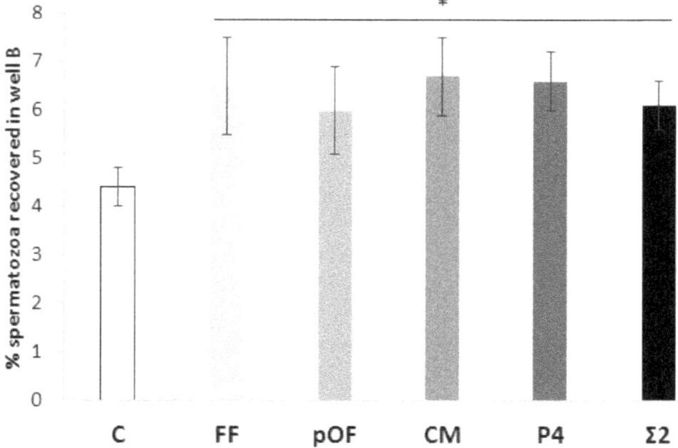

Figure 3. Chemotactic potential of chemoattractants. The percentage of spermatozoa attracted by the most effective concentration of each chemoattractant (0.25% FF, 0.25% pOF, 0.06% CM, 10.00 pM P4) was determined. The combination of the most effective concentration of chemoattractants (Σ2) according to experiment 1.1 was included in this study. A group in which there was not a supplementation of any chemoattractant in well B was included as control (C). The results are expressed as man ± SEM. Four replicates were performed. The asterisk (*) indicates significant statistical differences ($p < 0.05$).

3.4. Motility of Spermatozoa Selected by Chemoattractants

The motility of spermatozoa selected by the effective concentration of each chemoattractant is shown in Table 2. No statistical differences were found for any of the CASA parameters ($p > 0.05$).

3.5. Mechanism of Action of Chemoattractants through CatSper

Figure 4 shows the results of the concentrations of biofluids with higher chemotactic activity and it also shows that the chemotactic activity is related to CatSper. There was no difference in the % of spermatozoa attracted between chemoattractants or their combination (FF, pOF, CM, P4, Σ2; $p > 0.05$), or even when the spermatozoa were previously incubated with NNC (FF′, pOF′, CM′, P4′, Σ2′; $p > 0.05$). However, in both cases the % spermatozoa attracted were higher than their respective controls C and C′. The chemotactic action of all chemoattractants through CatSper was demonstrated by the reduction of the % of spermatozoa attracted by all chemoattractants until the point of being statistically similar to the control C. In other words, all chemoattractants in which spermatozoa had CatSper inhibited by NNC attracted the same % of spermatozoa as in the absence of chemoattractants. When comparing the effect of NNC on the controls (without chemoattractants), in C′ a significantly lower proportion of spermatozoa were recovered in well B as compared to C ($p < 0.05$). Additionally, Table S1 and Figure S3 showed that the motility and membrane integrity of spermatozoa is not affected by NNC.

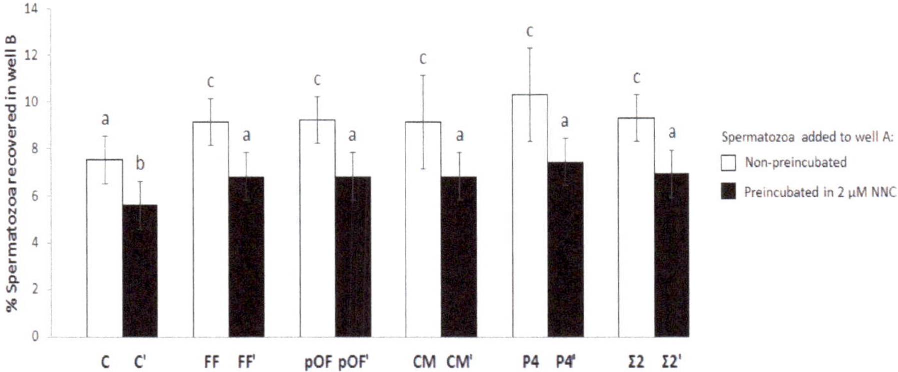

Figure 4. Chemotactic potential of chemoattractants and mechanism of action through CatSper. The percentage of spermatozoa attracted by the most effective concentration of each chemoattractant (0.25% FF, 0.25% pOF, 0.06% CM, 10.00 pM P4) was determined. That parameter was compared Table 10. min with 2 µM of the CatSper inhibitor NNC 55-0396 [28] (FF′, pOF′, CM′, P4′). The combination of the most effective concentration of chemoattractants according to experiment 1.1 (Figure S2; Σ2 and Σ2′) was included in this study. The groups in which there was not a supplementation of chemoattractant in well B for selecting spermatozoa were included as control (C and C′). The results are expressed as man ± SEM. Four replicates were performed. Different letters (a–c) indicate statistical differences between groups ($p < 0.05$).

Table 2. Motility of spermatozoa selected by chemotaxis to reproductive biofluids in the chemotaxis system. The effective concentration of chemoattractants were added to well B of the chemotaxis system: Follicular fluid (FF), periovulatory oviductal fluid (pOF), secretions of the cumulus cells (CM), progesterone (P4) and a combination of chemoattractants (Σ2 = 1/3 of 0.25% FF + 1/3 of 0.25% pOF + 1/3 of 0.06% CM). Spermatozoa (20 × 10^6 cells/mL) were added to well A, and the system was incubated for 20 min. Motility parameters of the spermatozoa recovered in well B were determined. A group in which spermatozoa were selected without chemoattractant was used as the control (C). Four replicates were performed.

	Mot (%)	MotPro (%)	VCL (µm/s)	VSL (µm/s)	VAP (µm/s)	LIN (%)	STR (%)	WOB (%)	ALH (µm)	BCF (Hz)
C	47.0 ± 8.0	21.0 ± 7.2	65.8 ± 8.1	44.8 ± 8.7	49.3 ±8.0	65.7 ± 7.0	87.5 ± 4.5	74.0 ± 4.3	1.8 ± 0.2	6.3 ± 0.7
FF	28.5 ± 3.2	12.0 ± 3.8	42.0 ± 5.6	29.8 ± 5.0	32.8 ± 4.9	70.0 ± 3.6	89.5 ± 2.0	77.5 ± 2.2	1.5 ± 0.2	6.7 ± 0.7
pOF	41.7 ± 5.3	20.8 ± 4.8	54.7 ± 3.7	39.3 ± 4.8	42.8 ± 4.3	70.5 ± 5.6	90.2 ± 2.9	77.2 ± 3.8	1.7 ± 0.2	7.0 ± 0.3
CM	45.2 ± 7.6	23.5 ± 6.9	52.3 ± 7.5	34.3 ± 6.2	39.2 ± 6.3	65.2 ± 3.7	86.8 ± 2.5	74.5 ± 2.3	1.3 ± 0.2	6.5 ± 0.4
P4	37.5 ± 9.4	22.7 ± 0.4	53.8 ± 9.9	37.2 ± 8.9	41.0 ± 9.0	66.3 ± 8.8	87.8 ± 4.4	73.7 ± 6.9	1.7 ± 0.3	7.0 ± 0.8
Σ2	29.8 ± 4.2	12.3 ± 4.9	51.7 ± 5.5	35.7 ± 4.7	38.7 ± 4.8	68.3 ± 3.9	90.3 ± 2.2	75.0 ± 2.9	1.5 ± 0.2	7.3 ± 0.3

Mot (%): percentage of total motile spermatozoa; MotPro (%): percentage of motile progressive spermatozoa; VCL (µm/s): curvilinear velocity; VSL (µm/s): straight-line velocity; VAP (µm/s): average path velocity; LIN (%): linearity of the curvilinear trajectory; STR (%): straightness; WOB (%): Wobble (VAP/VCL); ALH (µm): amplitude of lateral head displacement; BCF (Hz): beat cross-frequency. Two-way ANOVA and multiple pairwise Tukey test ($p < 0.05$) was carried out. Results are expressed as mean ± SEM.

3.6. Spermatozoa Selected by Chemotaxis to the Follicular Fluid (FF) Increases the Penetration of Oocytes In Vitro and the Rate and Quality of Blastocysts Production

During in vivo fertilization in the oviduct, the sperm must be guided towards the oocyte by different mechanisms, one of them is chemotaxis. Female reproductive biofluids can guide the spermatozoa and probably modulate their functionality, fertility, and potential in forming viable embryos. Table 3 shows the results of IVF with spermatozoa selected by chemotaxis. Spermatozoa selected by FF produced a higher % Pen ($p < 0.05$) but the same % Mon ($p > 0.05$) when compared to the rest of chemoattractants. In addition, spermatozoa selected by FF produced the highest proportion of spermatozoa bound to ZP (Spz/ZP), which was statistically the same to that obtained in C ($p > 0.05$). On the contrary, the spermatozoa selected by pOF and Σ2 showed the lowest % Pen, and Σ2 also produced the lowest Spz/ZP ($p < 0.05$).

Table 3. In vitro fertilization (IVF) with spermatozoa selected by chemotaxis to reproductive biofluids in the chemotaxis system. The effective concentration of chemoattractants were added to well B of the chemotaxis system: Follicular fluid (FF), periovulatory oviductal fluid (pOF), secretions of the cumulus cells (CM) progesterone (P4), and a combination of chemoattractants (Σ2 = 1/3 of 0.25% FF + 1/3 of 0.25% pOF + 1/3 of 0.06% CM). Spermatozoa (20×10^6 cells/mL) were added to well A, and the system was incubated for 20 min. A group in which spermatozoa were selected without chemoattractant was used as the control (C). Spermatozoa recovered in well B were used to perform the IVF at 25×10^3 spermatozoa/mL. Eighteen hours after insemination, putative zygotes were fixed with glutaraldehyde and stained with the bisbenzimide solution. Evaluation was carried out by fluorescence microscopy. Four replicates were performed.

	n	Pen (%)	Mon (%)	Spz/O (*n*)	Spz/ZP (*n*)
C	126	40.5 ± 4.4^{a}	33.3 ± 6.7	0.9 ± 0.1^{a}	9.2 ± 0.9^{a}
FF	139	66.2 ± 4.0^{b}	34.0 ± 4.9	1.4 ± 0.1^{b}	9.0 ± 0.5^{a}
pOF	121	$27.3 \pm 4.1^{a,c}$	45.4 ± 8.8	$0.5 \pm 0.1^{c,d}$	$2.6 \pm 0.2^{c,d}$
CM	95	39.0 ± 5.0^{a}	36.8 ± 7.9	$0.7 \pm 0.1^{b,c}$	5.3 ± 0.7^{b}
P4	107	37.4 ± 4.7^{a}	41.3 ± 8.0	$0.7 \pm 0.1^{b,c}$	$4.8 \pm 1.0^{b,c}$
Σ2	114	13.2 ± 3.2^{c}	60.0 ± 13.1	0.2 ± 0.1^{d}	1.2 ± 0.2^{d}

Pen (%): percentage of oocytes penetrated; Mon (%): percentage of oocytes penetrate by one spermatozoon; Spz/O: mean number of spermatozoa penetrating one oocyte; Spz/ZP: mean number of spermatozoa bound to the zona pellucida. One-way ANOVA was performed and the Tukey test of multiple comparisons. Results are expressed as the mean ± SEM. Different superscript ($^{a-d}$) within the same column indicate significant differences between experimental groups ($p < 0.05$).

Since the chemoattractant with the higher % Pen in IVF was FF, this group was also tested for IVC and the results are shown in Table 4. Although spermatozoa selected by FF did not modify the % of cleaved putative zygotes (2-cells; $p > 0.05$), they produced a significantly higher ($p < 0.05$) % of blastocysts (38.5 ± 3.7%) and expanded blastocyst (36.0 ± 2.5%) compared to the control group (27.6 ± 3.8% and 27.0 ± 2.2%, respectively). Furthermore, the number of cells of those blastocysts and expanded blastocyst were also significantly higher ($p < 0.05$) in the FF group (40.4 ± 2.3 and 54.5 ± 2.4, respectively) than in the control (35.9 ± 2.4 and 49.6 ± 2.1, respectively).

Table 4. In vitro embryo development (IVC) of putative zygotes produced by in vitro fertilization (IVF) with spermatozoa selected by follicular fluid (FF) in the chemotaxis system. IVF was performed with 25 × 10^3 selected spermatozoa/mL. After fertilization, putative zygotes continued the incubation in IVC medium (NCSU23) for up to 7 days (168 h). The percentage of embryos reaching 2-cells stage (48 h after insemination) and blastocyst and expanded blastocyst stages were analyzed (at 168 h after insemination). A group in which the spermatozoa were selected without chemoattractant was used as the control (C). Embryos after 7 days of culture were fixed with glutaraldehyde and stained with bisbenzimide solution. Evaluation was carried out by fluorescence microscopy. As an indicator of embryo quality, the number of cells forming each blastocyst was determined. Six replicates were performed.

	n	2-Cells (%)	Blastocyst		Expanded Blastocyst	
			%	Num. Cells	%	Num. Cells
C	824	42.1 ± 7.4	27.6 ± 3.8 ^{a}	35.9 ± 2.4 ^{a}	27.0 ± 2.2 ^{a}	49.6 ± 2.1 ^{a}
FF	718	39.8 ± 7.5	38.5 ± 3.7 ^{b}	40.4 ± 2.3 ^{b}	36.0 ± 2.5 ^{b}	54.5 ± 2.4 ^{b}

2-cells (%): percentage of oocytes that cleaved to the 2-cell stage of embryo development were evaluated at 48 h after insemination. Blastocyst and expanded blastocysts: percentage of 2-cells embryos that developed to blastocyst or expanded blastocyst stage (according to the classification proposed by Bó and Mapleroft, [26]) after 7 days (168 h) of culture. The number of cells were determined by fluorescence microscopy in blastocyst and expanded blastocysts fixed in 0.5% glutaraldehyde and stained with bisbenzimide solution. One-way ANOVA was performed and the Tukey test of multiple comparisons. Results are expressed as the mean ± SEM. The different superscripts ($^{a–b}$) within the same column indicate the significant differences between groups ($p < 0.05$).

4. Discussion

After ovulation, spermatozoa that remain attached to the oviductal epithelial cells in the sperm reservoir are released and continue their journey to meet the oocyte. Since the oviduct is a very narrow and full of crypts duct, the process of gametes encountering cannot be random. Not all sperm will reach the oocyte, and a strict and selective process occurs during their path. It has been suggested that spermatozoa could be attracted by chemical factors that would be released by the oocyte, cumulus cells, and/or the epithelial oviductal cells. Some possible chemoattractants that facilitate spermatozoon-oocyte interaction are substances present in FF, OF [12,33], and other secretions surrounding the oocyte [12,13,34]. Chemotaxis has been described as a concentration-dependent [27] and species-specific phenomenon [12–14,17]. However, the chemical nature of chemoattractants, their receptors and the underlying signaling pathways are still a subject up for debate. In the porcine species, the main sperm-chemotactic role was attributed to P4 [14,32,35,36], but other components were also suggested as potential chemoattractants [37,38].

A typical chemotaxis response is represented by a bell curve, where at lower or higher attractant dilutions no chemotactic response is observed. However at intermediate attractant dilutions, the proportion of chemotactic cells is the greatest [39]. The chemoattractant concentrations obtained in this study were higher than those described in the literature for other species [12,13,17,40]. These differences can be attributed, on the one hand, to the working conditions, since the devices and volumes used by those authors are different compared to ours. On the other hand, those differences could be attributed to some molecular species-specific events during sperm capacitation. However, this last assumption cannot be corroborated with other studies since, to the best of our knowledge, there are not any published studies developed in the porcine species. Regarding the concentration of P4, there is an in-depth study, not only concerning its chemotactic activity [1] but also its participation in the activation of calcium channels involved in the hyperactive movement of the spermatozoa. The concentration of P4 that attracted the largest number of porcine spermatozoa to well B was 10.00 pM, and above this amount there was no greater response. Similar results were observed in human [36], rabbit [36] and mouse sperm [41].

Regarding the chemotactic potential, we did not find statistical differences between experimental groups, but the phenomenon of spermatozoa attraction was clear since there was a significantly lower % of spermatozoa attracted in their absence. This result was unexpected since, despite all biofluids having common components, the concentrations of those components are different. Thus, it would be reasonable to think that the chemoattraction potential can vary between biofluids. It is possible that the effective concentrations of chemoattractants in the biofluids used in the present work were too high. In that case, the saturation of receptors could be a plausible explanation for the lack of effect of the biofluids. Thus, the reduction in the concentration of these biofluids to elucidate their chemotactic potential would make sense. To our knowledge, there are not any studies that show the physiological concentration of chemoattractants in porcine reproductive biofluids. And those may also vary according to the different phase of the estrous cycle. In mice, Oliveira et al. [12] observed that pOF induced a high proportion of spermatozoa travelling longer distances toward the chemotactic gradient, while FF caused an increase in spermatozoa velocity. However, the reasons why the values of spermatozoa attraction obtained in this work with pOF were lower than those obtained with FF remains to be clarified.

It seems logical to think that a chemoattractant whose composition is closest to that found by the spermatozoa in vivo should be the one that attracts more spermatozoa. That is represented in this work as a combination of biofluids (Σ). Nevertheless, the results in this study did not show that. Considering that the in vivo contact of the sperm with chemoattractants occurs sequentially, one hypothesis could be that some receptors may be activated at a given moment and consequently induce the activation of other receptors later [13,36]. The exposure of spermatozoa to chemoattractants in this study was not sequential and the chemotactic receptors could have been activated at the same time preventing a greater chemotactic response.

Although Armon and Eisenbach [42] provided evidence that hyperactivation is part of the chemotactic response of the spermatozoa, the motility results found here by CASA showed no differences between the experimental groups. The results in other studies carried until now about whether motility is affected by chemotaxis are quite controversial. Other authors such as Fabbri et al. [43] observed, in humans, an increase in motility and hyperactivation, but Isobe et al. [44] did not observe variations of motility derived from chemotaxis. The main differences of the present study with previous studies can be that the sperm motility was analyzed after sperm migration towards the chemoattractant. Armon and Eisenbach [42] analyzed the sperm trajectory of swimming in a spatial chemoattractant gradient and observed that hyperactivation was significantly reduced by chemoattractants compared to controls. These authors suggested that with the increase in the chemoattractant concentration the capacitated sperm represses the hyperactivation but keeps swimming in favor of the concentration gradient of chemoattractant.

The importance of increased intracellular Ca^{2+} in spermatozoa has been demonstrated during chemotaxis [45]. In the chemotactic process the hyperactive movement is produced (regulated by the activation of CatSper channels) and the directionality of the spermatozoa towards the chemoattractant changes [29]. It has been shown in porcine species that the CatSper inhibitor NNC blocked the spermatozoa release from oviduct isthmic epithelial cells [46]. This study used the same concentration of NNC and showed a decrease in the percentage of spermatozoa that migrated to the chemoattractant in all groups, including the control. However, the sperm accumulation in well B appeared not to be completely abolished by NNC, since that reduction was also observed when no chemoattractant was added to the system. Other substances with chemotactic activity could be present in the biofluids. Some candidates could be the 8.6 kDa protein similar to apolipoprotein B2 [38], or maybe the antithrombin III [37] which is responsible for this migration when CatSper are inhibited. However, the mechanism by which these substances stimulate motility has not yet been elucidated. The inhibitor NNC is quite unspecific for CatSper, which means that it could be unsuitable for chemotaxis assays. Rennhack et al. [47] showed that NNC exhibit serious adverse reactions in human sperm and evoke a sizeable and sustained increase of $[Ca^{2+}]_i$ and pH_i [48–50] in addition to stimulating the acrosomal exocytosis [50].

Although P4 has been identified as the main chemoattractant in pOF, FF, and CM, other components with chemotactic activity are due to be present in biofluids. Therefore, we decided to determine the concentration of E2, P4, cAMP, and cGMP in the biofluids studied.

The FF showed the highest concentrations of all of them except for cAMP. Interestingly, the FF was not the biofluid that showed the highest chemoattraction power. Perhaps some components offset the effect of others found in lower concentrations. For instance, cells are less sensitive to cGMP than to cAMP, but cGMP produces a more intense response [51]. Biofluids exhibit complex activity and the spermatozoa are simultaneously exposed to multiple ligands. This can lead to multiple effects and/or separate interactions. In this sense, the estrogen pretreatment elevates Ca^{2+} in spermatozoa apparently by a CatSper independent mechanism [28,52]. The estrogen pretreatment also reduces the concentration of Ca^{2+} in response to the stimulation with P4. However, that inhibition could be not produced when the concentrations of P4 are high [53].

Other questions that still need to be solved include why second messengers, such as cAMP and cGMP, are present in biofluids and their roles in chemotaxis. Purinergic signaling has been found to be a key component in the physiology of several tissues. Through the self-production of nucleotides and nucleosides and their binding to specific receptors, a wide range of cellular responses are modulated such as cell growth, differentiation, and motility. A clear example of the importance of these nucleotides was shown by Osycka-Salut et al. [54]. These authors demonstrated that bicarbonate in the extracellular medium produced an early increase in cAMP dependent on the soluble adenylyl cyclase (sAC) in the intra- and extracellular space in spermatozoa. Thus, it suggests that if the existence of a cAMP flow from the spermatozoa to the extracellular space were blocked, that it could result in the inhibition of capacitation. Another possible function of these nucleotides could be as an indicator of the direction to which the spermatozoon must go to find the oocyte, as in the case of sea urchins [55].

The functionality of the spermatozoa selected by pOF, FF, CM, and P4 was analyzed through IVF. The highest oocyte penetration was achieved when FF was used as the chemoattractant but the number of spermatozoa per oocyte was not affected by the biofluids used. A possible explanation of this would be that the FF increases the penetration of oocytes by modulating capacitation and acrosome reaction (reviewed by Hong et al. [11]) which consequently increases the penetration of oocytes. It would be logical to think that the number of spermatozoa per oocyte should also increase [56]. In this sense, Funahashi and Day [57] observed that the pre-fertilization incubation of porcine spermatozoa in suitable concentrations of porcine FF effectively reduces the incidence of polyspermy. Their results indicated that polyspermy blocking is produced by an interaction between FF and spermatozoa and not with oocytes. Another plausible explanation would be that the FF was the attractant that selects the subpopulation of spermatozoa with the highest quality and, therefore, with the highest penetration rate. These hypotheses are supported by the results of Ralt et al. [40], who reported that the FF potential to attract human spermatozoa is strongly correlated with the potential of those spermatozoa to fertilize the oocyte. However, under in vivo conditions, the spermatozoon contacts with other biofluids and biomolecules while it is attracted to the oocyte.

The lowest penetration rate obtained when spermatozoa were selected by pOF or Σ could be attributed to the presence of some factors that prevent capacitation before the spermatozoa contact with the oocyte. Soriano-Úbeda et al. [4] showed that pOF and CM decreases spermatozoa protein kinase A substrates and that tyrosine residues phosphorylation. And later, Zapata-Carmona et al. [25] showed that this process is reversible. During capacitation, the decapacitation factors from the SP are lost and that the intracellular Ca^{2+} concentration rises [58]. After that, the spermatozoa acquire additional PMCA4 from the oviduct via exosomes [59] which provide an adequate Ca^{2+} efflux to promote spermatozoa viability and prevent a premature acrosome reaction. Therefore, it cannot be discarded that some factors of pOF decreased spermatozoa capacitation, and other factors produced by the oocyte attracted the spermatozoa and prepared it for fertilization [60].

Until a few years ago, the research on embryo quality focused mainly on evaluating the impact of maternal factors on the embryo. However, it has recently been determined that paternal factors also share responsibility for contributing to better embryo quality. In our work, we have analyzed the quality of embryos obtained with spermatozoa selected by a density gradient and then migrated towards the FF. The results showed that there were no differences in the percentage of embryos divided at 48 h. However, those differences appeared later at the blastocysts and expanded blastocyst stages as well as in the number of cells that form those blastocysts. In both cases, the best quality was higher for the group of spermatozoa that migrated towards the FF. Therefore, the FF somehow selected the best spermatozoon. Gatica et al. [1] obtained better quality embryos when using sperm selected by chemotaxis and argued that this was due to less DNA damage that usually is very low in AI doses. In a previous study (data not shown) we analyzed DNA damage using acridine orange and the Halomax® kit and observed that the % altered DNA was very close to 0%.

It is interesting to observe that the differences between the groups appear at 168 h (7 days) of embryo culture in the blastocyst stage. This fact leads us to think that not only selection by FF is important to obtain spermatozoa of higher quality, but this fluid can provide the spermatozoa with certain factors that improve subsequent embryonic development. Based on this data, we can hypothesize that the results obtained here could be regulated by the extracellular vesicles (EVs) present in the FF. It has been observed that a brief co-incubation is sufficient for transferring components from EVs to the spermatozoa. These EVs participate in this way in the sperm formation process while providing them different types of RNA [61]. All this could explain the improvement in the quality of the embryos obtained with spermatozoa selected by chemotaxis towards the FF [62]. In this work we have observed that the speed of the blastocysts to expand is greater when the spermatozoa are selected by FF. Supporting this theory, Fatehi et al. [63] reported that bovine embryos in the faster division stage are more likely to become blastocysts, and that embryos of rapid division are associated with higher gestation rates.

5. Conclusions

Under in vitro conditions, the FF selects the best spermatozoa for an optimum and more physiological interaction with the oocyte. The knowledge acquired with this work can be useful in improving the current ART performed in animal production and as model for human clinic assays. From this point on, more studies are necessary to deepen the knowledge of gamete interaction in the physiological environment.

Supplementary Materials: The following are available online at https://www.mdpi.com/2076-2615/11/1/53/s1, Figure S1: Chemotaxis chamber diagram. Figure S2: Chemoattractant potential of chemoattractant combinations. Figure S3: Effect of CatSper inhibition on the membrane integrity of spermatozoa. Table S1: Effect of CatSper inhibition on the motility of spermatozoa.

Author Contributions: Conceptualization, L.A.V. and C.M.; Methodology, L.A.V., A.D., C.S.-Ú. and C.M.; Software, L.A.V.; Validation, L.A.V., C.S.-Ú. and C.M.; Formal Analysis, L.A.V., A.D., C.S.-Ú. and C.M.; Investigation, L.A.V. and A.D.; Resources, L.A.V. and C.M.; Data Curation, L.A.V., C.S.-Ú. and C.M.; Writing—Original Draft Preparation, L.A.V. and A.D.; Writing—Review & Editing, C.S.-Ú. and C.M.; Visualization, L.A.V., A.D., C.S.-Ú. and C.M.; Supervision, C.M.; Project Administration, L.A.V. and C.M.; Funding Acquisition, L.A.V. and C.M. All authors have read and agreed to the published version of the manuscript.

Funding: This work was supported by grants 20040/GERM/16 and 20020/SF/16 from Fundación Séneca (Agencia de Ciencia y Tecnología de la Región de Murcia, Spain) and PID2019-106380RB-I00/AEI/10.13039/501100011033 from the Spanish Ministry of Science and Innovation.

Institutional Review Board Statement: Not applicable.

Informed Consent Statement: Not applicable.

Data Availability Statement: The data presented in this study are available within the article or supplementary material.

Acknowledgments: The authors would like to thank Alejandro Torrecillas for his technical assistance, Florentin Daniel Staicu for his valuable support, Joaquín Gadea for his scientific advice, Department of Research and Development of CEFUSA and El Pozo S.A. (Alhama, Murcia, Spain) for providing the biological samples, and Melinda Furse for editing the language.

Conflicts of Interest: The authors declare no conflict of interest.

References

1. Gatica, L.V.; Guidobaldi, H.A.; Montesinos, M.M.; Teves, M.E.; Moreno, A.I.; Uñates, D.R.; Molina, R.I.; Giojalas, L.C. Picomolar gradients of progesterone select functional human sperm even in subfertile samples. *Mol. Hum. Reprod.* **2013**, *19*, 559–569. [CrossRef]
2. Romar, R.; Cánovas, S.; Matás, C.; Gadea, J.; Coy, P. Pig in vitro fertilization: Where are we and where do we go? *Theriogenology* **2019**, *137*, 113–121. [CrossRef]
3. Hunter, R.H.F. Ovarian control of very low sperm/egg ratios at the commencement of mammalian fertilisation to avoid polyspermy. *Mol. Reprod. Dev.* **1996**, *44*, 417–422. [CrossRef]
4. Soriano-Úbeda, C.; García-Vázquez, F.A.; Romero-Aguirregomezcorta, J.; Matás, C. Improving porcine in vitro fertilization output by simulating the oviductal environment. *Sci. Rep.* **2017**, *7*, 1–12. [CrossRef]
5. Eisenbach, M. Mammalian sperm chemotaxis and its association with capacitation. *Dev. Genet.* **1999**, *25*, 87–94. [CrossRef]
6. Machado-Oliveira, G.; Lefièvre, L.; Ford, C.; Herrero, M.B.; Barratt, C.; Connolly, T.J.; Nash, K.; Morales-Garcia, A.; Kirkman-Brown, J.; Publicover, S.J. Mobilisation of Ca2+ stores and flagellar regulation in human sperm by S-nitrosylation: A role for NO synthesised in the female reproductive tract. *Development* **2008**, *135*, 3677–3686. [CrossRef] [PubMed]
7. Kaupp, U.B.; Solzin, J.; Hildebrand, E.; Brown, J.E.; Helbig, A.; Hagen, V.; Beyerman, M.; Pampaloni, F.; Weyand, I. The signal flow and motor response controling chemotaxis of sea urchin sperm. *Nat. Cell Biol.* **2003**, *5*, 109–117. [CrossRef]
8. Matsumoto, M.; Solzin, J.; Helbig, A.; Hagen, V.; Ueno, S.; Kawase, O.; Maruyama, Y.; Ogiso, M.; Godde, M.; Minakata, H.; et al. A sperm-activating peptide controls a cGMP-signaling pathway in starfish sperm. *Dev. Biol.* **2003**, *260*, 314–324. [CrossRef]
9. Orihuela, P.A.; Parada-Bustamante, A.; Cortés, P.P.; Gatica, C.; Croxatto, H.B. Estrogen receptor, cyclic adenosine monophosphate, and protein kinase A are involved in the nongenomic pathway by which estradiol accelerates oviductal oocyte transport in cyclic rats. *Biol. Reprod.* **2003**, *68*, 1225–1231. [CrossRef]
10. Villanueva-Diaz, C.; Vadillo-Ortega, F.; Kably-Ambe, A.; Diaz-Perez, M.A.; Krivitzky, S.K. Evidence that human follicular fluid contains a chemoattractant for spermatozoa. *Fertil. Steril.* **1990**, *54*, 1180–1182. [CrossRef]
11. Hong, C.Y.; Chao, H.T.; Lee, S.L.; Wei, Y.H. Modification of human sperm function by human follicular fluid: A review. *Int. J. Androl.* **1993**, *16*, 93–96. [CrossRef] [PubMed]
12. Oliveira, R.G.; Tomasi, L.; Rovasio, R.A.; Giojalas, L.C. Increased velocity and induction of chemotactic response in mouse spermatozoa by follicular and oviductal fluids. *J. Reprod. Fertil.* **1999**, *115*, 23–27. [CrossRef] [PubMed]
13. Sun, F.; Bahat, A.; Gakamsky, A.; Girsh, E.; Katz, N.; Giojalas, L.C.; Tur-Kaspa, I.; Eisenbach, M. Human sperm chemotaxis: Both the oocyte and its surrounding cumulus cells secrete sperm chemoattractants. *Hum. Reprod.* **2005**, *20*, 761–767. [CrossRef] [PubMed]
14. Oren-Benaroya, R.; Orvieto, R.; Gakamsky, A.; Pinchasov, M.; Eisenbach, M. The sperm chemoattractant secreted from human cumulus cells is progesterone. *Hum. Reprod.* **2008**, *23*, 2339–2345. [CrossRef]
15. Guidobaldi, H.A.; Teves, M.E.; Uñates, D.R.; Anastasía, A.; Giojalas, L.C. Progesterone from the cumulus cells is the sperm chemoattractant secreted by the rabbit oocyte cumulus complex. *PLoS ONE* **2008**, *3*, e3040. [CrossRef]
16. Fabro, G.; Rovasio, R.A.; Civalero, S.; Frenkel, A.; Caplan, S.R.; Eisenbach, M.; Giojalas, L.C. Chemotaxis of capacitated rabbit spermatozoa to follicular fluid revealed by a novel directionality-based assay. *Biol. Reprod.* **2002**, *67*, 1565–1571. [CrossRef]
17. Gil, P.I.; Guidobaldi, H.A.; Teves, M.E.; Uñates, D.R.; Sanchez, R.; Giojalas, L.C. Chemotactic response of frozen-thawed bovine spermatozoa towards follicular fluid. *Anim. Reprod. Sci.* **2008**, *108*, 236–246. [CrossRef]
18. Rath, D.; Long, C.R.; Dobrinsky, J.R.; Welch, G.R.; Schreier, L.L.; Johnson, L.A. In vitro production of sexed embryos for gender preselection: High-speed sorting of X-chromosome-bearing sperm to produce pigs after embryo transfer. *J. Anim. Sci.* **1999**, *77*, 3346–3352. [CrossRef]
19. Petters, R.M.; Wells, K.D. Culture of pig embryos. *J. Reprod. Fertil.* **1993**, *48*, 61–73.
20. Matás, C.; Coy, P.; Romar, R.; Marco, M.; Gadea, J.; Ruiz, S. Effect of sperm preparation method on in vitro fertilization in pigs. *Reproduction* **2003**, *125*, 133–141. [CrossRef]
21. Matás, C.; Vieira, L.; García-Vázquez, F.A.; Avilés-López, K.; López-Úbeda, R.; Carvajal, J.A.; Gadea, J. Effects of centrifugation through three different discontinuous Percoll gradients on boar sperm function. *Anim. Reprod. Sci.* **2011**, *127*, 62–72. [CrossRef] [PubMed]
22. Funahashi, H.; Cantley, T.; Day, B. Different hormonal requirements of pig oocyte-cumulus complexes during maturation in vitro. *J. Reprod. Fertil.* **1994**, *101*, 159–165. [CrossRef] [PubMed]
23. Carrasco, L.C.; Romar, R.; Avilés, M.; Gadea, J.; Coy, P. Determination of glycosidase activity in porcine oviductal fluid at the different phases of the estrous cycle. *Reproduction* **2008**, *136*, 833–842. [CrossRef] [PubMed]

24. Soriano-Úbeda, C.; Romero-Aguirregomezcorta, J.; Matás, C.; Visconti, P.E.; García-Vázquez, F.A. Manipulation of bicarbonate concentration in sperm capacitation media improves in vitro fertilisation output in porcine species. *J. Anim. Sci. Biotecnol.* **2019**, *10*, 1–15. [CrossRef] [PubMed]
25. Zapata-Carmona, H.; Soriano-Úbeda, C.; París-Oller, E.; Matás, C. Periovulatory oviductal fluid decreases sperm protein kinase A activity, tyrosine phosphorylation, and in vitro fertilization in pig. *Andrology* **2019**, *3*, 1–13. [CrossRef] [PubMed]
26. Bó, G.; Mapletoft, R. Evaluation and classification of bovine embryos. *Anim. Reprod.* **2013**, *10*, 344–348.
27. Adler, J. A method for measuring chemotaxis and use of the method to determine optimum conditions for chemotaxis by Escherichia coli. *J. Gen. Microbiol.* **1973**, *74*, 77–91. [CrossRef]
28. Lishko, P.V.; Botchkina, I.L.; Kirichok, Y. Progesterone activates the principal Ca2+ channel of human sperm. *Nature* **2011**, *471*, 387–391. [CrossRef]
29. Eisenbach, M.; Giojalas, L.C. Sperm guidance in mammals-An unpaved road to the egg. *Nat. Rev. Mol. Cell Biol.* **2006**, *7*, 276–285. [CrossRef]
30. Harper, C.V.; Kirkman-Brown, J.C.; Barratt, C.L.R.; Publicover, S.J. Encoding of progesterone stimulus intensity by intracellular [Ca2+] ([Ca2+]i) in human spermatozoa. *Biochem. J.* **2003**, *372*, 407–417. [CrossRef]
31. Suarez, S.S. Control of hyperactivation in sperm. *Hum. Reprod. Update* **2008**, *14*, 647–657. [CrossRef] [PubMed]
32. Teves, M.E.; Guidobaldi, H.A.; Uñates, D.R.; Sanchez, R.; Miska, W.; Publicover, S.J.; Garcia, A.A.M.; Giojalas, L. Molecular mechanism for human sperm chemotaxis mediated by progesterone. *PLoS ONE* **2009**, *4*, e8211. [CrossRef] [PubMed]
33. Cohen-Dayag, A.; Tur-Kaspa, I.; Dor, J.; Mashiach, S.; Eisenbach, M. Sperm capacitation in humans is transient and correlates with chemotactic responsiveness to follicular factors. *Proc. Natl. Acad. Sci. USA* **1995**, *92*, 11039–11043. [CrossRef] [PubMed]
34. Giojalas, L.C.; Rovasio, R.A. Mouse spermatozoa modify their motility parameters and chemotactic response to factors from the oocyte microenvironment. *Int. J. Androl.* **1998**, *21*, 201–206. [CrossRef]
35. Villanueva-Diaz, C.; Arias-Martinez, J.; Bermejo-Martinez, L.; Vadillo-Ortega, F. Progesterone induces human sperm chemotaxis. *Fertil. Steril.* **1995**, *64*, 1183–1188. [CrossRef]
36. Teves, M.E.; Barbano, F.; Guidobaldi, H.A.; Sanchez, R.; Miska, W.; Giojalas, L.C. Progesterone at the picomolar range is a chemoattractant for mammalian spermatozoa. *Fertil. Steril.* **2006**, *86*, 745–749. [CrossRef]
37. Lee, S.L.; Wei, Y.H. The involvement of extracellular proteinases and proteinase inhibitors in mammalian fertilization. *Biotechnol. Appl. Biochem.* **1994**, *19*, 31.
38. Serrano, H.; Canchola, E.; García-Suárez, M.D. Sperm-attracting activity in follicular fluid associated to an 8.6-kDa protein. *Biochem. Biophys. Res. Commun.* **2001**, *283*, 782–784. [CrossRef]
39. Eisenbach, M. Towards understanding the molecular mechanism of sperm chemotaxis. *J. Gen. Physiol.* **2004**, *124*, 105–108. [CrossRef]
40. Ralt, D.; Goldenberg, M.; Fetterolf, P.; Thompson, D.; Dor, J.; Mashiach, S.; Garbers, D.L.; Eisenbach, M. Sperm attraction to a follicular factor(S) correlates with human egg fertilizability. *Obstet. Gynecol. Surv.* **1991**, *46*, 648–650. [CrossRef]
41. Guidobaldi, H.A.; Hirohashi, N.; Cubilla, M.; Buffone, M.G.; Giojalas, L.C. An intact acrosome is required for the chemotactic response to progesterone in mouse spermatozoa. *Mol. Reprod. Dev.* **2017**, *84*, 310–315. [CrossRef] [PubMed]
42. Armon, L.; Eisenbach, M. Behavioral mechanism during human sperm chemotaxis: Involvement of hyperactivation. *PLoS ONE* **2011**, *6*, e28359. [CrossRef]
43. Fabbri, R.; Porcu, E.; Lenzi, A.; Gandini, L.; Marsella, T.; Flamigni, C. Follicular fluid and human granulosa cell cultures: Influence on sperm kinetic parameters, hyperactivation, and acrosome reaction. *Fertil. Steril.* **1998**, *69*, 112–117. [CrossRef]
44. Isobe, T.; Minoura, H.; Tanaka, K.; Shibahara, T.; Hayashi, N.; Toyoda, N. The effect of RANTES on human sperm chemotaxis. *Hum. Reprod.* **2002**, *17*, 1441–1446. [CrossRef] [PubMed]
45. Seifert, R.; Flick, M.; Bönigk, W.; Alvarez, L.; Trötschel, C.; Poetsch, A.; Müller, A.; Goodwin, N.; Pelzer, P.; Kashikar, N.D.; et al. The CatSper channel controls chemosensation in sea urchin sperm. *EMBO J.* **2015**, *34*, 379–392. [CrossRef] [PubMed]
46. Machado, S.A.; Sharif, M.; Wang, H.; Bovin, N.; Miller, D.J. Release of porcine sperm from oviduct cells is stimulated by progesterone and requires CatSper. *Sci. Rep.* **2019**, *9*, 1–11. [CrossRef]
47. Rennhack, A.; Schiffer, C.; Brenker, C.; Fridman, D.; Nitao, E.T.; Cheng, Y.M.; Tamburrino, L.; Balbach, M.; Stölting, G.; Berger, T.K.; et al. A novel cross-species inhibitor to study the function of CatSper Ca2+ channels in sperm. *Br. J. Pharmacol.* **2018**, *175*, 3144–3161. [CrossRef]
48. Strünker, T.; Goodwin, N.; Brenker, C.; Kashikar, N.D.; Weyand, I.; Seifert, R.; Kaupp, U.B. The CatSper channel mediates progesterone-induced Ca2+ influx in human sperm. *Nature* **2011**, *471*, 382–387. [CrossRef]
49. Brenker, C.; Goodwin, N.; Weyand, I.; Kashikar, N.D.; Naruse, M.; Krähling, M.; Müller, A.; Benjamin Kaupp, U.; Strünker, T. The CatSper channel: A polymodal chemosensor in human sperm. *EMBO J.* **2012**, *31*, 1654–1665. [CrossRef]
50. Chávez, J.C.; De la Vega-Beltrán, J.L.; José, O.; Torres, P.; Nishigaki, T.; Treviño, C.L.; Darszon, A. Acrosomal alkalization triggers Ca2+ release and acrosome reaction in mammalian spermatozoa. *J. Cell. Physiol.* **2018**, *233*, 4735–4747. [CrossRef]
51. Gakamsky, A.; Armon, L.; Eisenbach, M. Behavioral response of human spermatozoa to a concentration jump of chemoattractants or intracellular cyclic nucleotides. *Hum. Reprod.* **2009**, *24*, 1152–1163. [CrossRef]
52. Luconi, M.; Porazzi, I.; Ferruzzi, P.; Marchiani, S.; Forti, G.; Baldi, E. Tyrosine phosphorylation of the a kinase anchoring protein 3 (AKAP3) and soluble adenylate cyclase are involved in the increase of human sperm motility by bicarbonate. *Biol. Reprod.* **2005**, *72*, 22–32. [CrossRef] [PubMed]

53. Mannowetz, N.; Miller, M.R.; Lishko, P.V. Regulation of the sperm calcium channel CatSper by endogenous steroids and plant triterpenoids. *Proc. Natl. Acad. Sci. USA* **2017**, *114*, 5743–5748. [CrossRef] [PubMed]
54. Osycka-Salut, C.; Diez, F.; Burdet, J.; Gervasi, M.G.; Franchi, A.; Bianciotti, L.G.; Davio, C.; Perez-Martinez, S. Cyclic AMP efflux, via MRPs and A1 adenosine receptors, is critical for bovine sperm capacitation. *Mol. Hum. Reprod.* **2014**, *20*, 89–99. [CrossRef] [PubMed]
55. Trötschel, C.; Hamzeh, H.; Alvarez, L.; Pascal, R.; Lavryk, F.; Bönigk, W.; Körschen, H.G.; Müller, A.; Poetsch, A.; Rennhack, A.; et al. Absolute proteomic quantification reveals design principles of sperm flagellar chemosensation. *EMBO J.* **2020**, *39*, 1–18. [CrossRef]
56. Funahashi, H. Polyspermic penetration in porcine IVM-IVF systems. *Reprod. Fertil. Dev.* **2003**, *15*, 167–177. [CrossRef]
57. Funahashi, H.; Day, B. Effects of follicular fluid at fertilization in vitro on sperm penetration in pig oocytes. *J. Reprod. Fertil.* **1993**, *99*, 97–103. [CrossRef]
58. Matás, C.; Sansegundo, M.; Ruiz, S.; García-Vázquez, F.A.; Gadea, J.; Romar, R.; Coy, P. Sperm treatment affects capacitation parameters and penetration ability of ejaculated and epididymal boar spermatozoa. *Theriogenology* **2010**, *74*, 1327–1340. [CrossRef]
59. Al-Dossary, A.A.; Strehler, E.E.; Martin-DeLeon, P.A. Expression and secretion of plasma membrane Ca2+-ATPase 4a (PMCA4a) during murine estrus: Association with oviductal exosomes and uptake in sperm. *PLoS ONE* **2013**, *8*, e80181. [CrossRef]
60. Armon, L.; Ben-Ami, I.; Ron-El, R.; Eisenbach, M. Human oocyte-derived sperm chemoattractant is a hydrophobic molecule associated with a carrier protein. *Fertil. Steril.* **2014**, *102*, 885–890. [CrossRef]
61. Martin-DeLeon, P.A. Uterosomes: Exosomal cargo during the estrus cycle and interaction with sperm. *Front. Biosci.* **2016**, *8*, 115–122. [CrossRef] [PubMed]
62. Franchi, A.; Moreno-Irusta, A.; Domínguez, E.M.; Adre, A.J.; Giojalas, L.C. Extracellular vesicles from oviductal isthmus and ampulla stimulate the induced acrosome reaction and signaling events associated with capacitation in bovine spermatozoa. *J. Cell. Biochem.* **2019**, *121*, 2877–2888. [CrossRef] [PubMed]
63. Fatehi, A.N.; Bevers, M.M.; Schoevers, E.; Roelen, B.A.J.; Colenbrander, B.; Gadella, B.M. DNA damage in bovine sperm does not block fertilization and early embryonic development but induces apoptosis after the first cleavages. *J. Androl.* **2006**, *27*, 176–188. [CrossRef] [PubMed]

Review

Oocyte Selection for In Vitro Embryo Production in Bovine Species: Noninvasive Approaches for New Challenges of Oocyte Competence

Luis Aguila [1,*], Favian Treulen [2], Jacinthe Therrien [1], Ricardo Felmer [3], Martha Valdivia [4] and Lawrence C Smith [1]

1 Centre de Recherche en Reproduction et Fértilité (CRRF), Université de Montréal, St-Hyacinthe, QC J2S 2M2, Canada; jacinthe.therrien@umontreal.ca (J.T.); lawrence.c.smith@umontreal.ca (L.C.S.)
2 School of Medical Technology, Faculty of Science, Universidad Mayor, Temuco 4801043, Chile; faviantreulen@gmail.com
3 Laboratory of Reproduction, Centre of Reproductive Biotechnology (CEBIOR-BIOREN), Faculty of Medicine, Universidad de La Frontera, Temuco 4811322, Chile; ricardo.felmer@ufrontera.cl
4 Laboratory of Animal Reproductive Physiology, Biological Sciences Faculty, Universidad Nacional Mayor de San Marcos, Lima 15088, Peru; mvaldiviac@unmsm.edu.pe
* Correspondence: luis.aguila.paredes@gmail.com

Received: 20 October 2020; Accepted: 19 November 2020; Published: 24 November 2020

Simple Summary: The efficiency of producing embryos using in vitro technologies in cattle species remains lower when compared to mice, indicating that the proportion of female gametes that fail to develop after in vitro manipulation is considerably large. Considering that the intrinsic quality of the oocyte is one of the main factors affecting embryo production, the precise identification of noninvasive markers that predict oocyte competence is of major interest. The aim of this review was to explore the current literature on different noninvasive markers associated with oocyte quality in the bovine model. Apart from some controversial findings, the presence of cycle-related structures in ovaries, a follicle size between 6 and 10 mm, a large slightly expanded investment without dark areas, large oocyte diameter (>120 microns), dark cytoplasm, and the presence of a round and smooth first polar body have been associated with better embryonic development. In addition, the combination of oocyte and zygote selection, spindle imaging, and the anti-Stokes Raman scattering microscopy together with studies decoding molecular cues in oocyte maturation have the potential to further optimize the identification of oocytes with better developmental competence for in vitro technologies in livestock species.

Abstract: The efficiency of producing embryos using in vitro technologies in livestock species rarely exceeds the 30–40% threshold, indicating that the proportion of oocytes that fail to develop after in vitro fertilization and culture is considerably large. Considering that the intrinsic quality of the oocyte is one of the main factors affecting blastocyst yield, the precise identification of noninvasive cellular or molecular markers that predict oocyte competence is of major interest to research and practical applications. The aim of this review was to explore the current literature on different noninvasive markers associated with oocyte quality in the bovine model. Apart from some controversial findings, the presence of cycle-related structures in ovaries, a follicle size between 6 and 10 mm, large number of surrounding cumulus cells, slightly expanded investment without dark areas, large oocyte diameter (>120 microns), dark cytoplasm, and the presence of a round and smooth first polar body have been associated with better competence. In addition, the combination of oocyte and zygote selection via brilliant cresyl blue (BCB) test, spindle imaging, and the anti-Stokes Raman scattering microscopy together with studies decoding molecular cues in oocyte maturation have the potential to further optimize the identification of oocytes with better developmental competence for in-vitro-derived technologies in livestock species.

Keywords: oocyte competence; livestock production; assisted reproductive technology; embryo development; micromanipulation; in vitro production

1. Introduction

In recent years, new knowledge in the field of assisted reproductive technologies (ART, has allowed researchers and practitioners to reach new hallmarks in oocyte and sperm in vitro competence. Gamete competence is the ability to undergo successful fertilization and develop a normal blastocyst that is capable of implanting in the uterus and generate viable offspring [1]. Many researchers are focused on identifying cellular and molecular markers to select the most competent oocyte and spermatozoon to produce embryos with higher implantation potential [2].

Although it is well known that the most common applications of ARTs in livestock species are for research purposes, some techniques, particularly in vitro embryo production (IVP), have become commercially viable and are extensively used for animal breeding [3]. Nonetheless, the efficiency of IVP technologies in livestock species, such as bovine, equine, and porcine, measured as the proportion of immature oocytes that reach the blastocyst stage, rarely exceeds the 30–40% threshold [4], which means that the proportion of oocytes that fail to develop following in vitro maturation, fertilization, and culture is considerably large. Contrary to humans, where eggs are mainly collected at the MII stage, in livestock species, the oocytes have to be matured in vitro due to the difficulty of obtaining a sufficient number of in vivo matured oocytes [5]. Additionally, given that the most frequent source of ovaries is slaughterhouse-derived animals, many important factors that influence oocyte quality, such as age of the donor, the stage of the estrous cycle, nutritional status, genetic potential, presence of a reproductive disorder, and others, are often unknown [6]. Therefore, it is almost impossible to avoid the retrieval of a heterogeneous population of oocytes that have a distinct ability to undergo maturation and support early embryonic development after fertilization, which is known as developmental competence or oocyte quality [7].

Considering that the intrinsic quality of the oocyte is one of the major factors affecting early embryonic development [8], and that embryo culture conditions have a crucial role in determining blastocyst quality [9], the precise selection of competent oocytes is vital for IVP technologies in livestock. Recently, the new arrival of bovine embryonic stem cells (ESCs) [10,11] emphasizes the already existing challenge in the selection of competent oocytes for the production of high-quality embryos through in vitro fertilization (IVF), intracytoplasmic sperm injection (ICSI) or somatic cell nuclear transfer (SCNT), and derivation of pluripotent stem cell lines, with promising applications in research or industry, such as in vitro breeding programs [12]. Usually, for IVP and micromanipulation procedures (ICSI and SCNT), the choice of the oocytes lie in morphological features that are easily assessed with light microscopy [13]. The major difference and/or advantage of conventional IVF compared to micromanipulation procedures is that fertilization can occur during gamete co-incubation when the oocyte has reached or is close to nuclear and cytoplasmic maturity [14]. Conversely, during micromanipulation procedures, the operator must accurately assess the maturity of the oocyte and, therefore, its competence [15]. Because the criteria used for grading and selecting oocytes vary among researchers, could be easily misinterpreted, and depend on the expert's evaluation and experience, the identification of noninvasive cellular or molecular markers that predict oocyte competence is a major research goal [16,17]. Despite efforts for finding molecular factors associated with oocyte quality, it is still challenging to find a visual marker that accurately predicts embryonic competence. Thus, this article reviews the current literature on different noninvasive markers that have been correlated with oocyte quality in cattle and explores the utility of each grading system.

2. Morphological and Visual Markers for the Selection of the Best Oocytes

2.1. Ovarian Morphology

During the retrieval of oocytes from slaughterhouse material, the collection of ovaries based on the presence or absence of estrus cycle structures, i.e., presence or absence of follicles and corpus luteum (CL), has been used as a straightforward noninvasive criterion to access developmentally competent oocytes. However, there are discrepancies among different studies in this regard. Early studies indicated that the presence of a dominant follicle (>10 mm) in one or both ovaries had a negative effect on in vitro developmental competence of oocytes derived from the subordinate follicles [18–20]. Manjunatha et al. [21] reported that embryonic development was higher in oocytes coming from ovaries with a CL and no dominant follicle, whereas gametes coming from ovaries that had a CL and a dominant follicle showed higher competence only when oocytes were derived from the dominant follicle. In agreement with this notion, Pirestani et al. [22] reported that oocytes derived from ovaries containing a large follicle (~20 mm) were less competent compared to those derived from ovaries containing a CL. Similarly, Penitente-Filho et al. [23] classified cumulus–oocyte complexes (COCs) under the stereomicroscope and indicated that ovaries with CL yielded a larger number of competent oocytes than ovaries without CL. However, the oocytes used in the latter study were not subjected to IVP to confirm their developmental competence. Overall, these studies indicate that the presence of a dominant follicle in the bovine ovary would negatively influence the subsequent embryo development, while the presence of a CL favors oocyte competence. In contrast, more recent studies indicated that the presence of a CL has negative effects on the developmental competence of ipsilateral oocytes [24,25]. However, this "negative" effect does not influence the competence of oocytes originated from large follicles (10–20 mm) as much as those derived from small and medium follicles (<9 mm) [25].

Ovaries without structures indicative of estrus cyclicity have less competent oocytes than others [21,26], as indicated by the presence of fewer than 10 follicles 2–5 mm in diameter and no large follicles [27]. In addition, other authors indicated that the developmental competence of bovine oocytes from antral follicles (2 to 8 mm) is not affected by either the presence of a dominant follicle or the phase of folliculogenesis [27–31]. Thus, despite the few discrepancies, it seems that the selection of ovaries based on the presence of cycle-related structures could help optimize access to oocytes with better developmental competence for in-vitro-derived technologies. Nevertheless, the positive or negative effects of ovarian structures on oocyte competence require further investigation to determine more precisely how these ovarian structures impact subsequent in vitro embryonic development.

2.2. Follicle Size

One of the most used criteria to obtain competent oocytes is the size of the follicle. Research over the past decades indicates that bovine oocytes gain competence at late stages of the follicular phase, when signs of atresia are observed for the first time, such as a slight expansion in the outer cumulus layers and some cytoplasmic granulations [7,32]. Therefore, the recommendation is that oocytes recovered from follicles between 6 and 10 mm develop more frequently to more advanced embryonic stages [7,33–36]. Although the acquisition of competence begins when the follicle reaches 3 mm and the effect of size becomes more important at 8 mm [19,37,38], success is not guaranteed even if the oocytes come from larger follicles [39].

The acquisition of oocyte competence seems to be due to the substrate support received and to the developmental phase at the time of removal from the follicle [7,32,34]. Recent reports indicate that the follicular fluid (FF) microenvironment of large follicles has higher levels of electrolytes, glucose, reactive oxygen species, glutathione, superoxide dismutase activity, lipids, cholesterol, pyruvate, and estradiol [33,40,41]. Moreover, oocytes derived from larger follicles also show a different transcriptional pattern for chromatin remodeling and metabolic pathways, such as lipid metabolism, cellular stress, and cell signaling, with respect to those coming from smaller sizes, which would favor their developmental potential [41,42]. Therefore, these findings indicate that large follicles (>6 mm) provide an appropriate microenvironment for the oocyte leading to better embryonic development.

2.3. Morphology of the Cumulus–Oocyte Complexes

The quality of COCs can be influenced by multiple factors, both intrinsic and extrinsic. Intrinsic factors include breed, age, reproductive status, metabolic and nutritional status, hormonal levels, and stage of the estrous cycle [43], whereas key extrinsic factors include the timing between slaughter and oocyte withdrawal from the ovary, morphology and methods of collecting the COCs, storage temperature of the ovaries, collection media, and micromanipulation skills of the operator [44].

Since intrinsic factors are more difficult to control when using slaughterhouse ovaries from cows of unknown origin, the morphology of the COC is relatively easy to evaluate and is often the most common criterion used to select and classify a standard collection of bovine oocytes [45–47]. Morphological criteria include the number and appearance of cumulus layers and the cytoplasmic features of the oocyte, such as the texture or brightness of its cytoplasm. Basically, the healthiest COC quality (Class I) relates to a complete cumulus cover with several compact cell layers; medium quality (Class II) has only partial cumulus cover and/or slightly expanded cumulus containing fewer than five cell layers; lastly, the worst quality (Class III) has a darker cytoplasm and the presence of dark spots with expanded cumulus, all indicative of follicular atresia (Figure 1). However, such classification criteria vary among laboratories.

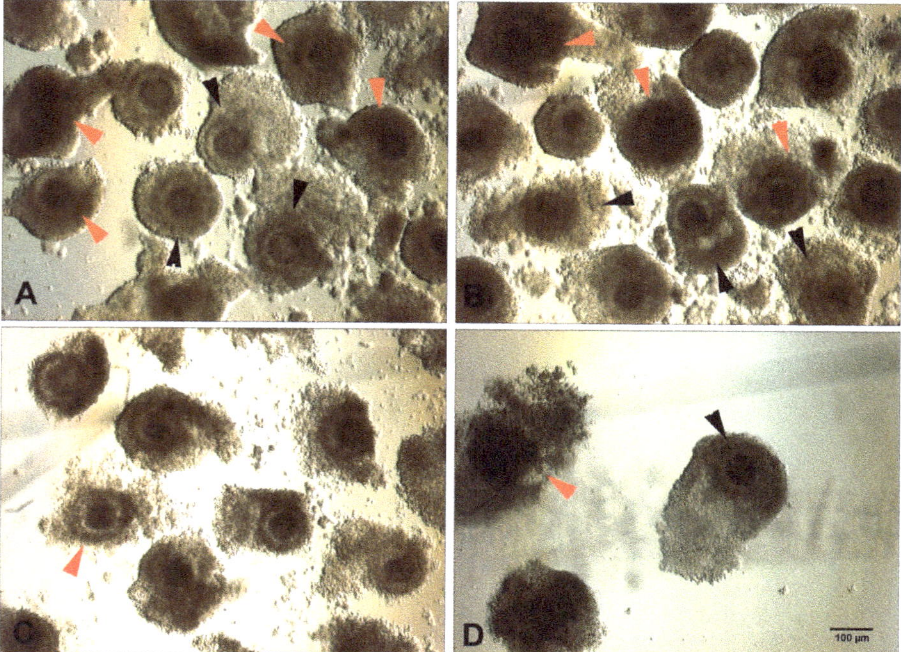

Figure 1. Representative images of bovine cumulus–oocyte complexes (COCs) after ovary withdrawal classified according to COC morphology. (**A**,**B**) Complete cumulus cover with several compacts (red arrows) and slightly loose (black arrows) cell layers; (**C**) partial cumulus cover and loose cell layers with signs of early atresia (red arrow); (**D**) COC showing clear signs of atresia (red arrow) and a black-punctate cytoplasm (black arrow).

The study by Wit et al. [30] classified COCs into three groups: (i) compact and bright, (ii) less compact and dark, and (iii) strongly expanded cumulus with dark spots, where developmental capacity, measured by in vitro embryo production, was correlated with COC appearance. Moreover, less compact and darker COCs showed faster meiotic resumption. Another study using similar categories reported that COCs with darker cumulus and ooplasm were the most competent in terms of cleavage and blastocyst yield after IVF and parthenogenetic activation [48]. In addition, this study showed that developmental competence was related to calcium currents in the plasma membrane and calcium stores in the cytoplasm of immature oocytes [48]. The report by Bilodeau-Goeseels et al. [49] divided COCs into six classes on the basis of their cumulus and ooplasm features. These authors found that, although oocytes with fewer than five layers of cumulus cells (CC) showed lower cleavage rates, their developmental potential to the blastocyst stage was similar to oocytes with more than five layers of CC. More recently, De Bem et al. [37] found that class III COCs, considered to be of poor morphological quality, were superior in terms of blastocyst development to the intermediate class II group, but similar to class I COCs, albeit without differences in blastocyst quality. Emanuelli et al. [50] indicated that COCs with partial (fewer than five cell layers) and expanded cumulus had higher levels of DNA fragmentation after in vitro maturation (IVM) and lower competence compared to healthier ones, in accordance with the report by Yuan et al. [51]. However, blastocysts derived from COCs with varied morphologies exhibited no variations in terms of quality assessed by the number of cells. In addition, Emanuelli et al. [50] further concluded that these differences were due to better nuclear maturation through enhanced maintenance of metaphase II (MII) block by COCs showing full cumulus coverage.

Thus, despite these contradictory results, most studies agree that COCs showing signs of early atresia yield high blastocyst rates compared to morphologically healthy COCs. Nonetheless, advanced atresia, with signs such as cytoplasmic granulations, fewer than five cumulus layers, and expanded cumulus with dark cellular masses or, strictly, its complete absence, show lower in vitro potential as measured by cleavage rates and blastocyst formation [30] (Figure 1C). Additionally, although morphological classification seems to influence the proportion of blastocysts formed, such criteria may not influence their quality. Therefore, when selecting COCs according to their cumulus investment and ooplasm texture, the ideal would be to target COCs with several cumulus cell layers (more than or at least five layers), compact and/or slightly expanded, with or without dark areas in the oocyte and cumulus.

2.4. Lipid Content

The morphological appearance of the ooplasm commonly assessed to select the oocytes [52,53] is influenced by lipid content in livestock species, such as cattle, pigs, and horses [54–56]. Lipids, in the form of lipid droplets (LDs), are signaling molecules with important roles in oocyte maturation and competence acquisition [57]. In the late stage of oocyte maturation and during preimplantation development, endogenous oocyte lipids work as an energy source [58,59] and as a lipid factory for energy reserve [60]. Failure to use lipids in oocytes has been shown to be related to inadequate nuclear maturation [61,62]. The number of LDs present in the cytoplasm increases as the oocyte grows [63] and, although the ooplasm organization does not undergo major changes during in vitro maturation to MII [56], the type and number of lipids in the LDs seem to be more dynamic and to undergo changes during meiotic progression to MII [59,64].

LDs aggregate in the form of dark clusters that can be seen in the ooplasm as a cytoplasmic darkness [55,65] (Figure 2). Cytoplasmic darkness can be homogeneous, affecting the entire cytoplasm or concentrated in the center, with a clear peripheral ring that gives the cytoplasm a darkened appearance (Figure 2B,D). This opaque appearance is more intense in pigs and domestic cats, followed by cows and finally sheep and goats, whose ooplasm is lighter. In the case of horses, lipid polarization is commonly observed, which facilitates the visualization of the spermatozoon within the oocyte [55,66].

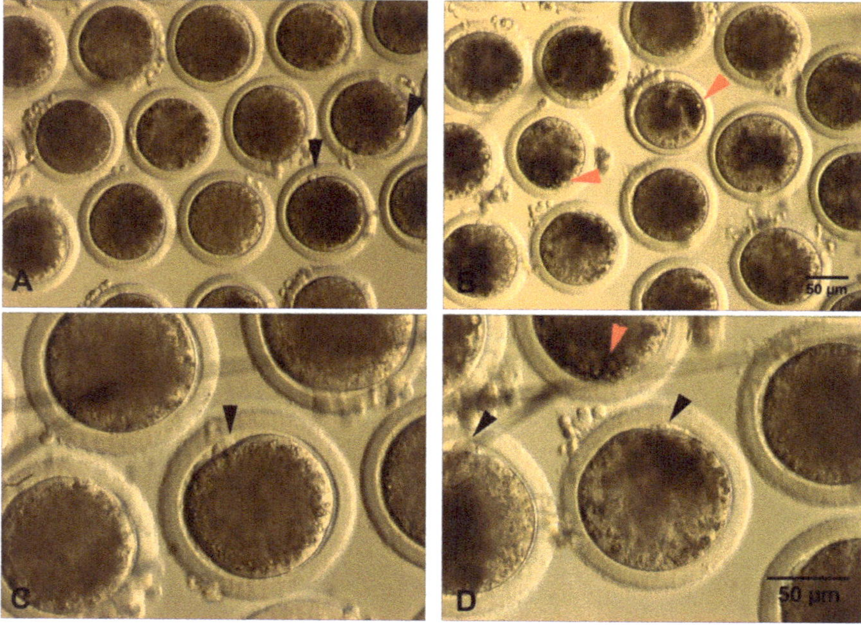

Figure 2. Denuded MII bovine oocytes after 24 h of IVM. (**A**,**C**) oocytes showing a homogeneous dark cytoplasm. Black arrows depict the first polar body; (**B**,**D**) oocytes showing a heterogeneous pale and punctuated cytoplasm. Black arrows indicate the first polar body, while red arrows depict dark areas of intense lipid accumulation (cytoplasmic granulations).

Several studies investigated the relationship between oocyte lipid content and competence. For instance, cytoplasm color can be used as a marker of lipid content and as predictive of the embryonic potential [67], as oocytes with a uniform and brown or dark cytoplasm contain more intracellular lipids than oocytes with a granular or pale cytoplasm [65]. Most studies demonstrated that oocytes with rough granulations or very pale ooplasm yield a lower preimplantation development [49,53,67]. Jeong et al. [68] classified the ooplasm in three categories: dark, brown, and pale. In this study, the content of mitochondria and the proportion of oocytes that reached the blastocyst stage were higher in darker oocytes. Moreover, Nagano et al. [67] reported that sperm penetration, monospermic fertilization, cleavage, and blastocyst rates were higher in oocytes with a brown ooplasm compared to those with pale or very dark ones. Moreover, brown oocytes with a dark edge or with dark spots showed, under electron microscopy, an organelle arrangement similar to in vivo matured oocytes, and pale or black oocytes appeared to be degenerating and/or aging [67]. The authors concluded that a dark ooplasm is associated with a lipid accumulation and better developmental competence, while a pale ooplasm would indicate fewer organelles and poor developmental potential [69]. Interestingly, a study by Prates et al. [70] distinguished fat areas of different color shades using the Nomarski interference differential contrast (NIC) as the fat gray value of porcine oocytes, reflecting alterations in lipid content, and proposed this tool as an appropriate and noninvasive technique to evaluate the lipid content of a single oocyte before or after in vitro maturation. Recently, the study of Jasensky et al. [71] reported the use of anti-Stokes Raman scattering (CARS) microscopy as a new non-invasive tool for the quantification of lipid content in mammalian oocytes. This study showed that the ~2 min of laser exposure was enough for a quantitative comparison of lipid content in mice oocytes at different developmental stages, as well as in oocytes of others mammalian species, and, more importantly, without detrimental effects (without the need to attach fluorescence labels) for subsequent preimplantation development. Thus, its application in live-cell imaging of oocytes is promising to provide alternative and/or additional information in order to improve the accuracy of subjective morphometric measurements.

Taken together, as stated by the review of Nagano and colleagues [69], a dark ooplasm indicates an accumulation of lipids and good developmental potential, a light-colored ooplasm indicates a deficiency of lipid stores and poor developmental potential, and a black ooplasm indicates aging and low developmental potential (Figure 2). Finally, the use of NIC and CARS should be further investigated as a potential noninvasive tool to evaluate the lipid content of single oocytes in livestock species.

2.5. Cumulus Expansion and Oocyte Size

Another parameter that is often used as an indirect indicator of oocyte quality is the degree of cumulus expansion following maturation, typically after 20 to 24 h of culture in an in vitro maturation environment. Grades 1 to 3 (sometimes 4) are attributed to increasing degrees of expansion (1: modest expansion, characterized by few morphologic changes compared to before maturation, 2: partial expansion, and 3: complete or almost complete expansion) [72–74].

Although the expansion of CCs has been described as the basis for oocyte maturation [75] and early reports supported the idea that quantity and quality of the expanded cumulus mass were correlated with developmental capacity [76], its usefulness as an indicator of developmental potential in bovine seems to be modest [77]. For instance, studies by Anchordoquy et al. [78], Dovolou et al. [79], and Rosa et al. [80] reported that, under different experimental conditions, the cumulus expansion index was not indicative of blastocyst yield or quality. Similarly, another study indicated that inhibition of cumulus expansion by enzymatic hyaluronidase degradation did not affect cleavage or blastocyst development [81]. Nonetheless, as shown by Fukui et al. [82], more than an indicator of developmental competence, CCs and their expansion play an important role in fertilization by inducing the acrosome reaction and, therefore, promoting higher fertilization rates.

In addition to follicle size, oocyte size has been used as a noninvasive quality parameter. Although it is difficult to measure the precise diameter of the oocyte during IVF, oocyte selection based on diameter can be used as a routine step during micromanipulation protocols. The study of Fair et al. [83] classified oocytes recovered from slaughterhouse ovaries into four groups (<100 microns, 100 to 110 microns, 110 to 120 microns, and >120 microns). Rates of resumption of meiosis to MII were higher for oocytes >110 microns. Moreover, oocytes <110 microns were transcriptionally active, suggesting that they were still in the growth phase of oogenesis [83,84]. Similarly, Anguita et al. [85] reported that cleavage and blastocyst rates were higher in oocytes >110 microns. Moreover, Otoi et al. [86] and Arlotto et al. [29] found that oocytes >115 microns had better rates of nuclear maturation and a lower incidence of polyspermy after IVF, but cleavage rates and development to the blastocyst stage were optimal in oocytes >120 microns. Huang et al. [87] and Yang et al. [88] compared oocytes collected from initial antral follicles (0.5–1 mm in diameter) cultured in vitro for 14–16 days with oocytes collected from antral follicles (2–8 mm in diameter), cultured, and submitted to IVM. The authors reported better maturation rate for oocytes >115 microns, optimal for oocytes >120 microns, but developmental competence was only high for oocytes collected from antral follicles and of size >120 microns.

These results suggest that bovine oocytes acquire meiotic competence with a diameter of 115 microns, but full developmental competence is acquired around 120 microns, possibly because smaller oocytes have not yet completed their growth phase [46]. Thus, the selection of follicles between 6 and 10 mm, with oocyte diameters >115 and <130 microns, has the potential to optimize developmental outcomes.

2.6. First Polar Body Assessment

At the end of IVM and after the removal of CCs, it is easy to perform a detailed observation of morphological features [13], including the assessment of oocyte shape, cytoplasm color and granulation, regularity and thickness of the zona pellucida, size of the perivitelline space, presence of vacuoles, and presence or absence of the first polar body (PB1) and its morphology. Extrusion of PB1 in mammalian oocytes is a cellular landmark of meiotic maturation, and its assessment is frequently used as an indicator of nuclear maturation [89]. Thus, its absence indicates that the oocyte is immature or that it has degraded due to aging; however, its presence does not guarantee that the oocytes have completed their maturation process, and some of them remain incompetent despite exhibiting morphologic features of nuclear maturation [90].

In bovine species, extrusion of PB1 begins at 16–18 h after IVM [91–94]. Nonetheless, oocytes acquire the highest developmental competence at around 5–10 h after PB1 extrusion [14,95]. Dominko and First [95] indicated that oocytes that extruded their PB1 after 16 h of IVM were only capable of reaching higher developmental competence after 24 h of in vitro culture. Thus, cytoplasmic maturation in cattle occurs several hours after nuclear maturation, probably between 24 and 30 h after the beginning of IVM.

Unfortunately, there are no studies that analyzed the influence of the first PB morphology on oocyte competence in cattle. However, one study using porcine oocytes indicated that PB1 with a smooth or intact surface was indicative of a more advanced cytoplasmic maturation and better embryonic development in vitro than those with a fragmented or rough surface [96]. Despite lacking studies in domestic species, studies in humans investigated the association between PB1 morphology and oocyte competence [97,98]. Ebner et al. [99] conducted a retrospective study using 70 consecutive ICSI cases in which oocyte classification based on PB1 morphology revealed that oocytes with intact, well-shaped PB1 yield better fertilization and high embryonic quality. Later, Ebner et al. [97] confirmed the relationship among PB1 morphology, fertilization, and blastocyst quality, as well as a positive effect on implantation and pregnancy rates. Similarly, Rose et al. [100] reported that oocytes with an intact PB1 show better fertilization and embryonic development, whereas those displaying a PB1 with morphological abnormalities such as a larger size, irregularities, coarse surface, or fragmentation are less competent during an IVF protocol, having poor implantation capabilities after embryo transfer. In contrast, others did not report any correlation [101–103]. Thus, there is a lack of consensus on the impact of PB1 morphology on oocyte competence and embryonic development in humans. It is also important to note that some PB1 abnormalities may be an artefact of oocyte manipulation (mainly during the denudation process) or aging [104].

In summary, although the selection of oocytes with PB1 of a homogeneous, round shape with a smooth or intact surface may be indicative of a better oocyte, the usefulness of this selection criterion in livestock requires further research to establish its real predictive value for oocyte competence.

2.7. Polarized Light Microscopy

Polarized light microscopy (PLM) has been used in different mammalian oocytes since it allows the noninvasive assessment of subcellular features such as the meiotic spindle and zona pellucida birefringence (ZPB). To learn about the principles and equipment required for PLM in detail, readers are directed to excellent reviews on the subject [105,106].

2.7.1. Evaluation of the Meiotic Spindle and Zona Pellucida Birefringence

Using PLM, it is possible to locate and evaluate the morphology of the meiotic spindle to confirm egg maturation, which has been positively correlated with developmental competence [90,107–109]. This method avoids damaging the spindle during the ICSI procedure, considering that the position of the PB1 can be altered when CCs are removed during preparation for ICSI [110]. Furthermore, PLM has been successfully used to remove the meiotic spindle and chromosomes (enucleation) in mice [111], bovines [112], and pigs [113], with an average efficiency of 90% and, more importantly, avoiding the exposure to ultraviolet (UV) rays and their detrimental effect on embryonic development.

In livestock species, the dark appearance of the ooplasm, attributed to high lipid contents, is known to interfere with spindle imaging [113] and, as in humans, precludes the detection of meiotic spindle abnormalities [102,113,114]. Therefore, spindle birefringence should be carefully considered as an index of gamete quality and chromosome alignment in some species. In pigs, a negative PLM signal was associated with to reduced maturation and poor development potential [113]. In the same study, when the PLM system was used for spindle removal, the overall enucleation efficiency was 92.6%, indicating that PLM is an effective tool for performing enucleation in pigs. A few years later, the same group evaluated the use of PLM to assess the meiotic spindle of in vitro matured bovine oocytes after vitrification and warming [115]. They were able to confirm the presence of the meiotic spindle in 99% of the analyzed eggs. Moreover, after vitrification and warming, meiotic spindles were detected in 79% of oocytes. Interestingly, thawed oocytes that displayed a positive PLM signal showed better competence in terms of cleavage and blastocyst rates after parthenogenetic activation, indicating that PLM can be a useful tool for assessing post-warming viability in vitrified bovine oocytes.

Overall, these studies demonstrate that PLM efficiently detects the meiotic spindle of livestock oocytes and does not affect early embryonic development. However, the selection of cattle oocytes on the basis of the presence of a PLM signal does not seem to offer improvement in IVP outcomes yet.

2.7.2. Assessment of the Zona Pellucida Birefringence

In addition, PLM has been used for the evaluation of the ZPB, which in humans has been associated with oocyte quality [116–118], although this is still under debate [119,120]. The few studies in cattle showed that a lower ZPB is related to high-quality oocytes and improved blastocyst development [121,122], whereas two studies in horses reported conflicting results, indicating beneficial effects of both low ZPB [123] and high ZPB [124]. Because most of the studies with PLM were carried out in mice and humans with conflicting results, its potential application and practical use in cattle and other livestock species needs further assessment. Contrary to humans, where the number of highly valuable oocytes from donors is relatively low, livestock oocytes obtained from slaughterhouse ovaries allow a more stringent selection. Furthermore, assessment of the meiotic spindle can be a laborious procedure, which delays the overall process of in vitro manipulation and embryo production. Thus, its application will require showing a clear advantage over conventional approaches using the morphological criterion mentioned above for oocyte selection. However, PLM might be beneficial when individual oocytes are of high value, such as oocytes recovered from elite cows by ovum pick-up (OPU) [111,113].

2.8. Brilliant Cresyl Blue (BCB) Staining

Another approach that demonstrated predictive potential is the evaluation of glucose-6-phosphate dehydrogenase (G6PDH) activity via brilliant cresyl blue (BCB) staining. BCB is a dye that determines the intracellular activity of G6PDH. Activity of G6PDH is observed during the oocyte growth phase (BCB^-: colorless cytoplasm, increased G6PDH) due to the demand of ribose-6-phosphate for nucleotide synthesis. This activity is low (BCB^+: colored cytoplasm, low G6PDH) in oocytes that have completed their growth phase [125]. This technique has been successfully employed in various species, including cattle [125–127].

Although previous reports found that the developmental competence of oocytes with low G6PDH activity (BCB^+) was higher than that of oocytes with a high G6PDH activity (BCB^-), the absence of differences in terms of embryonic development between BCB^+ and the untreated control group decreases the utility of the BCB test in IVP technology [128]. However, it is unquestionable that BCB^+ oocytes have statistically higher developmental competence than BCB^- oocytes, both in IVF and somatic cell nuclear transfer (SCNT) [128].

Later studies continued to show only a trend of BCB^+ oocytes toward greater developmental potential. Better blastocyst rates at day 7 were reported by Silva et al. [129], and a study by Fakruzzaman et al. [130] reported higher blastocyst quality on the basis of total apoptotic cells and mitochondria numbers. Similarly, Castaneda et al. [131] indicated that the higher lipid content of BCB^+ bovine oocytes might be associated with their better developmental competence. Interestingly, another article indicated that co-culture with BCB^- oocytes during IVM affects negatively the capacity of BCB^+ oocytes to undergo embryonic development [132]. However, other authors suggested that the BCB test is not sufficient for identification of the most competent gametes [133]. Nonetheless, the combination of oocyte and zygote selection using BCB staining would improve the efficiency of embryo selection [134]. Therefore, the BCB test can be a valuable tool when used together with classical morphological classification and could be useful for the selection of oocytes with a higher implantation potential. Nonetheless, an assessment of the effects of BCB staining on post-implantation development is necessary to elucidate its usefulness for IVP technologies, not only for research but also in the industry of animal production. A summary of the morphological and visual indicators associated with oocyte competence is shown in Table 1.

Table 1. Summary of the morphological and visual indicators of oocyte competence.

Reference	Criteria	Recommendation
[28,30,31]	Ovarian morphology	Presence of cycle-related structures
[7,33–35]	Follicle size	>5 mm
[30,49,50]	Morphology of the cumulus–oocyte complexes (COCs)	COCs with at least five layers of cumulus cells (CC), compact and/or slightly expanded cumulus, with or without dark spots in the oocyte and cumulus
[53,67,68]	Lipid content	Dark ooplasm indicates high competence, light-colored indicates lacking lipids and poor competence, and black ooplasm indicates aging
[77–80]	Cumulus expansion and oocyte size	Not associated to oocyte quality; important role in fertilization
[29,83,86,135]	Oocyte size	Diameters >115 and <130 microns
[96–99]	First polar body (PB1) morphology	PB1 of a homogeneous, round shape with a smooth or intact surface
[112,114,115]	Meiotic spindle and zona pellucida birefringence	Useful tool for micromanipulation procedures (intracytoplasmic sperm injection (ICSI) or somatic cell nuclear transfer (SCNT)) and for assessing post-warming integrity of meiotic spindle of vitrified bovine oocytes
[121,122]	Zona pellucida birefringence (ZPB)	Lower ZPB is related to high quality oocytes and improved blastocyst development
[115,128–130,134]	Brilliant cresyl blue staining	BCB^+ oocytes have higher developmental competence than BCB^- oocytes

3. Non-invasive Molecular Approaches

Many studies are being performed in mammals in order to find molecular markers predictive of oocyte quality. So far, most of the data show considerable variations, perhaps due to different experimental conditions and/or the criterion of quality/competence, resulting in varied scientific views.

3.1. Cell Death (Apoptosis) in Cumulus Cells

Because morphological evaluation prior to maturation does not allow to discriminate the atretic oocytes from healthier ones [135], one of the earlier noninvasive markers of oocyte competence was the level of apoptosis in CC, seen as DNA fragmentation, externalization of phosphatidylserine (EP), and/or the expression ratio of anti-apoptotic (Bcl-2) and pro-apoptotic (Bax) genes (BCL-2/BAX). Early studies found that the CC of bovine COCs undergo progressive apoptosis during IVM [136], and this was negatively correlated with the oocyte developmental capacity [51]. However, results reported by Janowski et al. [137] supported the notion that follicular cells surrounding the more competent oocytes have a higher degree of apoptosis. Later, Warzych et al. [138] showed that the level of apoptosis in CC was not associated with morphology or the oocyte meiotic stage, suggesting that the extent of apoptosis in CC is not a reliable quality marker for gamete competence. Similarly, the study of Anguita et al. [135] showed that embryonic developmental potential increased together with oocyte diameter, but this developmental competence was not related to the incidence of apoptosis. Recently, another study indicated that optimum control of the meiosis block, nuclear maturation, and developmental potential were associated with less DNA fragmentation in CC [50].

Similarly, in the human model, the majority of related studies have focused on granulosa cells (GC) isolated from FF during oocyte collection. Apoptosis, evaluated by EP, of GC was negatively associated with egg and embryo numbers in IVF/ICSI cycles, pregnancy rate, and live birth rate after IVF [139,140]. However, contrarily, it was also reported that the EP in GC is not related to follicular quality and oocyte competence during ICSI [141]. Thus, in the bovine and human models, it is still controversial whether apoptosis of GC and/or CC can impact the developmental potential of the oocyte.

3.2. *Transcriptomic and Proteomic of Cumulus Cells*

Many new genomic tools helped to deepen the understanding in the area of oocyte–cumulus communication, as well as molecular pathways required for the acquisition of competence in mammalian gametes and embryos. For instance, recent advances in RNA-Seq technology offer a global transcriptomic approach for identifying differentially expressed genes associated with competence and embryonic development.

Among the molecular approaches, study of the transcriptomic profile of the surrounding cumulus is one of the most popular attempts at finding molecular markers associated with gamete competence in mammals. The "noninvasive" strategy is based on profiling the gene expression of a small biopsy before IVM, maintaining COC integrity, and following the embryonic development of the respective oocyte. This is also called "oocyte fate" [142]. Although several studies in cattle already found several genes in CCs from germinal vesicle (GV) [16,35,143–151] and MII oocytes [144,152] to be associated with oocyte competence, only a few reports matched the oocyte fate with the transcriptomic profile obtained from the CCs or granulosa cells (Table 2). There is some consensus regarding pathways correlated positively with oocyte competence, including the cell cycle (CCND1, CCNB2, and CCNA2 genes) [143,145,153], cell growth and proliferation, (CD44, TGFB1, EGF, FGF11, PRL, and GH genes) [35,147–149,154], and steroidogenesis (HSD3B2 and CYP11A1 genes) [16,154]. On the contrary, genes related to cellular apoptosis would be associated with a low competence (ATRX, KRT8, ANGPT2, KCNJ8, and ANKRD1 genes) [142,147,152,155].

Table 2. Summary of studies performing transcriptomic and proteomic analysis of CC and/or GC.

Reference	Oocyte Stage	Criterion of Developmental Competence	Technique Used	Genes and/or Pathways Associated with High Competence	Genes and/or Pathways Associated with Low Competence
Transcriptomic					
[16]	GV	GC collected 2 h before and 6 h after LH surge	qPCR and microarray analysis	TNFAIP6, HAS2, HSD3B2, PLOD2, CHSY1 (differentiation, cell growth, protein translation, apoptosis-related, lipid and glucose metabolism, ECM formation)	ENO1, DNAJB6, GJA1, SYNPO, ZNF330, MYO1D (protein synthesis cellular movement, cell signaling, molecular transport, nucleic acid metabolism)
[35]	GV	follicle size (1.0–3.0, 3.1–6.0, 6.1–8.0, and ≥8.1 mm)	qPCR	FSHR (follicle stimulant hormone receptor), GH (cell growth), and EGF (cell growth and differentiation)	N.A
[142]	GV	Cell arrest and oocyte fate	qPCR and microarray analysis	GATM (post-translational modification, amino-acid metabolism, and free-radical scavenging)	AGPAT9 (lipid metabolism), CLIC3 (chloride ion concentration control, cell volume regulation, and apoptosis), KRT8 (cellular assembly and organization, apoptosis)
[143]	GV	Follicle size (>5 mm vs. <2mm)	qPCR and SSH	Oct4, Msx1 (transcription factors), Znf198 (TFGb and activin signaling), NDFIP1(posttranslational modification), CCNA2 (cell cycle), SLB (stabilization and translation of mRNAs encoding histones)	N.A
[144]	GV	Adult vs. prepuberal donors	qPCR and microarray analysis	N.A	CTSB, CTSK, CTSS, and CTSZ (cathepsin family of lysosomal cysteine proteinases)
[145]	GV	OPU 6 h post LH vs. slaughterhouse oocytes after 6 h IVM	qPCR and microarray analysis	PTTG1, CDC5L, CKS1B, CCNB2 (cell cycle), PSMB2, PRDX1 (cell metabolism), RGS16 (cell signaling), SKIIP (gene expression), and chromatin support H2A	BMP15, GDF9, CCNB1, and STK6 (follicle–oocyte interaction and cell cycle)
[146]	GV	Brilliant cresyl blue staining	qPCR	N.A	CTSB, CTSK, CTSS, and CTSZ (cathepsin family of lysosomal cysteine proteinases)
[147]	GV	GC collected after FSH withdrawal	qPCR and microarray analysis	SMAD7, STAT1 (transcription), PRL and GH (cell growth, proliferation), BMPR1B, IGF2, RELN, and TFPI2 (follicle growth), NRP1 (angiogenesis), GFPT2, TF, and VNN1 (oxidative stress response)	KCNJ8 and ANKRD1 (apoptosis and inflammation)
[148]	GV	Follicle size (>8 mm vs. <3mm)	qPCR and microarray analysis	FGF11 (cell growth, and differentiation), IGFBP4 and SPRY1 (cell cycle, DNA repair)	ARHGAP22, COL18A1, and GPC4 (cell cycle, signaling)
[149]	GV	IVM plus FSH or phorbol myristate acetate (PMA) treatment	qPCR and microarray analysis	HAS2, INHBA, EGFR, GREM1, CD44, TNFAIP6, PTGS2, HSP90B1, SERPINE2, PTX3 (differentiation, cell growth, protein translation, apoptosis, lipid and glucose metabolism, ECM formation)	N.A
[150]	GV	Follicle size and oocyte fate	qPCR	GPC4 (regulation of growth factors, adhesion, signaling, proliferation, and differentiation)	N.A

Table 2. *Cont.*

Reference	Oocyte Stage	Criterion of Developmental Competence	Technique Used	Genes and/or Pathways Associated with High Competence	Genes and/or Pathways Associated with Low Competence
[151]	GV	COCs morphology and oocyte fate	qPCR	N.A	FSHR, IGF1R, CYP11al, and HSD3β (cell growth, cell differentiation, steroidogenesis)
[153]	GV	Maturation outcome and oocyte fate	RNA-seq	CCND1, BMP15, GDF9, H19, KLF4, GPC1, SYCP3, and CTSB (cell cycle, meiosis, cell signaling, metabolism, and apoptosis)	N.A
[154]	GV	FSH withdrawals; follicles from 5 mm aspirated by OPU	qPCR and microarray analysis	CYP11A1 (steroidogenesis), NSDHL (cholesterol synthesis), GATM (creatine biosynthesis), MAN1A1 (functional gap junction-mediated communication), VNN1 (oxidative stress response), NRP1 (angiogenesis), TGFB1 (cell growth and differentiation)	N.A
[155]	GV	Chromatin compaction, follicle size, and BCB staining	qPCR and microarray analysis	GATM (posttranslational modification, amino-acid metabolism, and free-radical scavenging), MAN1A1 (functional gap junction-mediated COC communication), ZIP8 (zinc transporter)	ANGPT2 (cell death, apoptosis)
Proteomic					
[156]	MII	Matured in vivo vs. IVM	MALDI TOF	KEGG pathways of the complement and coagulation cascade, ECM–receptor interactions, steroid biosynthesis, glucose and carbohydrate metabolism	N.A
[157]	GV	COC morphology and follicle size (>2 mm to 8 mm)	2-DLCMS	Integrin signaling, actin cytoskeleton signaling, ephrin receptor signaling, PI3K signaling, MAPK signaling	N.A
[158]	GV	COC morphology and follicle size (>2 mm to 8 mm)	2-DLCMS	4395 proteins were expressed in the CCs; 858 proteins were common to both CCs and oocytes	N.A

MII: meiosis II, GV: germinal vesicle, CC: cumulus cells, GC: granulosa cells, qPCR: quantitative reverse transcription PCR, RNA-seq: RNA sequencing, IF: immunofluorescence, SSH: suppressive subtractive hybridization, BCB: Brilliant cresyl blue, 2-DLCMS: two-dimensional liquid chromatography-tandem mass spectrometry, MALDI TOF: matrix-assisted laser desorption/ionization-time of flight, ECM: extracellular matrix, MAPK: mitogen-activated protein kinases, PI3K: phosphatidylinositol 3-kinase, IVM: in vitro *maturation*. * N.A = not available.

On the other hand, studies analyzing the proteomic profile of the cumulus–oocyte complex (COC) are scarce. Moreover, most of them have done invasive analysis in a pool of oocytes; thus, oocyte fate could not be followed (Table 2). Nonetheless, the few studies described many proteins involved in cell signaling that may have a role in cumulus–oocyte communication and competence. Most of the proteins are involved in components of integrin, actin cytoskeleton, mitogen-activated protein kinases (MAPK) and phosphatidylinositol 3-kinase (PI3K) signaling pathways, extracellular matrix (ECM) receptor interactions, steroid biosynthesis, and glucose and carbohydrate metabolism, which may have implications in various reproductive processes such as oocyte development and maturation [156–158] (Table 2). A recent study reported a highly sensitive approach to characterize the CC proteome from a single COC after in vivo or in vitro maturation [156]. This method shows the potential to directly connect the cumulus proteome to the developmental potential of the corresponding oocyte, as already performed at the gene expression level.

3.3. Follicular Fluid Analysis

It is well known that the composition of FF has an impact on the developmental capacity of the oocyte and, thus, the resulting embryo. Excellent articles reviewed the importance of FF on oocyte physiology and fertility [159–161]. This fluid contains proteins, cytokines, growth factors, steroids, metabolites, and other indeterminate factors [159]. Therefore, by studying its composition, it should be possible to predict oocyte competence and fertilization outcomes [162–164]. Metabolites in the FF, such as glucose and potassium, have already been positively associated with oocyte quality in cattle [41,165]. However, studies linking the FF features with the respective oocyte fate in bovines have not been performed yet. Reports in humans have positively associated the presence of anti-Müllerian hormone (AMH) in FF with competence of the respective oocyte [166,167], although with some contradictory results [168,169]. Conversely, a recent study that used a large population of transferred embryos matching FF samples indicated that the AMH level in FF following withdrawal from the ovarian follicle is closely linked to the oocyte's competence, and it is a suitable predictor of a live birth after single embryo transfer [170]. In the cow, it was already reported that AMH concentrations can be predictive of the number of ovulations and embryos produced in response to ovarian stimulation by FSH [171–173], making it a suitable molecule to be related to the oocyte competence.

In addition, other molecules in FF of cattle that show promising results are microRNAs (miRNAs). The bovine FF contains free miRNAs, as well as some associated with exosomes [174,175]. Recently, the study of Pasquariello et al. [176] showed, for the first time, the miRNA content of different populations of oocytes categorized according to their competence. Interestingly, they discovered that the most differentially expressed miRNAs (miR-24, miR-10a, and miR-320a) in FF found in highly competent follicles were part of the regulation of the neurotrophin signaling pathway, which supports follicle formation and development, as well as the TGF-βsignaling pathway that controls the production of ovarian peptide hormones. Therefore, linking FF molecules such as AMH or miRNAs with gamete competence is an encouraging strategy in the field of oocyte selection. However, we have to consider that it will be applicable only when the fast collection and analysis of FF from individual follicles become practicable.

4. Conclusions and Future Perspectives

The classification and selection of oocytes in livestock species for in vitro embryo production and for micromanipulation techniques, such as ICSI and SCNT, can be one of the most important steps to reach superior embryonic development and quality. Although more sophisticated methods (qRT-PCR, global transcriptomic, and proteomic analysis) have been studied since a few decades ago, the lack of a quick enough method producing reliable results hinders the implementation of these technologies. Moreover, molecular analysis requires high-tech equipment and technical staff that would be cost-ineffective in most research laboratories. Thus, although oocyte selection based on morphologic criteria appears to be insufficient to distinguish more competent gametes, in real practice, when 100–300 oocytes are waiting to be processed during micromanipulation experiments, it seems to be the only available strategy so far. Furthermore, studies that perform embryo transfers are also important to effectively evaluate developmental potential, as successful embryo implantation is highly dependent on the quality of the embryo and the intricate relationship it establishes with the uterine endometrium. Ultimately, with the advent of bovine embryonic stem cells, greater scrutiny of oocytes with high developmental potential is necessary, for the production of stable pluripotent stem cell lines to be used in basic science, forward and reverse genetics, epigenetics, gene imprinting, and the production of animal models with applications in animal production. Thus, in addition to improving the conditions to support in vitro maturation, the implementation of new tools for the assessment of gamete competence, together with studies decoding molecular cues in oocyte maturation, will improve our understanding of this complex process and will more precisely identify the synchrony between nuclear and cytoplasmic maturation in livestock species.

Author Contributions: All authors contributed equally to reviewing the literature, as well as writing and editing the review. All authors read and agreed to the published version of the manuscript.

Funding: This article received no external funding.

Conflicts of Interest: The authors declare no conflict of interest.

References

1. Conti, M.; Franciosi, F. Acquisition of oocyte competence to develop as an embryo: Integrated nuclear and cytoplasmic events. *Hum. Reprod. Update* **2018**, *24*, 245–266. [CrossRef] [PubMed]
2. Ruvolo, G.; Fattouh, R.R.; Bosco, L.; Brucculeri, A.M.; Cittadini, E. New molecular markers for the evaluation of gamete quality. *J. Assist Reprod. Genet.* **2013**, *30*, 207–212. [CrossRef]
3. Van Wagtendonk-de Leeuw, A.M. Ovum pick up and in vitro production in the bovine after use in several generations: A 2005 status. *Theriogenology* **2006**, *65*, 914–925. [CrossRef] [PubMed]
4. Rizos, D.; Clemente, M.; Bermejo-Alvarez, P.; de La Fuente, J.; Lonergan, P.; Gutierrez-Adan, A. Consequences of in vitro culture conditions on embryo development and quality. *Reprod. Domest. Anim.* **2008**, *43* (Suppl. 4), 44–50. [CrossRef] [PubMed]
5. Telfer, E.E.; Sakaguchi, K.; Clarkson, Y.L.; McLaughlin, M. In vitro growth of immature bovine follicles and oocytes. *Reprod. Fertil. Dev.* **2019**, *32*, 1–6. [CrossRef] [PubMed]
6. Lonergan, P.; Fair, T. Maturation of Oocytes in Vitro. *Annu. Rev. Anim. Biosci.* **2016**, *4*, 255–268. [CrossRef]
7. Blondin, P.; Sirard, M.A. Oocyte and follicular morphology as determining characteristics for developmental competence in bovine oocytes. *Mol. Reprod. Dev.* **1995**, *41*, 54–62. [CrossRef]
8. Krisher, R.L. The effect of oocyte quality on development. *J. Anim. Sci.* **2004**, *82* (E-Suppl), E14–E23. [CrossRef]
9. Rizos, D.; Ward, F.; Duffy, P.; Boland, M.P.; Lonergan, P. Consequences of bovine oocyte maturation, fertilization or early embryo development in vitro versus in vivo: Implications for blastocyst yield and blastocyst quality. *Mol. Reprod. Dev.* **2002**, *61*, 234–248. [CrossRef]
10. Bogliotti, Y.S.; Wu, J.; Vilarino, M.; Okamura, D.; Soto, D.A.; Zhong, C.; Sakurai, M.; Sampaio, R.V.; Suzuki, K.; Izpisua Belmonte, J.C.; et al. Efficient derivation of stable primed pluripotent embryonic stem cells from bovine blastocysts. *Proc. Natl. Acad. Sci. USA* **2018**, *115*, 2090–2095. [CrossRef]
11. Navarro, M.; Soto, D.A.; Pinzon, C.A.; Wu, J.; Ross, P.J. Livestock pluripotency is finally captured in vitro. *Reprod. Fertil. Dev.* **2019**, *32*, 11–39. [CrossRef] [PubMed]
12. Goszczynski, D.E.; Cheng, H.; Demyda-Peyras, S.; Medrano, J.F.; Wu, J.; Ross, P.J. In vitro breeding: Application of embryonic stem cells to animal productiondagger. *Biol. Reprod.* **2019**, *100*, 885–895. [CrossRef] [PubMed]
13. Ebner, T.; Moser, M.; Sommergruber, M.; Tews, G. Selection based on morphological assessment of oocytes and embryos at different stages of preimplantation development: A review. *Hum. Reprod. Update* **2003**, *9*, 251–262. [CrossRef] [PubMed]
14. Koyama, K.; Kang, S.S.; Huang, W.; Yanagawa, Y.; Takahashi, Y.; Nagano, M. Estimation of the optimal timing of fertilization for embryo development of in vitro-matured bovine oocytes based on the times of nuclear maturation and sperm penetration. *J. Vet. Med. Sci.* **2014**, *76*, 653–659. [CrossRef]

15. Coticchio, G.; Sereni, E.; Serrao, L.; Mazzone, S.; Iadarola, I.; Borini, A. What criteria for the definition of oocyte quality? *Ann. N. Y. Acad. Sci.* **2004**, *1034*, 132–144. [CrossRef]
16. Assidi, M.; Montag, M.; Van der Ven, K.; Sirard, M.A. Biomarkers of human oocyte developmental competence expressed in cumulus cells before ICSI: A preliminary study. *J. Assist. Reprod. Genet.* **2011**, *28*, 173–188. [CrossRef]
17. Goovaerts, I.G.; Leroy, J.L.; Jorssen, E.P.; Bols, P.E. Noninvasive bovine oocyte quality assessment: Possibilities of a single oocyte culture. *Theriogenology* **2010**, *74*, 1509–1520. [CrossRef]
18. Varisanga, M.D.; Sumantri, C.; Murakami, M.; Fahrudin, M.; Suzuki, T. Morphological classification of the ovaries in relation to the subsequent oocyte quality for IVF-produced bovine embryos. *Theriogenology* **1998**, *50*, 1015–1023. [CrossRef]
19. Hagemann, L.J.; Beaumont, S.E.; Berg, M.; Donnison, M.J.; Ledgard, A.; Peterson, A.J.; Schurmann, A.; Tervit, H.R. Development during single IVP of bovine oocytes from dissected follicles: Interactive effects of estrous cycle stage, follicle size and atresia. *Mol. Reprod. Dev.* **1999**, *53*, 451–458. [CrossRef]
20. Hagemann, L.J. Influence of the dominant follicle on oocytes from subordinate follicles. *Theriogenology* **1999**, *51*, 449–459. [CrossRef]
21. Manjunatha, B.M.; Gupta, P.S.; Ravindra, J.P.; Devaraj, M.; Ramesh, H.S.; Nandi, S. In vitro developmental competence of buffalo oocytes collected at various stages of the estrous cycle. *Theriogenology* **2007**, *68*, 882–888. [CrossRef] [PubMed]
22. Pirestani, A.; Hosseini, S.M.; Hajian, M.; Forouzanfar, M.; Moulavi, F.; Abedi, P.; Gourabi, H.; Shahverdi, A.; Taqi Dizaj, A.V.; Nasr Esfahani, M.H. Effect of ovarian cyclic status on in vitro embryo production in cattle. *Int. J. Fertil. Steril.* **2011**, *4*, 172–175. [PubMed]
23. Penitente-Filho, J.M.; Jimenez, C.R.; Zolini, A.M.; Carrascal, E.; Azevedo, J.L.; Silveira, C.O.; Oliveira, F.A.; Torres, C.A. Influence of corpus luteum and ovarian volume on the number and quality of bovine oocytes. *Anim. Sci. J.* **2015**, *86*, 148–152. [CrossRef] [PubMed]
24. Hajarian, H.; Shahsavari, M.H.; Karami-shabankareh, H.; Dashtizad, M. The presence of corpus luteum may have a negative impact on in vitro developmental competency of bovine oocytes. *Reprod. Biol.* **2016**, *16*, 47–52. [CrossRef] [PubMed]
25. Karami Shabankareh, H.; Shahsavari, M.H.; Hajarian, H.; Moghaddam, G. In vitro developmental competence of bovine oocytes: Effect of corpus luteum and follicle size. *Iran J. Reprod. Med.* **2015**, *13*, 615–622.
26. Chohan, K.R.; Hunter, A.G. Effect of reproductive status on in vitro developmental competence of bovine oocytes. *J. Vet. Sci.* **2003**, *4*, 67–72. [CrossRef]
27. Gandolfi, F.; Luciano, A.M.; Modina, S.; Ponzini, A.; Pocar, P.; Armstrong, D.T.; Lauria, A. The in vitro developmental competence of bovine oocytes can be related to the morphology of the ovary. *Theriogenology* **1997**, *48*, 1153–1160. [CrossRef]
28. Smith, L.C.; Olivera-Angel, M.; Groome, N.P.; Bhatia, B.; Price, C.A. Oocyte quality in small antral follicles in the presence or absence of a large dominant follicle in cattle. *J. Reprod. Fertil.* **1996**, *106*, 193–199. [CrossRef]
29. Arlotto, T.; Schwartz, J.L.; First, N.L.; Leibfried-Rutledge, M.L. Aspects of follicle and oocyte stage that affect in vitro maturation and development of bovine oocytes. *Theriogenology* **1996**, *45*, 943–956. [CrossRef]
30. De Wit, A.A.; Wurth, Y.A.; Kruip, T.A. Effect of ovarian phase and follicle quality on morphology and developmental capacity of the bovine cumulus-oocyte complex. *J. Anim. Sci.* **2000**, *78*, 1277–1283. [CrossRef]
31. Chian, R.C.; Chung, J.T.; Downey, B.R.; Tan, S.L. Maturational and developmental competence of immature oocytes retrieved from bovine ovaries at different phases of folliculogenesis. *Reprod. Biomed. Online* **2002**, *4*, 127–132. [CrossRef]
32. Lonergan, P.; Monaghan, P.; Rizos, D.; Boland, M.P.; Gordon, I. Effect of follicle size on bovine oocyte quality and developmental competence following maturation, fertilization, and culture in vitro. *Mol. Reprod. Dev.* **1994**, *37*, 48–53. [CrossRef] [PubMed]
33. Iwata, H.; Hashimoto, S.; Ohota, M.; Kimura, K.; Shibano, K.; Miyake, M. Effects of follicle size and electrolytes and glucose in maturation medium on nuclear maturation and developmental competence of bovine oocytes. *Reproduction* **2004**, *127*, 159–164. [CrossRef] [PubMed]
34. Lequarre, A.S.; Vigneron, C.; Ribaucour, F.; Holm, P.; Donnay, I.; Dalbies-Tran, R.; Callesen, H.; Mermillod, P. Influence of antral follicle size on oocyte characteristics and embryo development in the bovine. *Theriogenology* **2005**, *63*, 841–859. [CrossRef] [PubMed]

35. Caixeta, E.S.; Ripamonte, P.; Franco, M.M.; Junior, J.B.; Dode, M.A. Effect of follicle size on mRNA expression in cumulus cells and oocytes of Bos indicus: An approach to identify marker genes for developmental competence. *Reprod. Fertil. Dev.* **2009**, *21*, 655–664. [CrossRef]
36. Nivet, A.L.; Bunel, A.; Labrecque, R.; Belanger, J.; Vigneault, C.; Blondin, P.; Sirard, M.A. FSH withdrawal improves developmental competence of oocytes in the bovine model. *Reproduction* **2012**, *143*, 165–171. [CrossRef]
37. De Bem, T.; Adona, P.R.; Bressan, F.F.; Mesquita, L.G.; Chiaratti, M.R.; Meirelles, F.V.; Leal, C. The influence of morphology, follicle size and Bcl-2 and Bax transcripts on the developmental competence of bovine oocytes. *Reprod. Domest. Anim.* **2014**, *49*, 576–583. [CrossRef]
38. Kauffold, J.; Amer, H.A.; Bergfeld, U.; Weber, W.; Sobiraj, A. The in vitro developmental competence of oocytes from juvenile calves is related to follicular diameter. *J. Reprod. Dev.* **2005**, *51*, 325–332. [CrossRef]
39. Hendriksen, P.J.; Vos, P.L.; Steenweg, W.N.; Bevers, M.M.; Dieleman, S.J. Bovine follicular development and its effect on the in vitro competence of oocytes. *Theriogenology* **2000**, *53*, 11–20. [CrossRef]
40. Annes, K.; Muller, D.B.; Vilela, J.A.P.; Valente, R.S.; Caetano, D.P.; Cibin, F.W.S.; Milazzotto, M.P.; Mesquita, F.S.; Belaz, K.R.A.; Eberlin, M.N.; et al. Influence of follicle size on bovine oocyte lipid composition, follicular metabolic and stress markers, embryo development and blastocyst lipid content. *Reprod. Fertil. Dev.* **2019**, *31*, 462–472. [CrossRef]
41. Alves, G.P.; Cordeiro, F.B.; Bruna de Lima, C.; Annes, K.; Cristina Dos Santos, E.; Ispada, J.; Fontes, P.K.; Nogueira, M.F.G.; Nichi, M.; Milazzotto, M.P. Follicular environment as a predictive tool for embryo development and kinetics in cattle. *Reprod. Fertil. Dev.* **2019**, *31*, 451–461. [CrossRef] [PubMed]
42. Labrecque, R.; Fournier, E.; Sirard, M.A. Transcriptome analysis of bovine oocytes from distinct follicle sizes: Insights from correlation network analysis. *Mol. Reprod. Dev.* **2016**, *83*, 558–569. [CrossRef] [PubMed]
43. Moussa, M.; Shu, J.; Zhang, X.H.; Zeng, F. Maternal control of oocyte quality in cattle "a review". *Anim. Reprod. Sci.* **2015**, *155*, 11–27. [CrossRef] [PubMed]
44. Tello, M.F.; Lorenzo, M.S.; Luchetti, C.G.; Maruri, A.; Cruzans, P.R.; Alvarez, G.M.; Gambarotta, M.C.; Salamone, D.F.; Cetica, P.D.; Lombardo, D.M. Apoptosis in porcine cumulus-oocyte complexes: Relationship with their morphology and the developmental competence. *Mol. Reprod. Dev.* **2020**, *87*, 274–283. [CrossRef] [PubMed]
45. Leibfried, L.; First, N.L. Characterization of bovine follicular oocytes and their ability to mature in vitro. *J. Anim. Sci.* **1979**, *48*, 76–86. [CrossRef] [PubMed]
46. Hazeleger, N.L.; Hill, D.J.; Stubbing, R.B.; Walton, J.S. Relationship of morphology and follicular fluid environment of bovine oocytes to their developmental potential in vitro. *Theriogenology* **1995**, *43*, 509–522. [CrossRef]
47. Madison, V.; Avery, B.; Greve, T. Selection of immature bovine oocytes for developmental potential in vitro. *Anim. Reprod. Sci.* **1992**, *27*, 1–11. [CrossRef]
48. Boni, R.; Cuomo, A.; Tosti, E. Developmental potential in bovine oocytes is related to cumulus-oocyte complex grade, calcium current activity, and calcium stores. *Biol. Reprod.* **2002**, *66*, 836–842. [CrossRef]
49. Bilodeau-Goeseels, S.; Panich, P. Effects of oocyte quality on development and transcriptional activity in early bovine embryos. *Anim. Reprod. Sci.* **2002**, *71*, 143–155. [CrossRef]
50. Emanuelli, I.P.; Costa, C.B.; Rafagnin Marinho, L.S.; Seneda, M.M.; Meirelles, F.V. Cumulus-oocyte interactions and programmed cell death in bovine embryos produced in vitro. *Theriogenology* **2019**, *126*, 81–87. [CrossRef]
51. Yuan, Y.Q.; Van Soom, A.; Leroy, J.L.; Dewulf, J.; Van Zeveren, A.; de Kruif, A.; Peelman, L.J. Apoptosis in cumulus cells, but not in oocytes, may influence bovine embryonic developmental competence. *Theriogenology* **2005**, *63*, 2147–2163. [CrossRef]
52. De Loos, F.; van Vliet, C.; van Maurik, P.; Kruip, T.A. Morphology of immature bovine oocytes. *Gamete Res.* **1989**, *24*, 197–204. [CrossRef] [PubMed]
53. Hawk, H.; Wall, R. Improved yields of bovine blastocysts from in vitro-produced oocytes. I. Selection of oocytes and zygotes. *Theriogenology* **1994**, *41*, 1571–1583. [CrossRef]
54. Dunning, K.R.; Russell, D.L.; Robker, R.L. Lipids and oocyte developmental competence: The role of fatty acids and beta-oxidation. *Reproduction* **2014**, *148*, R15–R27. [CrossRef] [PubMed]
55. Genicot, G.; Leroy, J.L.; Soom, A.V.; Donnay, I. The use of a fluorescent dye, Nile red, to evaluate the lipid content of single mammalian oocytes. *Theriogenology* **2005**, *63*, 1181–1194. [CrossRef]

56. Van Blerkom, J.; Bell, H.; Weipz, D. Cellular and developmental biological aspects of bovine meiotic maturation, fertilization, and preimplantation embryogenesis in vitro. *J. Electron. Microsc. Tech.* **1990**, *16*, 298–323. [CrossRef]
57. McKeegan, P.J.; Sturmey, R.G. The role of fatty acids in oocyte and early embryo development. *Reprod. Fertil. Dev.* **2011**, *24*, 59–67. [CrossRef]
58. Sturmey, R.G.; Reis, A.; Leese, H.J.; McEvoy, T.G. Role of fatty acids in energy provision during oocyte maturation and early embryo development. *Reprod. Domest. Anim.* **2009**, *44* (Suppl. 3), 50–58. [CrossRef]
59. Kim, J.Y.; Kinoshita, M.; Ohnishi, M.; Fukui, Y. Lipid and fatty acid analysis of fresh and frozen-thawed immature and in vitro matured bovine oocytes. *Reproduction* **2001**, *122*, 131–138. [CrossRef]
60. Pavani, K.C.; Rocha, A.; Oliveira, E.; da Silva, F.M.; Sousa, M. Novel ultrastructural findings in bovine oocytes matured in vitro. *Theriogenology* **2020**, *143*, 88–97. [CrossRef]
61. Dadarwal, D.; Honparkhe, M.; Dias, F.C.; Alce, T.; Lessard, C.; Singh, J. Effect of superstimulation protocols on nuclear maturation and distribution of lipid droplets in bovine oocytes. *Reprod. Fertil. Dev.* **2015**, *27*, 1137–1146. [CrossRef] [PubMed]
62. Auclair, S.; Uzbekov, R.; Elis, S.; Sanchez, L.; Kireev, I.; Lardic, L.; Dalbies-Tran, R.; Uzbekova, S. Absence of cumulus cells during in vitro maturation affects lipid metabolism in bovine oocytes. *Am. J. Physiol. Endocrinol. Metab.* **2013**, *304*, E599–E613. [CrossRef] [PubMed]
63. Fair, T.; Hulshof, S.C.; Hyttel, P.; Greve, T.; Boland, M. Oocyte ultrastructure in bovine primordial to early tertiary follicles. *Anat. Embryol. (Berl)* **1997**, *195*, 327–336. [CrossRef] [PubMed]
64. Prates, E.G.; Nunes, J.T.; Pereira, R.M. A role of lipid metabolism during cumulus-oocyte complex maturation: Impact of lipid modulators to improve embryo production. *Mediators Inflamm.* **2014**, *2014*, 692067. [CrossRef] [PubMed]
65. Leroy, J.L.; Genicot, G.; Donnay, I.; Van Soom, A. Evaluation of the lipid content in bovine oocytes and embryos with nile red: A practical approach. *Reprod. Domest. Anim.* **2005**, *40*, 76–78. [CrossRef]
66. Salamone, D.F.; Canel, N.G.; Rodriguez, M.B. Intracytoplasmic sperm injection in domestic and wild mammals. *Reproduction* **2017**, *154*, F111–F124. [CrossRef]
67. Nagano, M.; Katagiri, S.; Takahashi, Y. Relationship between bovine oocyte morphology and in vitro developmental potential. *Zygote* **2006**, *14*, 53–61. [CrossRef]
68. Jeong, W.J.; Cho, S.J.; Lee, H.S.; Deb, G.K.; Lee, Y.S.; Kwon, T.H.; Kong, I.K. Effect of cytoplasmic lipid content on in vitro developmental efficiency of bovine IVP embryos. *Theriogenology* **2009**, *72*, 584–589. [CrossRef]
69. Nagano, M. Acquisition of developmental competence and in vitro growth culture of bovine oocytes. *J. Reprod. Dev.* **2019**, *65*, 195–201. [CrossRef]
70. Prates, E.G.; Marques, C.C.; Baptista, M.C.; Vasques, M.I.; Carolino, N.; Horta, A.E.; Charneca, R.; Nunes, J.T.; Pereira, R.M. Fat area and lipid droplet morphology of porcine oocytes during in vitro maturation with trans-10, cis-12 conjugated linoleic acid and forskolin. *Animal* **2013**, *7*, 602–609. [CrossRef]
71. Jasensky, J.; Boughton, A.P.; Khmaladze, A.; Ding, J.; Zhang, C.; Swain, J.E.; Smith, G.W.; Chen, Z.; Smith, G.D. Live-cell quantification and comparison of mammalian oocyte cytosolic lipid content between species, during development, and in relation to body composition using nonlinear vibrational microscopy. *Analyst* **2016**, *141*, 4694–4706. [CrossRef] [PubMed]
72. Machado, M.F.; Caixeta, E.S.; Sudiman, J.; Gilchrist, R.B.; Thompson, J.G.; Lima, P.F.; Price, C.A.; Buratini, J. Fibroblast growth factor 17 and bone morphogenetic protein 15 enhance cumulus expansion and improve quality of in vitro-produced embryos in cattle. *Theriogenology* **2015**, *84*, 390–398. [CrossRef] [PubMed]
73. Zhang, K.; Hansen, P.J.; Ealy, A.D. Fibroblast growth factor 10 enhances bovine oocyte maturation and developmental competence in vitro. *Reproduction* **2010**, *140*, 815–826. [CrossRef] [PubMed]
74. Kobayashi, K.; Yamashita, S.; Hoshi, H. Influence of epidermal growth factor and transforming growth factor-alpha on in vitro maturation of cumulus cell-enclosed bovine oocytes in a defined medium. *J. Reprod. Fertil.* **1994**, *100*, 439–446. [CrossRef] [PubMed]
75. Allworth, A.E.; Albertini, D.F. Meiotic maturation in cultured bovine oocytes is accompanied by remodeling of the cumulus cell cytoskeleton. *Dev. Biol.* **1993**, *158*, 101–112. [CrossRef]
76. Furnus, C.C.; de Matos, D.G.; Moses, D.F. Cumulus expansion during in vitro maturation of bovine oocytes: Relationship with intracellular glutathione level and its role on subsequent embryo development. *Mol. Reprod. Dev.* **1998**, *51*, 76–83. [CrossRef]

77. Choi, Y.H.; Carnevale, E.M.; Seidel, G.E., Jr.; Squire, E.L. Effects of gonadotropins on bovine oocytes matured in TCM-199. *Theriogenology* **2001**, *56*, 661–670. [CrossRef]
78. Anchordoquy, J.P.; Anchordoquy, J.M.; Sirini, M.A.; Testa, J.A.; Peral-Garcia, P.; Furnus, C.C. The importance of manganese in the cytoplasmic maturation of cattle oocytes: Blastocyst production improvement regardless of cumulus cells presence during in vitro maturation. *Zygote* **2016**, *24*, 139–148. [CrossRef]
79. Dovolou, E.; Messinis, I.E.; Periquesta, E.; Dafopoulos, K.; Gutierrez-Adan, A.; Amiridis, G.S. Ghrelin accelerates in vitro maturation of bovine oocytes. *Reprod. Domest. Anim.* **2014**, *49*, 665–672. [CrossRef]
80. Rosa, D.E.; Anchordoquy, J.M.; Anchordoquy, J.P.; Sirini, M.A.; Testa, J.A.; Mattioli, G.A.; Furnus, C.C. Analyses of apoptosis and DNA damage in bovine cumulus cells after in vitro maturation with different copper concentrations: Consequences on early embryo development. *Zygote* **2016**, *24*, 869–879. [CrossRef]
81. Marei, W.F.; Ghafari, F.; Fouladi-Nashta, A.A. Role of hyaluronic acid in maturation and further early embryo development of bovine oocytes. *Theriogenology* **2012**, *78*, 670–677. [CrossRef] [PubMed]
82. Fukui, Y. Effect of follicle cells on the acrosome reaction, fertilization, and developmental competence of bovine oocytes matured in vitro. *Mol. Reprod. Dev.* **1990**, *26*, 40–46. [CrossRef]
83. Fair, T.; Hyttel, P.; Greve, T. Bovine oocyte diameter in relation to maturational competence and transcriptional activity. *Mol. Reprod. Dev.* **1995**, *42*, 437–442. [CrossRef]
84. Fair, T.; Hyttel, P.; Greve, T.; Boland, M. Nucleus structure and transcriptional activity in relation to oocyte diameter in cattle. *Mol. Reprod. Dev.* **1996**, *43*, 503–512. [CrossRef]
85. Anguita, B.; Jimenez-Macedo, A.R.; Izquierdo, D.; Mogas, T.; Paramio, M.T. Effect of oocyte diameter on meiotic competence, embryo development, p34 (cdc2) expression and MPF activity in prepubertal goat oocytes. *Theriogenology* **2007**, *67*, 526–536. [CrossRef] [PubMed]
86. Otoi, T.; Yamamoto, K.; Koyama, N.; Tachikawa, S.; Suzuki, T. Bovine oocyte diameter in relation to developmental competence. *Theriogenology* **1997**, *48*, 769–774. [CrossRef]
87. Huang, W.; Nagano, M.; Kang, S.S.; Yanagawa, Y.; Takahashi, Y. Effects of in vitro growth culture duration and prematuration culture on maturational and developmental competences of bovine oocytes derived from early antral follicles. *Theriogenology* **2013**, *80*, 793–799. [CrossRef]
88. Yang, Y.; Kanno, C.; Huang, W.; Kang, S.S.; Yanagawa, Y.; Nagano, M. Effect of bone morphogenetic protein-4 on in vitro growth, steroidogenesis and subsequent developmental competence of the oocyte-granulosa cell complex derived from bovine early antral follicles. *Reprod. Biol. Endocrinol.* **2016**, *14*, 3. [CrossRef]
89. Cavalera, F.; Zanoni, M.; Merico, V.; Sacchi, L.; Bellazzi, R.; Garagna, S.; Zuccotti, M. Chromatin organization and timing of polar body I extrusion identify developmentally competent mouse oocytes. *Int. J. Dev. Biol.* **2019**, *63*, 245–251. [CrossRef]
90. Holubcova, Z.; Kyjovska, D.; Martonova, M.; Paralova, D.; Klenkova, T.; Otevrel, P.; Stepanova, R.; Kloudova, S.; Hampl, A. Egg maturity assessment prior to ICSI prevents premature fertilization of late-maturing oocytes. *J. Assist. Reprod. Genet.* **2019**, *36*, 445–452. [CrossRef]
91. Nandi, S.; Ravindranatha, B.M.; Gupta, P.S.; Sarma, P.V. Timing of sequential changes in cumulus cells and first polar body extrusion during in vitro maturation of buffalo oocytes. *Theriogenology* **2002**, *57*, 1151–1159. [CrossRef]
92. Van der Westerlaken, L.A.; van der Schans, A.; Eyestone, W.H.; de Boer, H.A. Kinetics of first polar body extrusion and the effect of time of stripping of the cumulus and time of insemination on developmental competence of bovine oocytes. *Theriogenology* **1994**, *42*, 361–370. [CrossRef]
93. Park, Y.S.; Kim, S.S.; Kim, J.M.; Park, H.D.; Byun, M.D. The effects of duration of in vitro maturation of bovine oocytes on subsequent development, quality and transfer of embryos. *Theriogenology* **2005**, *64*, 123–134. [CrossRef] [PubMed]
94. Ward, F.; Enright, B.; Rizos, D.; Boland, M.; Lonergan, P. Optimization of in vitro bovine embryo production: Effect of duration of maturation, length of gamete co-incubation, sperm concentration and sire. *Theriogenology* **2002**, *57*, 2105–2117. [CrossRef]
95. Dominko, T.; First, N.L. Timing of meiotic progression in bovine oocytes and its effect on early embryo development. *Mol. Reprod. Dev.* **1997**, *47*, 456–467. [CrossRef]
96. Hu, J.; Jin, C.; Zheng, H.; Liu, Q.; Zhu, W.; Zeng, Z.; Wu, J.; Wang, Y.; Li, J.; Zhang, X.; et al. First polar body morphology affects potential development of porcine parthenogenetic embryo in vitro. *Zygote* **2015**, *23*, 615–621. [CrossRef]

97. Ebner, T.; Moser, M.; Sommergruber, M.; Yaman, C.; Pfleger, U.; Tews, G. First polar body morphology and blastocyst formation rate in ICSI patients. *Hum. Reprod.* **2002**, *17*, 2415–2418. [CrossRef]
98. Zhou, W.; Fu, L.; Sha, W.; Chu, D.; Li, Y. Relationship of polar bodies morphology to embryo quality and pregnancy outcome. *Zygote* **2016**, *24*, 401–407. [CrossRef]
99. Ebner, T.; Yaman, C.; Moser, M.; Sommergruber, M.; Feichtinger, O.; Tews, G. Prognostic value of first polar body morphology on fertilization rate and embryo quality in intracytoplasmic sperm injection. *Hum. Reprod.* **2000**, *15*, 427–430. [CrossRef]
100. Rose, B.I.; Laky, D. Polar body fragmentation in IVM oocytes is associated with impaired fertilization and embryo development. *J. Assist. Reprod. Genet.* **2013**, *30*, 679–682. [CrossRef]
101. Halvaei, I.; Khalili, M.A.; Soleimani, M.; Razi, M.H. Evaluating the Role of First Polar Body Morphology on Rates of Fertilization and Embryo Development in ICSI Cycles. *Int. J. Fertil. Steril.* **2011**, *5*, 110–115. [PubMed]
102. De Santis, L.; Cino, I.; Rabellotti, E.; Calzi, F.; Persico, P.; Borini, A.; Coticchio, G. Polar body morphology and spindle imaging as predictors of oocyte quality. *Reprod. Biomed. Online* **2005**, *11*, 36–42. [CrossRef]
103. Ciotti, P.M.; Notarangelo, L.; Morselli-Labate, A.M.; Felletti, V.; Porcu, E.; Venturoli, S. First polar body morphology before ICSI is not related to embryo quality or pregnancy rate. *Hum. Reprod.* **2004**, *19*, 2334–2339. [CrossRef] [PubMed]
104. Verlinsky, Y.; Lerner, S.; Illkevitch, N.; Kuznetsov, V.; Kuznetsov, I.; Cieslak, J.; Kuliev, A. Is there any predictive value of first polar body morphology for embryo genotype or developmental potential? *Reprod. Biomed. Online* **2003**, *7*, 336–341. [CrossRef]
105. Caamano, J.N.; Munoz, M.; Diez, C.; Gomez, E. Polarized light microscopy in mammalian oocytes. *Reprod. Domest. Anim.* **2010**, *45* (Suppl. 2), 49–56. [CrossRef] [PubMed]
106. Montag, M.; Koster, M.; van der Ven, K.; van der Ven, H. Gamete competence assessment by polarizing optics in assisted reproduction. *Hum. Reprod. Update* **2011**, *17*, 654–666. [CrossRef]
107. Yu, Y.; Yan, J.; Liu, Z.C.; Yan, L.Y.; Li, M.; Zhou, Q.; Qiao, J. Optimal timing of oocyte maturation and its relationship with the spindle assembly and developmental competence of in vitro matured human oocytes. *Fertil. Steril.* **2011**, *96*, 73–78. [CrossRef]
108. Kilani, S.; Cooke, S.; Tilia, L.; Chapman, M. Does meiotic spindle normality predict improved blastocyst development, implantation and live birth rates? *Fertil. Steril.* **2011**, *96*, 389–393. [CrossRef]
109. Tomari, H.; Honjo, K.; Kunitake, K.; Aramaki, N.; Kuhara, S.; Hidaka, N.; Nishimura, K.; Nagata, Y.; Horiuchi, T. Meiotic spindle size is a strong indicator of human oocyte quality. *Reprod. Med. Biol.* **2018**, *17*, 268–274. [CrossRef]
110. Rienzi, L.; Ubaldi, F.; Martinez, F.; Iacobelli, M.; Minasi, M.G.; Ferrero, S.; Tesarik, J.; Greco, E. Relationship between meiotic spindle location with regard to the polar body position and oocyte developmental potential after ICSI. *Hum. Reprod.* **2003**, *18*, 1289–1293. [CrossRef]
111. Liu, L.; Oldenbourg, R.; Trimarchi, J.R.; Keefe, D.L. A reliable, noninvasive technique for spindle imaging and enucleation of mammalian oocytes. *Nat. Biotechnol.* **2000**, *18*, 223–225. [CrossRef] [PubMed]
112. Lu, F.; Shi, D.; Wei, J.; Yang, S.; Wei, Y. Development of embryos reconstructed by interspecies nuclear transfer of adult fibroblasts between buffalo (Bubalus bubalis) and cattle (Bos indicus). *Theriogenology* **2005**, *64*, 1309–1319. [CrossRef] [PubMed]
113. Caamano, J.N.; Maside, C.; Gil, M.A.; Munoz, M.; Cuello, C.; Diez, C.; Sanchez-Osorio, J.R.; Martin, D.; Gomis, J.; Vazquez, J.M.; et al. Use of polarized light microscopy in porcine reproductive technologies. *Theriogenology* **2011**, *76*, 669–677. [CrossRef] [PubMed]
114. Rienzi, L.; Martinez, F.; Ubaldi, F.; Minasi, M.G.; Iacobelli, M.; Tesarik, J.; Greco, E. Polscope analysis of meiotic spindle changes in living metaphase II human oocytes during the freezing and thawing procedures. *Hum. Reprod.* **2004**, *19*, 655–659. [CrossRef]
115. Caamano, J.N.; Diez, C.; Trigal, B.; Munoz, M.; Morato, R.; Martin, D.; Carrocera, S.; Mogas, T.; Gomez, E. Assessment of meiotic spindle configuration and post-warming bovine oocyte viability using polarized light microscopy. *Reprod. Domest Anim.* **2013**, *48*, 470–476. [CrossRef]
116. Montag, M.; Schimming, T.; Koster, M.; Zhou, C.; Dorn, C.; Rosing, B.; van der Ven, H.; Ven der Ven, K. Oocyte zona birefringence intensity is associated with embryonic implantation potential in ICSI cycles. *Reprod. Biomed. Online* **2008**, *16*, 239–244. [CrossRef]

117. Madaschi, C.; de Souza Bonetti, T.C.; de Almeida Ferreira Braga, D.P.; Pasqualotto, F.F.; Iaconelli, A., Jr.; Borges, E., Jr. Spindle imaging: A marker for embryo development and implantation. *Fertil. Steril.* **2008**, *90*, 194–198. [CrossRef]
118. De Almeida Ferreira Braga, D.P.; de Cassia Savio Figueira, R.; Queiroz, P.; Madaschi, C.; Iaconelli, A., Jr.; Borges, E., Jr. Zona pellucida birefringence in in vivo and in vitro matured oocytes. *Fertil. Steril.* **2010**, *94*, 2050–2053. [CrossRef]
119. Petersen, C.G.; Vagnini, L.D.; Mauri, A.L.; Massaro, F.C.; Silva, L.F.; Cavagna, M.; Baruffi, R.L.; Oliveira, J.B.; Franco, J.G., Jr. Evaluation of zona pellucida birefringence intensity during in vitro maturation of oocytes from stimulated cycles. *Reprod. Biol. Endocrinol.* **2011**, *9*, 53. [CrossRef]
120. Ashourzadeh, S.; Khalili, M.A.; Omidi, M.; Mahani, S.N.; Kalantar, S.M.; Aflatoonian, A.; Habibzadeh, V. Noninvasive assays of in vitro matured human oocytes showed insignificant correlation with fertilization and embryo development. *Arch. Gynecol. Obstet.* **2015**, *292*, 459–463. [CrossRef]
121. Held, E.; Mertens, E.M.; Mohammadi-Sangcheshmeh, A.; Salilew-Wondim, D.; Besenfelder, U.; Havlicek, V.; Herrler, A.; Tesfaye, D.; Schellander, K.; Holker, M. Zona pellucida birefringence correlates with developmental capacity of bovine oocytes classified by maturational environment, COC morphology and G6PDH activity. *Reprod. Fertil. Dev.* **2012**, *24*, 568–579. [CrossRef] [PubMed]
122. Koester, M.; Mohammadi-Sangcheshmeh, A.; Montag, M.; Rings, F.; Schimming, T.; Tesfaye, D.; Schellander, K.; Hoelker, M. Evaluation of bovine zona pellucida characteristics in polarized light as a prognostic marker for embryonic developmental potential. *Reproduction* **2011**, *141*, 779–787. [CrossRef] [PubMed]
123. Bertero, A.; Ritrovato, F.; Evangelista, F.; Stabile, V.; Fortina, R.; Ricci, A.; Revelli, A.; Vincenti, L.; Nervo, T. Evaluation of equine oocyte developmental competence using polarized light microscopy. *Reproduction* **2017**, *153*, 775–784. [CrossRef] [PubMed]
124. Mohammadi-Sangcheshmeh, A.; Held, E.; Rings, F.; Ghanem, N.; Salilew-Wondim, D.; Tesfaye, D.; Sieme, H.; Schellander, K.; Hoelker, M. Developmental competence of equine oocytes: Impacts of zona pellucida birefringence and maternally derived transcript expression. *Reprod. Fertil. Dev.* **2014**, *26*, 441–452. [CrossRef]
125. Pujol, M.; Lopez-Bejar, M.; Paramio, M.T. Developmental competence of heifer oocytes selected using the brilliant cresyl blue (BCB) test. *Theriogenology* **2004**, *61*, 735–744. [CrossRef]
126. Alm, H.; Torner, H.; Lohrke, B.; Viergutz, T.; Ghoneim, I.M.; Kanitz, W. Bovine blastocyst development rate in vitro is influenced by selection of oocytes by brillant cresyl blue staining before IVM as indicator for glucose-6-phosphate dehydrogenase activity. *Theriogenology* **2005**, *63*, 2194–2205. [CrossRef]
127. Bhojwani, S.; Alm, H.; Torner, H.; Kanitz, W.; Poehland, R. Selection of developmentally competent oocytes through brilliant cresyl blue stain enhances blastocyst development rate after bovine nuclear transfer. *Theriogenology* **2007**, *67*, 341–345. [CrossRef]
128. Opiela, J.; Katska-Ksiazkiewicz, L. The utility of Brilliant Cresyl Blue (BCB) staining of mammalian oocytes used for in vitro embryo production (IVP). *Reprod. Biol.* **2013**, *13*, 177–183. [CrossRef]
129. Silva, D.S.; Rodriguez, P.; Galuppo, A.; Arruda, N.S.; Rodrigues, J.L. Selection of bovine oocytes by brilliant cresyl blue staining: Effect on meiosis progression, organelle distribution and embryo development. *Zygote* **2013**, *21*, 250–255. [CrossRef]
130. Fakruzzaman, M.; Bang, J.I.; Lee, K.L.; Kim, S.S.; Ha, A.N.; Ghanem, N.; Han, C.H.; Cho, K.W.; White, K.L.; Kong, I.K. Mitochondrial content and gene expression profiles in oocyte-derived embryos of cattle selected on the basis of brilliant cresyl blue staining. *Anim. Reprod. Sci.* **2013**, *142*, 19–27. [CrossRef]
131. Castaneda, C.A.; Kaye, P.; Pantaleon, M.; Phillips, N.; Norman, S.; Fry, R.; D'Occhio, M.J. Lipid content, active mitochondria and brilliant cresyl blue staining in bovine oocytes. *Theriogenology* **2013**, *79*, 417–422. [CrossRef] [PubMed]
132. Salviano, M.B.; Collares, F.J.; Becker, B.S.; Rodrigues, B.A.; Rodrigues, J.L. Bovine non-competent oocytes (BCB-) negatively impact the capacity of competent (BCB+) oocytes to undergo in vitro maturation, fertilisation and embryonic development. *Zygote* **2016**, *24*, 245–251. [CrossRef] [PubMed]
133. Karami Shabankareh, H.; Azimi, G.; Torki, M. Developmental competence of bovine oocytes selected based on follicle size and using the brilliant cresyl blue (BCB) test. *Iran. J. Reprod. Med.* **2014**, *12*, 771–778. [PubMed]

134. Mirshamsi, S.M.; Karamishabankareh, H.; Ahmadi-Hamedani, M.; Soltani, L.; Hajarian, H.; Abdolmohammadi, A.R. Combination of oocyte and zygote selection by brilliant cresyl blue (BCB) test enhanced prediction of developmental potential to the blastocyst in cattle. *Anim. Reprod. Sci.* **2013**, *136*, 245–251. [CrossRef] [PubMed]
135. Anguita, B.; Vandaele, L.; Mateusen, B.; Maes, D.; Van Soom, A. Developmental competence of bovine oocytes is not related to apoptosis incidence in oocytes, cumulus cells and blastocysts. *Theriogenology* **2007**, *67*, 537–549. [CrossRef]
136. Ikeda, S.; Imai, H.; Yamada, M. Apoptosis in cumulus cells during in vitro maturation of bovine cumulus-enclosed oocytes. *Reproduction* **2003**, *125*, 369–376. [CrossRef]
137. Janowski, D.; Salilew-Wondim, D.; Torner, H.; Tesfaye, D.; Ghanem, N.; Tomek, W.; El-Sayed, A.; Schellander, K.; Holker, M. Incidence of apoptosis and transcript abundance in bovine follicular cells is associated with the quality of the enclosed oocyte. *Theriogenology* **2012**, *78*, 656–669. [CrossRef]
138. Warzych, E.; Pers-Kamczyc, E.; Krzywak, A.; Dudzinska, S.; Lechniak, D. Apoptotic index within cumulus cells is a questionable marker of meiotic competence of bovine oocytes matured in vitro. *Reprod. Biol.* **2013**, *13*, 82–87. [CrossRef]
139. Lee, K.S.; Joo, B.S.; Na, Y.J.; Yoon, M.S.; Choi, O.H.; Kim, W.W. Cumulus cells apoptosis as an indicator to predict the quality of oocytes and the outcome of IVF-ET. *J. Assist. Reprod. Genet.* **2001**, *18*, 490–498. [CrossRef]
140. Fan, Y.; Chang, Y.; Wei, L.; Chen, J.; Li, J.; Goldsmith, S.; Silber, S.; Liang, X. Apoptosis of mural granulosa cells is increased in women with diminished ovarian reserve. *J. Assist. Reprod. Genet.* **2019**, *36*, 1225–1235. [CrossRef]
141. Clavero, A.; Castilla, J.A.; Nunez, A.I.; Garcia-Pena, M.L.; Maldonado, V.; Fontes, J.; Mendoza, N.; Martinez, L. Apoptosis in human granulosa cells after induction of ovulation in women participating in an intracytoplasmic sperm injection program. *Eur. J. Obstet. Gynecol. Reprod. Biol.* **2003**, *110*, 181–185. [CrossRef]
142. Bunel, A.; Jorssen, E.P.; Merckx, E.; Leroy, J.L.; Bols, P.E.; Sirard, M.A. Individual bovine in vitro embryo production and cumulus cell transcriptomic analysis to distinguish cumulus-oocyte complexes with high or low developmental potential. *Theriogenology* **2015**, *83*, 228–237. [CrossRef] [PubMed]
143. Donnison, M.; Pfeffer, P.L. Isolation of genes associated with developmentally competent bovine oocytes and quantitation of their levels during development. *Biol. Reprod.* **2004**, *71*, 1813–1821. [CrossRef]
144. Bettegowda, A.; Patel, O.V.; Lee, K.B.; Park, K.E.; Salem, M.; Yao, J.; Ireland, J.J.; Smith, G.W. Identification of novel bovine cumulus cell molecular markers predictive of oocyte competence: Functional and diagnostic implications. *Biol. Reprod.* **2008**, *79*, 301–309. [CrossRef] [PubMed]
145. Mourot, M.; Dufort, I.; Gravel, C.; Algriany, O.; Dieleman, S.; Sirard, M.A. The influence of follicle size, FSH-enriched maturation medium, and early cleavage on bovine oocyte maternal mRNA levels. *Mol. Reprod. Dev.* **2006**, *73*, 1367–1379. [CrossRef] [PubMed]
146. Ashry, M.; Lee, K.; Mondal, M.; Datta, T.K.; Folger, J.K.; Rajput, S.K.; Zhang, K.; Hemeida, N.A.; Smith, G.W. Expression of TGFbeta superfamily components and other markers of oocyte quality in oocytes selected by brilliant cresyl blue staining: Relevance to early embryonic development. *Mol. Reprod. Dev.* **2015**, *82*, 251–264. [CrossRef] [PubMed]
147. Nivet, A.L.; Vigneault, C.; Blondin, P.; Sirard, M.A. Changes in granulosa cells' gene expression associated with increased oocyte competence in bovine. *Reproduction* **2013**, *145*, 555–565. [CrossRef] [PubMed]
148. Melo, E.O.; Cordeiro, D.M.; Pellegrino, R.; Wei, Z.; Daye, Z.J.; Nishimura, R.C.; Dode, M.A. Identification of molecular markers for oocyte competence in bovine cumulus cells. *Anim. Genet.* **2017**, *48*, 19–29. [CrossRef]
149. Assidi, M.; Dufort, I.; Ali, A.; Hamel, M.; Algriany, O.; Dielemann, S.; Sirard, M.A. Identification of potential markers of oocyte competence expressed in bovine cumulus cells matured with follicle-stimulating hormone and/or phorbol myristate acetate in vitro. *Biol. Reprod.* **2008**, *79*, 209–222. [CrossRef]
150. Kussano, N.R.; Leme, L.O.; Guimaraes, A.L.; Franco, M.M.; Dode, M.A. Molecular markers for oocyte competence in bovine cumulus cells. *Theriogenology* **2016**, *85*, 1167–1176. [CrossRef]
151. Khurchabilig, A.; Sato, A.; Ashibe, S.; Hara, A.; Fukumori, R.; Nagao, Y. Expression levels of FSHR, IGF1R, CYP11al and HSD3beta in cumulus cells can predict in vitro developmental competence of bovine oocytes. *Zygote* **2020**, 1–7. [CrossRef]
152. O'Shea, L.C.; Daly, E.; Hensey, C.; Fair, T. ATRX is a novel progesterone-regulated protein and biomarker of low developmental potential in mammalian oocytes. *Reproduction* **2017**, *153*, 671–682. [CrossRef] [PubMed]

153. Xiong, X.R.; Lan, D.L.; Li, J.; Yin, S.; Xiong, Y.; Zi, X.D. Identification of differential abundances of mRNA transcript in cumulus cells and CCND1 associated with yak oocyte developmental competence. *Anim. Reprod. Sci.* **2019**, *208*, 106135. [CrossRef] [PubMed]
154. Bunel, A.; Nivet, A.L.; Blondin, P.; Vigneault, C.; Richard, F.J.; Sirard, M.A. Cumulus cell gene expression associated with pre-ovulatory acquisition of developmental competence in bovine oocytes. *Reprod. Fertil. Dev.* **2014**, *26*, 855–865. [CrossRef] [PubMed]
155. Dieci, C.; Lodde, V.; Labreque, R.; Dufort, I.: Tessaro, I.; Sirard, M.A.; Luciano, A.M. Differences in cumulus cell gene expression indicate the benefit of a pre-maturation step to improve in-vitro bovine embryo production. *Mol. Hum. Reprod.* **2016**, *22*, 882–897. [CrossRef] [PubMed]
156. Walter, J.; Monthoux, C.; Fortes, C.; Grossmann, J.; Roschitzki, B.; Meili, T.; Riond, B.; Hofmann-Lehmann, R.; Naegeli, H.; Bleul, U. The bovine cumulus proteome is influenced by maturation condition and maturational competence of the oocyte. *Sci. Rep.* **2020**, *10*, 9880. [CrossRef]
157. Peddinti, D.; Memili, E.; Burgess, S.C. Proteomics-based systems biology modeling of bovine germinal vesicle stage oocyte and cumulus cell interaction. *PLoS ONE* **2010**, *5*, e11240. [CrossRef]
158. Memili, E.; Peddinti, D.; Shack, L.A.; Nanduri, B.; McCarthy, F.; Sagirkaya, H.; Burgess, S.C. Bovine germinal vesicle oocyte and cumulus cell proteomics. *Reproduction* **2007**, *133*, 1107–1120. [CrossRef]
159. Sutton, M.L.; Gilchrist, R.B.; Thompson, J.G. Effects of in-vivo and in-vitro environments on the metabolism of the cumulus-oocyte complex and its influence on oocyte developmental capacity. *Hum. Reprod. Update* **2003**, *9*, 35–48. [CrossRef]
160. Revelli, A.; Delle Piane, L.; Casano, S.; Molinari, E.; Massobrio, M.; Rinaudo, P. Follicular fluid content and oocyte quality: From single biochemical markers to metabolomics. *Reprod. Biol. Endocrinol.* **2009**, *7*, 40. [CrossRef]
161. Wrenzycki, C.; Stinshoff, H. Maturation environment and impact on subsequent developmental competence of bovine oocytes. *Reprod. Domest. Anim.* **2013**, *48* (Suppl. 1), 38–43. [CrossRef] [PubMed]
162. Matoba, S.; Bender, K.; Fahey, A.G.; Mamo. S.; Brennan, L.; Lonergan, P.; Fair, T. Predictive value of bovine follicular components as markers of oocyte developmental potential. *Reprod. Fertil. Dev.* **2014**, *26*, 337–345. [CrossRef] [PubMed]
163. Bender, K.; Walsh, S.; Evans, A.C.; Fair, T.; Brennan, L. Metabolite concentrations in follicular fluid may explain differences in fertility between heifers and lactating cows. *Reproduction* **2010**, *139*, 1047–1055. [CrossRef] [PubMed]
164. Zachut, M.; Sood, P.; Levin, Y.; Moallem, U. Proteomic analysis of preovulatory follicular fluid reveals differentially abundant proteins in less fertile dairy cows. *J. Proteomics* **2016**, *139*, 122–129. [CrossRef] [PubMed]
165. Alves, B.G.; Alves, K.A.; Lucio, A.C.; Martins, M.C.; Silva, T.H.; Alves, B.G.; Braga, L.S.; Silva, T.V.; Viu, M.A.; Beletti, M.E.; et al. Ovarian activity and oocyte quality associated with the biochemical profile of serum and follicular fluid from Girolando dairy cows postpartum. *Anim. Reprod. Sci.* **2014**, *146*, 117–125. [CrossRef] [PubMed]
166. Takahashi, C.; Fujito, A.; Kazuka, M.; Sugiyama, R.; Ito, H.; Isaka, K. Anti-Mullerian hormone substance from follicular fluid is positively associated with success in oocyte fertilization during in vitro fertilization. *Fertil. Steril.* **2008**, *89*, 586–591. [CrossRef]
167. Kim, J.H.; Lee, J.R.; Chang, H.J.; Jee, B.C.; Suh, C.S.; Kim, S.H. Anti-Mullerian hormone levels in the follicular fluid of the preovulatory follicle: A predictor for oocyte fertilization and quality of embryo. *J. Korean Med. Sci.* **2014**, *29*, 1266–1270. [CrossRef]
168. Tramisak Milakovic, T.; Panic Horvat, L.; Cavlovic, K.; Smiljan Severinski, N.; Vlasic, H.; Vlastelic, I.; Ljiljak, D.; Radojcic Badovinac, A. Follicular fluid anti-Mullerian hormone: A predictive marker of fertilization capacity of MII oocytes. *Arch. Gynecol. Obstet.* **2015**, *291*, 681–687. [CrossRef]
169. Revelli, A.; Canosa, S.; Bergandi, L.; Skorokhod, O.A.; Biasoni, V.; Carosso, A.; Bertagna, A.; Maule, M.; Aldieri, E.; D'Eufemia, M.D.; et al. Oocyte polarized light microscopy, assay of specific follicular fluid metabolites, and gene expression in cumulus cells as different approaches to predict fertilization efficiency after ICSI. *Reprod. Biol. Endocrinol.* **2017**, *15*, 47. [CrossRef]
170. Ciepiela, P.; Duleba, A.J.; Kario, A.; Chelstowski, K.; Branecka-Wozniak, D.; Kurzawa, R. Oocyte matched follicular fluid anti-Mullerian hormone is an excellent predictor of live birth after fresh single embryo transfer. *Hum. Reprod.* **2019**, *34*, 2244–2253. [CrossRef]

171. Rico, C.; Fabre, S.; Medigue, C.; di Clemente, N.; Clement, F.; Bontoux, M.; Touze, J.L.; Dupont, M.; Briant, E.; Remy, B.; et al. Anti-mullerian hormone is an endocrine marker of ovarian gonadotropin-responsive follicles and can help to predict superovulatory responses in the cow. *Biol. Reprod.* **2009**, *80*, 50–59. [CrossRef] [PubMed]
172. Rico, C.; Drouilhet, L.; Salvetti, P.; Dalbies-Tran, R.; Jarrier, P.; Touze, J.L.; Pillet, E.; Ponsart, C.; Fabre, S.; Monniaux, D. Determination of anti-Mullerian hormone concentrations in blood as a tool to select Holstein donor cows for embryo production: From the laboratory to the farm. *Reprod. Fertil. Dev.* **2012**, *24*, 932–944. [CrossRef]
173. Monniaux, D.; Barbey, S.; Rico, C.; Fabre, S.; Gallard, Y.; Larroque, H. Anti-Mullerian hormone: A predictive marker of embryo production in cattle? *Reprod. Fertil. Dev.* **2010**, *22*, 1083–1091. [CrossRef] [PubMed]
174. Gebremedhn, S.; Salilew-Wondim, D.; Ahmad, I.; Sahadevan, S.; Hossain, M.M.; Hoelker, M.; Rings, F.; Neuhoff, C.; Tholen, E.; Looft, C.; et al. MicroRNA Expression Profile in Bovine Granulosa Cells of Preovulatory Dominant and Subordinate Follicles during the Late Follicular Phase of the Estrous Cycle. *PLoS ONE* **2015**, *10*, e0125912. [CrossRef] [PubMed]
175. Sohel, M.M.; Hoelker, M.; Noferesti, S.S.; Salilew-Wondim, D.; Tholen, E.; Looft, C.; Rings, F.; Uddin, M.J.; Spencer, T.E.; Schellander, K.; et al. Exosomal and Non-Exosomal Transport of Extra-Cellular microRNAs in Follicular Fluid: Implications for Bovine Oocyte Developmental Competence. *PLoS ONE* **2013**, *8*, e78505. [CrossRef] [PubMed]
176. Pasquariello, R.; Manzoni, E.F.M.; Fiandanese, N.; Viglino, A.; Pocar, P.; Brevini, T.A.L.; Williams, J.L.; Gandolfi, F. Implications of miRNA expression pattern in bovine oocytes and follicular fluids for developmental competence. *Theriogenology* **2020**, *145*, 77–85. [CrossRef] [PubMed]

Publisher's Note: MDPI stays neutral with regard to jurisdictional claims in published maps and institutional affiliations.

Article

Cellular and Molecular Events that Occur in the Oocyte during Prolonged Ovarian Storage in Sheep

Alicia Martín-Maestro, Irene Sánchez-Ajofrín *, Carolina Maside, Patricia Peris-Frau, Daniela-Alejandra Medina-Chávez, Beatriz Cardoso, José Carlos Navarro, María Rocío Fernández-Santos, José Julián Garde and Ana Josefa Soler *

SaBio IREC (CSIC-UCLM-JCCM), ETSIAM, Campus Universitario, s/n, 02071 Albacete, Spain; alicia.martinmaestro@uclm.es (A.M.-M.); carolina.maside@uclm.es (C.M.); patricia.peris@uclm.es (P.P.-F.); daniela.medina@uclm.es (D.-A.M.-C.); beacardoso_14@hotmail.com (B.C.); jnavarropedrosa@gmail.com (J.C.N.); mrocio.fernandez@uclm.es (M.R.F.-S.); julian.garde@uclm.es (J.J.G.)

* Correspondence: irene.ssanchez@uclm.es (I.S.-A.); anajosefa.soler@uclm.es (A.J.S.)

Received: 1 November 2020; Accepted: 13 December 2020; Published: 17 December 2020

Simple Summary: Establishing efficient in vitro embryo production (IVP) protocols in sheep usually requires prolonged transportation of post-mortem ovaries since adult animals are often slaughtered in abattoirs far from laboratories. In this study, different analyses were carried out to investigate important cellular and molecular aspects of hypoxic injury on excised ovaries over time in order to understand the factors jeopardizing the development of competent oocytes during prolonged transport times. We observed that, when ovaries were stored for more than 7 h, the quality and developmental potential of oocytes and cumulus cells were greatly reduced. Moreover, the use of medium TCM199 over saline solution also had deleterious effects. Beyond transport time, strategies aimed at reducing these damages may improve oocyte quality and developmental competence.

Abstract: For the past two decades, there has been a growing interest in the application of in vitro embryo production (IVP) in small ruminants such as sheep. To improve efficiency, a large number abattoir-derived ovaries must be used, and long distances from the laboratory are usually inevitable when adult animals are used. In that scenario, prolonged sheep ovary transportation may negatively affect oocyte developmental competence. Here, we evaluated the effect of ovary storage time (3, 5, 7, 9, 11 and 13 h) and the medium in which they were transported (TCM199 and saline solution) on oocyte quality. Thus, live/dead status, early apoptosis, DNA fragmentation, reduced glutathione (GSH) and reactive oxygen species (ROS) content, caspase-3 activity, mitochondrial membrane potential and distribution, and relative abundance of mRNA transcript levels were assessed in oocytes. After in vitro maturation (IVM), cumulus cell viability and quality, meiotic and fertilization competence, embryo rates and blastocyst quality were also evaluated. The results revealed that, after 7 h of storage, oocyte quality and developmental potential were significantly impaired since higher rates of dead oocytes and DNA fragmentation and lower rates of viable, matured and fertilized oocytes were observed. The percentage of cleavage, blastocyst rates and cumulus cell parameters (viability, active mitochondria and GSH/ROS ratio) were also decreased. Moreover, the preservation of ovaries in medium TCM199 had a detrimental effect on cumulus cells and oocyte competence. In conclusion, ovary transport times up to 5 h in saline solution are the most adequate storage conditions to maintain oocyte quality as well as developmental capacity in sheep. A strategy to rescue the poor developmental potential of stored oocytes will be necessary for successful production of high-quality embryos when longer ovarian preservation times are necessary.

Keywords: sheep; ovary storage; transport; oocyte; in vitro embryo production

1. Introduction

Assisted reproduction technologies (ARTs) in small ruminants, such as sheep, have great potential for genetic improvement and dissemination programs, since they allow for a rapid and sustainable increase in animals of great genetic merit. Furthermore, ARTs are effective tools in the preservation of endangered species or breeds as well as in disease eradication programs [1].

Though the main lines of investigation in small ruminants have focused on germplasm banks and artificial insemination [2], over the past few decades, slight advances have been made toward the use of in vitro embryo production (IVP) [1–3]. Generally, improving the efficiency of IVP protocols in these species entails the use of ovaries of dead animals because a large number of samples should be collected. Unlike their in vivo counterparts, oocytes retrieved from dead animals exhibit reduced developmental potential [4]. Moreover, the slaughterhouses where adult animals are slaughtered are usually located in strategic places and often far from research laboratories. Storing ovaries for long periods of time due to long distances could compromise the viability of the oocyte. Considering that the quality of oocytes determines the developmental potential of embryos after fertilization [5], the preservation of oocyte integrity from the moment the animal dies until the ovaries are processed is of critical importance.

Immediately following death, the lack of blood flow prevents oxygen and energy supply and places the ovaries under ischemic conditions [6]. The main mechanism of injury in ischemia is hypoxia, and cells with high metabolic rates, including the ones that form the ovarian tissue, tend to be damaged very rapidly [7,8]. Acute hypoxia results in ATP depletion that triggers a switch to glycolysis, the major anaerobic pathway for ATP production [9]. ATP is broken down without being resynthesized, and eventually, decreased energy efficiency and accumulation of lactic acid produced by glycolysis also reduce intracellular pH, resulting in additional cellular dysfunction [6,9]. Notwithstanding the relationship between hypoxia/ischemia and organ damage is being established, the physiological mechanisms by which oocyte quality is affected with increased ovary storage time remain to be completely understood. It is therefore necessary to elucidate the events occurring in the oocyte at the cellular and molecular levels for the purpose of taking the appropriate measures to reduce this damage.

The goal of this study, therefore, was to evaluate the cellular and molecular events related to oocyte quality and developmental competence that occur throughout storage of sheep oocytes within ischemic ovaries after the death of the animal. This may help to create a better understanding of the mechanism of oocyte injury obtained from ischemic ovaries and to identify the temporal window for successful fertilization in IVP, particularly when ovary transport times are inevitably long. Ultimately, it will contribute to developing a strategy to reverse the poor developmental potential of stored oocytes within ovaries, which will have a great significance for ARTs.

2. Materials and Methods

The adult sheep ovaries were collected from an authorized slaughterhouse ("Ovinos Manchegos"), and sperm samples were obtained from the Germplasm Bank of the "Reproduction Biology Group", which is officially authorized for collecting and storing semen from sheep (ES07RS02OC). All chemicals were acquired from Merck Life Sciences (Madrid, Spain) unless otherwise stated.

2.1. Oocyte Collection and In Vitro Maturation

Adult sheep ovaries ($n = 1420$) were obtained post-mortem and transported at 30 °C in physiological saline (8.9 gr/L NaCl) supplemented with penicillin (0.1 g/L) or at 38.5 °C in TCM199 medium supplemented with polyvinylpyrrolidone (PVA; 1 g/L), 4-(2-hydroxyethyl)-1-piperazineethanesulfonic acid (HEPES) (6.51 g/L), streptomycin (0.1 g/L), penicillin (0.1 g/L) and sodium bicarbonate (0.4 g/L) and were maintained for 13 h in the same media and at the same temperature. The mean age of animals was around 6 years old, and the sheep breeds were mainly Merino or mixed. Immature cumulus–oocyte complexes (COCs) were recovered by slicing the ovaries with a scalpel at 3, 5, 7, 9, 11 and 13 h post ovary collection. Then, a total of 4258 COCs from 8 replicates having a clear and homogeneous or moderate granular ooplasm and surrounded by at least three layers of tightly packed cumulus cells were selected and placed in TCM199 medium supplemented with HEPES (2.38 mg/mL), heparin (2 µL/mL) and gentamycin (4 µL/mL). In each replicate, the COCs of ovaries from the same treatments were mixed and homogeneously distributed. From those, 2327 COCs were mechanically denuded by vortex in phosphate-buffered saline (PBS) supplemented with 0.1% PVA (*w/v*; PBS-PVA) and oocytes were either directly analysed, fixed in 0.5% glutaraldehyde (*v/v*) and stored at 4 °C for terminal deoxynucleotidyl transferase mediated dUTP nick-end labelling (TUNEL) analysis or snap-frozen and stored at −80 °C for mRNA analysis. In addition, 1931 COCs, collected from the last 4 replicates, were matured, fertilized and cultured in vitro following the protocol by Sánchez-Ajofrín et al. [10]. Briefly, COCs were washed in TCM199-gentamycin (4 µL/mL) and randomly placed in four-well dishes containing 500 µL of TCM199 and 4 µL/mL gentamycin, 100 µM cysteamine, 10 ng/mL follicle stimulating hormone, 10 ng/mL luteinizing hormone and 10% fetal calf serum [11] under mineral oil (Nidacon, Gothenburg, Sweden) and an atmosphere of 5% CO_2 at 38.5 °C with maximal humidity.

2.2. *In Vitro Fertilization (IVF)*

After approximately 22 h, COCs were partially denuded by gentle pipetting, divided into groups of 40–45 oocytes and placed in four-well plates containing 450 µL of synthetic oviductal fluid (SOF, Table S1), as described by Takahashi and First [12], with 10% oestrous sheep serum (ESS). Frozen-thawed spermatozoa were separated using a Percoll® density gradient (45%/90%) from two rams and capacitated for 15 min at 38.5 °C in 5% CO_2 with SOF and 10% ESS. Spermatozoa were subsequently co-incubated with the oocytes at a final concentration of 10^6 spermatozoa/mL for 18 h at 38.5 °C in 5% CO_2.

2.3. *In Vitro Culture (IVC)*

After 18 h post-insemination (hpi), presumptive zygotes were transferred to 25 µL IVC droplets (approximately one embryo per µL) containing SOF supplemented with 3 mg/mL of bovine serum albumin and cultured in a humidified atmosphere of 5% CO_2, 5% O_2 and 90% N_2 in air until day 8 post-insemination (dpi). Cleavage rate and blastocyst yield were examined at 48 hpi and 6, 7 and 8 dpi, respectively. All expanded blastocysts were fixed in 0.5% glutaraldehyde (*v/v*) and stored for TUNEL analysis and cell-number evaluation.

2.4. *Early Apoptosis Assay*

To determine early apoptosis in oocytes, a total of 356 immature and denuded sheep oocytes were incubated in Annexin-V, fluorescein isothiocyanate (FITC) staining kit (Thermo Fisher Scientific, Barcelona, Spain) according to the manufacturer's instructions. Briefly, oocytes were stained for 15 min with Annexin-V/FITC and 100 µg/mL propidium iodide (PI) at 37 °C in the dark. After incubation, oocytes were washed thrice in PBS-PVA and mounted on slides. Samples were evaluated at ×20 augmentation by fluorescence microscopy (Eclipse 80i, Nikon Instruments Europe, Amsterdam, The Netherlands) with the Intensilight C-HGFI module. The filter for excitation and the emitted fluorescence were EX 450-490 nm (DM 505; BA 520). Oocytes were classified into the following groups: early apoptotic oocytes (Annexin-V positive signal and PI negative signal; Figure 1A-a), viable oocytes (Annexin-V and PI signals were both negative; Figure 1A-b) and dead oocytes (Annexin-V and PI positive signals; Figure 1A-c and negative Annexin-V signal in the membrane and positive PI signal; Figure 1A-d).

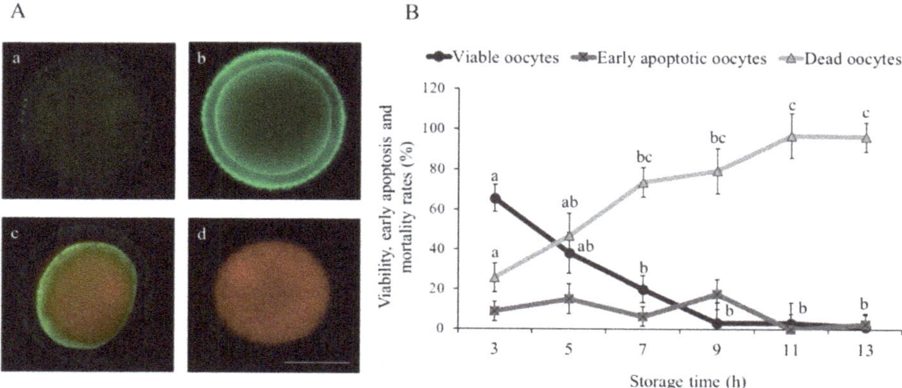

Figure 1. Effect of ovary storage time (3 to 13 h) on live/dead status and early apoptosis of immature sheep oocytes. (**A**) Representative images of sheep oocyte classification using Annexin-V staining: (a) viable oocyte, (b) early apoptotic oocyte and (c,d) dead oocytes. Scale bar = 50 µm. (**B**) Viability and early apoptosis rates (%): results are expressed as mean ± SEM. [a,b,c] Different letters indicate differences ($p \leq 0.05$) among storage times.

2.5. Measurement of Glutathione (GSH) and Reactive Oxygen Species (ROS)

A total of 350 immature oocytes were incubated in 50 µM Cell Tracker™ Blue (Thermo Fisher Scientific, Barcelona, Spain) and 10 µM of CM-H_2DCFDA (Thermo Fisher Scientific, Barcelona, Spain) for 30 min at 37 °C in the dark to detect intracellular glutathione (GSH) and reactive oxygen species (ROS) levels, respectively. Oocytes were subsequently washed thrice in PBS-PVA and then placed on glass slides under cover slips. The fluorescence intensity was observed using ×20 augmentation by fluorescence microscopy (Eclipse 80i, Nikon Instruments Europe, Amsterdam, The Netherlands) and quantified using ImageJ 1.45s software (National Institutes of Health, Bethesda, MD, USA; Figure 2A-a GSH oocyte level and 2A-b ROS oocyte level).

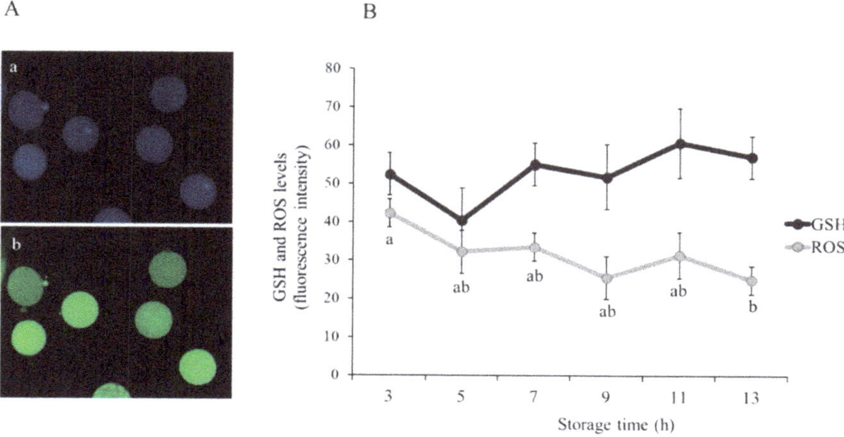

Figure 2. Effect of ovary storage time (3 to 13 h) on intracellular glutathione (GSH) and reactive oxygen species (ROS) levels of immature sheep oocytes: (**A**) representative images of intracellular (**a**) GSH and (**b**) ROS oocyte levels, scale bar = 100 µm, and (**B**) fluorescence intensity of GSH and ROS levels. Results are expressed as mean ± SEM. [a,b] Different letters indicate differences ($p \leq 0.05$) among storage times.

2.6. DNA Fragmentation Assay

The TUNEL method was used to detect DNA fragmentation combined with PI staining (oocytes) or Hoechst 33342 staining (blastocyst). Fixed immature sheep oocytes (n = 370) and blastocysts (n = 120) were permeabilized in 0.5% Triton X-100 in PBS for 1 h at room temperature. Next, In Situ Cell Death Detection Kit (Merck Life Sciences, Madrid, Spain) was used for the detection of DNA strand breaks in oocytes and blastomeres. According to the manufacturer's instructions, samples were placed in 30 µL drops of TUNEL reagent with fluorescein isothiocyanate conjugated deoxyuridine 5-triphosphate (dUTP) and the enzyme terminal deoxy-nucleotidyl transferase and were incubated for 1 h at 37 °C. The positive control was incubated with DNAse (0.2 U/µL) at 37 °C in the dark for 1 h, while the negative control was incubated in the absence of enzyme terminal deoxynucleotidyl transferase. Immediately after, immature oocytes and blastocysts were washed three times in PBS-PVA and transferred onto slides in a drop of Slowfade™ with 6.25 µg/mL PI and 5 µg/mL Hoechst 33342 fluorescent dye, respectively. Samples were evaluated using a ×20 augmentation by fluorescence microscopy (Eclipse 80i, Nikon Instruments Europe, Amsterdam, The Netherlands). The DNA damage in oocytes was classified as TUNEL-positive (Figure 3A-a) and -negative (Figure 3A-b) according to fragmented cell nuclei. The DNA fragmentation in blastocysts was determined by the number of cells with fragmented nuclei (TUNEL-positive) in relation to the total cell number.

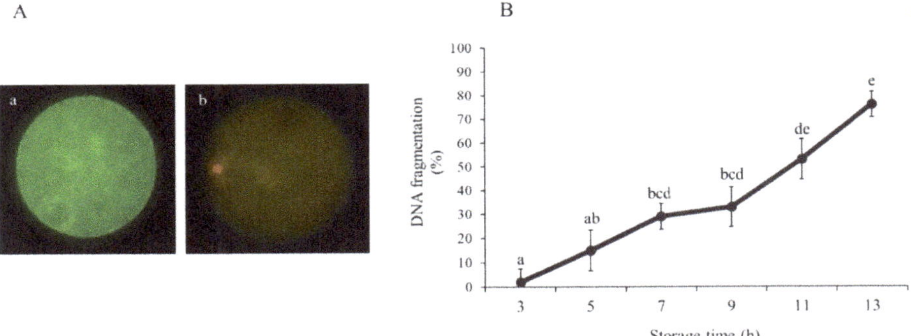

Figure 3. Effect of ovary storage time (3–13 h) on DNA fragmentation (positive terminal deoxynucleotidyl transferase mediated dUTP nick-end labelling (TUNEL) staining) of immature sheep oocytes: (**A**) representative images of (a) TUNEL-positive oocyte and (b) TUNEL-negative oocyte (scale bar = 50 µm) and (**B**) oocyte DNA fragmentation rates. Results are expressed as mean ± SEM. ^{a,b,c,d,e} Different letters indicate differences ($p \leq 0.05$) among storage times.

2.7. Measurement of Caspase-3 Activity

To monitor caspase-3 activity, 352 immature sheep oocytes were incubated for 30 min at 37 °C in 25 µL droplets of PBS-PVA containing 5 mM of PhiPhiLux-G1D2 (OncoImmunin Inc., Gaithersburg, MD, USA). After incubation, oocytes were washed twice in PBS-PVA and placed on slides under cover slips. Caspase activity was determined by fluorescence microscopy (Eclipse 80i, Nikon Instruments Europe, Amsterdam, The Netherlands), and intensity per unit area was quantified using ImageJ 1.45s software (National Institutes of Health, Bethesda, MD, USA; Figure S1).

2.8. Mitochondrial Membrane Potential Analysis

Membrane potential was determined by incubating 340 immature oocytes for 30 min at 37 °C in 0.5 µM of JC-1 dye (Thermo Fisher Scientific, Barcelona, Spain). After incubation, oocytes were washed twice for 5 min and then placed on glass slides. Oocytes were examined by ×20 augmentation by fluorescence microscopy (Eclipse 80i, Nikon Instruments Europe, Amsterdam, The Netherlands).

Relative mitochondrial membrane potential was determined as the ratio of J-aggregate to J-monomer staining intensity with ImageJ 1.45s software (National Institutes of Health, Bethesda, MD, USA; Figure S2).

2.9. Assessment of Mitochondrial Distribution

Mitochondrial distribution patterns were examined by MitoTracker® Red CMXRos (Thermo Fisher Scientific, Barcelona, Spain). At least 363 immature oocytes were incubated in PBS-PVA supplemented with 100 nM dye at 37 °C for 20 min. Oocytes were washed and then placed on glass slides and examined under ×20 augmentation by fluorescence microscopy (Eclipse 80i, Nikon Instruments Europe, Amsterdam, The Netherlands). Mitochondrial distribution was classified into two categories: abnormal mitochondrial distribution (Figure S3-d) in the cytoplasm and normal distribution (Figure S3a–c).

2.10. Quantification of Transcript Abundance

A total of 196 sheep oocytes were subjected to RNA extraction, complementary DNA (cDNA) synthesis and quantitative real-time PCR (qPCR) analysis as previously reported by Sánchez-Ajofrín et al. with minor modifications [13]. The RNA from groups of approximately 10 oocytes (3 replicates) was extracted using Dynabeads® (Invitrogen, California, CA, USA) following the protocol by [14]. Briefly, oocytes were lysed at room temperature in 50 µL binding buffer for 5 min and hybridized with 10 µL magnetic beads for another 5 min. Then, samples were washed twice in 50 µL buffer A and twice in buffer B. Next, mRNA samples were eluted with 28 µL Tris-HCl. Following this, reverse transcription was carried out using the Fermentas™ First Strand cDNA Synthesis Kit (Thermo Scientific, Barcelona, Spain) in a total volume of 40 µL. After heating the samples at 65 °C for 5 min, cDNA was synthesized by adding 2 µL of reaction buffer (5×), 2 µL of dNTP Mix, 1 µL of RiboLock RNase Inhibitor, 1 µL of M-MuLV Reverse Transcriptase and 2.5 µL of nuclease-free water. Subsequently, reverse transcription reaction was performed by incubating for 5 min at 25 °C, followed by 60 min at 37 °C and 5 min at 70 °C.

After cDNA synthesis, PowerUp™SYBR® Green Master Mix (Thermo Fisher Scientific, Barcelona, Spain) and a LightCycler 480 II system (Roche, Barcelona, Spain) were employed to determine the relative abundance of mRNA transcripts by qPCR. A final volume of 20 µL was reached by adding 10 µL master mix, 400 nM each of forward and reverse primers, 2 µL of cDNA template and nuclease-free water. The following PCR amplification conditions were used: 50 °C for 2 min, 95 °C for 2 min, 40 cycles of 95 °C for 15 s and 60 °C for 1 min. Immediately after, a melting curve analysis was performed to eliminate contamination by heating the samples to 95 °C for 5 s in a ramp rate of 4.4 °C/s, followed by 65 °C for 1 min with a heating rate of 2.2 °C/s and continuous fluorescence measurement. Each sample was analysed in duplicate, and reactions without any cDNA template (2 µL nuclease-free water) were used as the negative control.

The comparative cycle threshold and $2^{-\Delta\Delta CT}$ methods [15,16] were used to calculate the relative transcript abundances of candidate genes: BCL2-associated X protein (*BAX*), BCL2 apoptosis regulator (*BCL2*), bone morphogenetic protein 15 (*BMP15*), caspase-3 (*CASP3*), fibroblast growth factor 16 (*FGF16*) and growth differentiation factor 9 (*GDF9*). Quantification was normalized against that of the endogenous control (Peptidylprolyl Isomerase A (*PPIA*)). Information on the qPCR primers is provided in Table S2.

2.11. In Vitro Maturation and Fertilization Assessment

After maturation, oocytes were stripped of the surrounding cumulus cells by gentle pipetting. To examine oocyte maturation and sperm penetration, cells were stained with Hoechst 33342 (1 µg/mL) for 10 min at room temperature, washed in PBS-PVA and then analysed with ×20 augmentation by fluorescence microscopy (Eclipse 80i, Nikon Instruments Europe, Amsterdam, The Netherlands). Maturation rate was defined as the number of oocytes with an evident polar body and metaphase II

(MII) plate relative to the total number of oocytes analysed. Oocytes containing both female and male pronuclei (regardless of stage of decondensation) relative to the total number of oocytes matured were considered fertilized and were classified as normal (2PN) according to the number of swollen sperm heads and pronuclei in the cytoplasm.

2.12. Flow Cytometry Analysis of Cumulus Cells

Cumulus cells were collected from in vitro-matured oocytes and centrifuged at 12,000 rpm during 5 min. Pellet was resuspended in 125 µL PBS-PVA preequilibrated at 37 °C, and samples were stained and analysed by flow cytometry. To quantify cell live/dead status and apoptosis, samples were incubated with 10 µM YO-PRO-1 and 0.5 µM PI; for mitochondrial activity, cells were incubated with 200 mM of MitoTracker™ Deep Red (Thermo Fisher Scientific, Barcelona, Spain) for 20 min at 38.5 °C in the dark and then stained with 10 µM YO-PRO-1 and 0.5 µM PI; and for oxidative status (GSH and ROS levels), cells were incubated with 10 µM of Cell Tracker™ Blue (Thermo Fisher Scientific, Barcelona, Spain) and 10 µM of CM-H_2DCFDA (Thermo Fisher Scientific, Barcelona, Spain) for 30 min at 38.5 °C to assess GSH and ROS levels, respectively, and subsequently stained with 0.5 µM PI. The percentage of YO-PRO-1−/PI− showed the proportion of viable cells, while YO-PRO-1+/PI− showed that of apoptotic cells. Viable cells with active mitochondria were represented by the percentage of MitoTracker+/YO-PRO-1−. Finally, oxidative status was measured by GSH and ROS production only in viable cells (PI−).

Cumulus cells analyses were conducted using a FlowSight® imaging flow cytometer (Amnis, Merck-Millipore, Germany) equipped with violet, blue and red excitation lasers (405, 488 and 642 nm), 12 channels of detection and 10 available fluorescent channels. The system was controlled using INSPIRE® software (v.3). The flow cytometer was calibrated daily using calibration beads according to the manufacturer's instructions. A compensation overlap was performed before each experiment, and 1000 events were acquired per sample. In all cases, dot plots with aspect ratio and area were employed to exclude debris from cumulus cell populations and regions used to quantify cells subpopulation depended on the particular assay. The raw data were analysed using IDEAS® software (AMNIS) and out of focus cells, debris and cell clumps were excluded from the analysis.

2.13. Statistical Analysis

After determining that data were normally distributed and that variances were not heterogeneous, live/dead status, early apoptosis, DNA fragmentation, caspase-3 activity, GSH and ROS content, mitochondrial distribution and membrane potential, meiotic and fertilization competence, embryo production and total cell number were analysed by factorial ANOVA using the SPSS software (IBM, Armonk, NY, USA). For that, time of ovary storage (3, 5, 7, 9, 11 and 13 h) or type of ovary transport medium (TCM199 and saline solution) and the replicate (for IVM, IVF, embryo production, total number of cells and DNA fragmentation in blastocysts and cumulus cell analyses, four replicates were performed; for evaluation of oocyte viability and early apoptosis, DNA fragmentation, caspase-3 activity, GSH and ROS content, and mitochondrial membrane potential and distribution, eight replicates were performed) were considered fixed effects. Additionally, another factorial ANOVA was conducted to examine the relative abundances of mRNA transcripts, with time of ovary storage or type of medium and qPCR technical replicate (two replicates) as the fixed effects and the different target genes as the dependent variable. When a significant effect was observed, post hoc comparisons with Bonferroni correction were carried out. There was no evidence of statistically significant interactions between storage time and medium composition. Results are presented as mean ± S.E.M.

3. Results

3.1. Effect of Ovarian Transport Time on Oocyte Viability and Quality

As shown in Figure 1B, the percentage of viable immature sheep oocytes decreased ($p < 0.05$) from 7 h onwards (3 h = 65.62 ± 6.67% vs. 7 h = 20.00 ± 6.67%, 9 h = 3.12 ± 10.20%, 11 h = 3.12 ± 10.20% and 13 h = 1.25 ± 6.67%). Moreover, the lowest percentages ($p < 0.05$) of dead oocytes were observed at 3 and 5 h (25.62 ± 7.32% and 46.87 ± 11.18%, respectively) compared to 11 and 13 h (96.87 ± 11.18% and 96.25 ± 7.32%, respectively). Early apoptosis detected by phosphatidylserine localization using Annexin-V staining was not statistically different among groups ($p > 0.05$; Figure 1B).

A significantly higher level ($p < 0.05$) of ROS was recorded in immature sheep oocytes recovered from ovaries stored for 3 h (42.35 ± 3.69) compared to oocytes obtained from ovaries stored for 13 h (24.93 ± 3.69; Figure 2B). As also shown in Figure 2B, the different treatments did not exhibit different ($p > 0.05$) levels of GSH.

The number of immature sheep oocytes with TUNEL-positive fragmented DNA was lower ($p < 0.05$) at 3 h (1.94 ± 5.48%) of ovary storage, increased from 7 h (28.91 ± 5.48%) and recorded a maximum value at 13 h (75.88 ± 5.48%; Figure 3B).

The duration of ovary storage did not affect ($p > 0.05$) immature sheep oocyte caspase-3 intracellular activity (Figure S1), mitochondrial membrane potential (Figure S2) and mitochondrial distribution (Figure S3). Moreover, as shown in Figure S4, the relative abundance of mRNA transcripts of genes related to apoptosis (*BAX*, *BCL2* and *CASP3*) and oocyte quality (*BMP15*, *GDF9* and *FGF16*) did not show differences ($p > 0.05$) between 3, 7 and 13 h of ovary storage in immature sheep oocytes.

3.2. Effect of Ovarian Transport Time on the In Vitro Maturation and Fertilization Potential of Oocytes

The percentage of MII sheep oocytes recovered from ovaries stored for 3 h was higher ($p < 0.05$) compared to 7 and 13 h (Table 1. As expected, fertilization rate (2PN) was significantly increased after IVF in 3 h-derived oocytes compared to 7 and 13 h, with no significant differences between the latter groups (Table 1).

Table 1. In vitro maturation and fertilization of sheep oocytes retrieved from ovaries stored for 3, 7 and 13 h.

Storage Time (h)	Total Oocyte (*n*)	Maturation MII (%)	Fertilization 2PN (%)
3	187	68.33 ± 6.20 ^{a}	44.49 ± 4.47 ^{a}
7	144	30.07 ± 6.53 ^{b}	15.31 ± 4.70 ^{b}
13	140	4.69 ± 6.53 ^{c}	1.56 ± 4.70 ^{b}

Data are expressed as mean ± SEM. The results represent four replicates. ^{a,b,c} Different letters indicate differences ($p \leq 0.05$) among storage times.

3.3. Effect of Ovarian Transport Time on Cumulus Cells from Cumulus–Oocyte Complexes (COCs)

After maturation, COCs collected from ovaries stored for 3, 7 and 13 h were gently pipetted and detached cumulus cells were analysed by flow cytometry. Live/dead status, apoptosis, active mitochondria, and GSH and ROS levels were assessed. The results showed reduced ($p < 0.05$) cell viability as time increased (3 h = 82.06 ± 3.87% vs. 7 h = 59.16 ± 3.87% vs. 13 h = 26.53 ± 3.87%; Figure 4). Moreover, there was a lower ($p < 0.05$) percentage of dead cells at 3 h compared to 13 h (13.72 ± 6.08% and 58.50 ± 6.08%, respectively), although apoptosis did not show significant differences between storage time ($p > 0.05$; Figure 4).

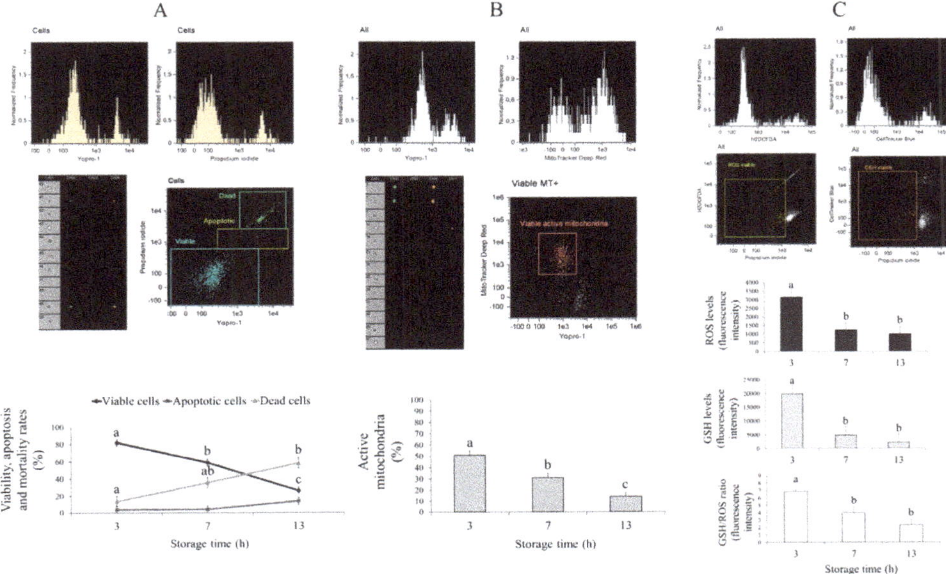

Figure 4. Effect of ovary storage time on (**A**) live/dead status and apoptosis, (**B**) percentage of active mitochondria and (**C**) intracellular GSH and ROS levels and the GSH/ROS ratio of cumulus cells collected from mature oocytes retrieved from ovaries stored for 3, 7 and 13 h: the results are expressed as mean ± SEM. [a,b,c] Different letters indicate differences ($p \leq 0.05$) among treatments.

Our results also showed reduced ($p < 0.05$) mitochondrial activity with increased storage time (3 h = 51.08 ± 3.78%, 7 h = 31.45 ± 3.45% and 13 h = 14.14 ± 3.45%; Figure 4). Furthermore, GSH and ROS levels and the ratio of GSH/ROS were higher after 3 h (GSH = 19890.59 ± 3044.95, ROS = 3153.65 ± 418.98 and GSH/ROS = 6.89 ± 0.78) of ovary storage compared to 7 h (GSH = 4813.30 ± 3044.95, ROS = 1234.67 ± 418.98 and GSH/ROS = 3.98 ± 0.78) and 13 h (GSH = 2213.70 ± 3044.95, ROS = 1009.18 ± 418.98 and GSH/ROS = 2.31 ± 0.78; Figure 4).

3.4. Effect of Ovarian Transport Time on In Vitro Embryo Development and Blastocyst Quality

The proportion of sheep oocytes that progressed to the first cleavage stage after IVF was significantly lower ($p < 0.05$) with increasing ovary storage time (Table 2. The percentage of total expanded blastocysts and percentage of blastocysts relative to the number of cleaved embryos was drastically decreased ($p < 0.05$) after 7 h of storage compared 3 h. Sheep blastocysts were not produced after 13 h of ovary storage (Table 2). However, the total cell number (3 h = 134.30 ± 6.01 and 7 h = 150.16 ± 12.02) and proportion of TUNEL-positive blastomeres (3 h = 13.02 ± 0.68% and 7 h = 12.34 ± 1.36%) in sheep blastocysts were similar ($p > 0.05$) between 3 and 7 h.

Table 2. The effect of ovary storage time on rates of cleavage and blastocyst development in sheep.

Storage Time (h)	Total Oocyte (*n*)	Cleaved Embryo at 48 hpi (%)	Expanded Blastocyst (%)	
			Total	Cleaved
3	555	65.96 ± 5.23 [a]	26.68 ± 2.19 [a]	40.82 ± 4.58 [a]
7	519	18.16 ± 5.23 [b]	4.66 ± 2.19 [b]	13.23 ± 4.58 [b]
13	386	2.59 ± 5.23 [b]	0.25 ± 2.19 [b]	6.25± 4.58 [b]

Data are expressed as mean ± SEM. The results represent four replicates. [a,b,c] Different letters indicate differences ($p \leq 0.05$) among storage times.

3.5. Effect of Medium Type During Ovary Transport on Oocyte Developmental Competence and Quality

The storage of sheep ovaries in TCM199 medium or saline solution did not show significant differences ($p > 0.05$) between the oocyte quality parameters studied, including oocyte live/dead status, apoptosis, caspase-3 intracellular activity, GSH and ROS levels, DNA fragmentation, mitochondrial distribution, mitochondrial membrane potential and relative mRNA transcript abundance (Figure S5).

The storage of ovaries with TCM199 resulted in lower ($p < 0.05$) rates of IVM, IVF, cleavage and blastocysts relative to the number of cleaved embryos (Table 3|). Nevertheless, the total cell number of blastocysts was decreased ($p < 0.05$) as a result of storing the ovaries in saline solution (125.46 ± 6.72%) compared to TCM199 (159.0 ± 10.83%). Moreover, the number of total expanded blastocysts and TUNEL-positive blastomeres was similar ($p > 0.05$) in both media (Table 3).

Cumulus cells were also affected by the type of ovary storage medium. Thus, cumulus cells from ovaries stored in saline solution showed a greater percentage ($p < 0.05$) of viable cells and active mitochondria, while the TCM199 medium exhibited higher rates ($p < 0.05$) of dead cells (Table 4). The proportion of apoptotic cells and the GSH and ROS levels did not show significant differences ($p > 0.05$) between media (Table 4).

4. Discussion

In the present study, storage of sheep ovaries beyond 7 h had a detrimental effect on oocyte quality and subsequent development to the blastocyst stage. Similar results were obtained in rat ovaries where apparent histological changes were observed after 3 h of ischemia [17]. Notably, an in-depth analysis revealed that, after 7 h, oocyte live/dead status was dramatically reduced along with increasing storage time. After organ removal, an immediate consequence of the cessation of blood supply is the deprivation of oxygen and nutrients as well as the accumulation of metabolic waste, which may lead to cellular damage [18]. The crucial event is ATP depletion, which occurs within the first few minutes of oxygen stoppage [19]. This early event results in a transition from aerobic to anaerobic metabolism, which in turn leads to a rise in lactate and H^+ levels that contribute in many mechanisms to cell injury related to ischemia [19]. In addition, the susceptibility of different types of cells to ischemic damage varies according to the degree of metabolic activity, and those with higher rates require a greater ongoing production of ATP [19]. For this reason, cells that form the ovarian tissue tend to be injured very rapidly by hypoxia [7,8].

Reduced viability of oocytes in post-mortem ovaries has been linked to degeneration of protein and DNA in horses [20] and domestic cats [21]. Remarkably, in the current study, we observed that the number of oocytes with fragmented DNA started to greatly increase after an ischemic time of 7 h, with no evidence of other oocyte apoptosis markers being significantly different (caspase-3 activity, phosphatidylserine binding by Annexin-V and mRNA transcripts). Moreover, there were no evident differences between storage times regarding mitochondrial membrane potential and distribution, which have been previously linked to many apoptotic stimuli [22–24]. Traditionally, apoptosis has been characterized by DNA damage as visualized by the TUNEL assay [25]. However, identification of terminal deoxy-nucleotidyl transferase (Tdt)–mediated deoxyuridine 5-triphosphate (dUTP) labelling in the nucleus of dying cells is not sufficient to demonstrate that cells are undergoing apoptosis, as the chromosomal DNA degradation and resulting DNA strand breaks also occur in necrotic cells [25,26]. Moreover, a pattern of TUNEL staining typical of necrotic cells was noticed in the present study with extensive staining of the cytosol, which may be due to the formation of large DNA fragments during karyorrhexis that are released into the cytosol upon nuclear disintegration [25,27].

Table 3. The effect of medium type during ovary transport on oocyte developmental competence and blastocyst quality in sheep.

Treatment	Total Oocyte (n)	Maturation MII (%)	Fertilization 2PN (%)	Cleaved Embryo at 48 hpi (%)	Total Expanded Blastocyst (%)	Expanded Blastocyst/ Cleaved (%)	TUNEL-Positive Blastomeres (%)
TCM199	934	28.27 ± 5.30^{b}	13.64 ± 3.83^{b}	23.81 ± 4.27^{b}	9.14 ± 1.78	13.15 ± 3.74^{b}	13.94 ± 1.22
Saline solution	997	43.46 ± 5.15^{a}	27.27 ± 3.71^{a}	34.00 ± 4.27^{a}	11.93 ± 1.78	27.05 ± 3.74^{a}	11.42 ± 0.76

Data are expressed as mean ± SEM. The results represent four replicates. ^{a,b} Different letters indicate differences ($p \leq 0.05$) among media.

Table 4. The effect of medium type during ovary transport on cumulus cells in sheep.

Treatment	Viable Cells (%)	Apoptotic Cells (%)	Dead Cells (%)	Active Mitochondria (%)	GSH Levels (Fluorescence Intensity)	ROS Levels (Fluorescence Intensity)
TCM199	44.97 ± 3.87^{b}	9.59 ± 3.29	45.07 ± 5.31^{b}	26.42 ± 3.46^{b}	8347.83 ± 2338.17	1591.14 ± 324.92
Saline solution	66.87 ± 3.87^{a}	5.07 ± 3.29	27.08 ± 5.31^{a}	38.97 ± 3.61^{a}	9597.23 ± 2338.17	2007.19 ± 324.92

Data are expressed as mean ± SEM. The results represent four replicates. ^{a,b} Different letters indicate differences ($p \leq 0.05$) among media.

Interestingly, intracellular ROS levels at 3 h were higher than at 13 h. It has been suggested that ROS quickly accumulate at the onset of ischemia despite limited O_2 supply [28,29]. In the presence of xanthine oxidase (XOD) and O_2, hypoxanthine can be converted to xanthine, which simultaneously produces superoxide anion ($O_2^{\bullet-}$) [30]. During ischemia, the accumulation of XOD and hypoxanthine results in increased $O_2^{\bullet-}$ production [31]. Considering that molecular O_2 is the limited substrate in this reaction and that its availability decreases over time, it was not surprising to find that, after 13 h of ovary storage, intracellular ROS levels were significantly lower.

Successful embryo development will largely depend on the optimal accumulation of organelles, metabolites and maternal RNAs during oocyte growth [32]. In our study, the duration of ovary storage did not affect mitochondrial membrane potential, mitochondrial distribution and mRNA transcript of genes related to oocyte quality, although the embryonic yield was lower beyond 7 h of storage. A possible explanation could be that the assessments were performed on immature oocytes and that, in this cellular state, fluorescent dyes displayed less sensitivity to intracellular changes. Thus, different studies have shown contradictory data in relation to mitochondrial activity between immature and mature oocytes [33,34].

Besides, our results revealed that the number of oocytes showing MII and 2 PN after fertilization were lower when ovaries were stored longer than 7 h. Gaulden [35] suggested that hypoxic conditions could reduce oocyte intracellular pH, influencing the organization and stability of the meiotic metaphase spindle. Moreover, other researchers found that normal oocytes collected from under-oxygenated follicles showed chromosomal defects such as a compact arrangement on the MII spindle [36]. In our case, rather than chromosomal alterations, it is likely that the low rates of MII and 2PN are due to the high oocyte mortality following acute deprivation of oxygen after ovary collection.

To gain further awareness of the biological consequences of prolonged transport time in stored oocytes, we examined cumulus cells and oocyte development in vitro. Cumulus cells play a critical role in oocyte maturation because they supply ions, metabolites and regulatory molecules that are necessary for meiotic progression, normal nuclear and cytoplasmic maturation of oocytes, and subsequent embryonic development after fertilization [37,38]. As expected, cumulus cells lost their supportive and protective functions when subjected to storage times longer than 7 h, since they showed decreased viability and impaired redox status and mitochondrial activity. Likewise, 7 and 13 h of ovarian storage resulted in drastic reduction in oocyte maturation, fertilization, cleavage and blastocyst rates. In fact, after 13 h of preservation, blastocysts were unable to develop. Therefore, reduced developmental potential of in vitro matured oocytes may also be related to impaired cumulus cell functions. Similar to our results, other studies have shown that the length of time that ovaries are held before oocyte recovery also affected developmental potential in several species. In sheep and pig, a delay of only 5–7 h reduced the maturation rate compared to that of oocytes placed immediately into maturation culture [39] or after 3 h of storage [6], respectively; in horses, 5–9 h had the same effects [40]. In addition, long-term storage (7–8 h) of ovaries reduced blastocyst formation rates after IVF in cattle [41] and intracytoplasmic sperm injection (ICSI) in horses [42].

Besides duration of ovarian storage, the type of medium where these organs are held plays an important role in determining appropriate transport conditions for oocyte survival and in vitro embryo development. Because TCM199 has more components (glucose, vitamins, amino acids and adenine sulphate) than saline solution and fully grown follicles are more metabolically active, we speculated that both follicles and oocytes may be better supported by the more complex medium. Although there were no differences between media for parameters used as indicators of oocyte quality, reduced oocyte developmental competence and cumulus cell quality were evidenced when ovaries were preserved in TCM199 medium. One possible explanation may be that ovaries from this group were kept at a higher temperature (38.5 °C) than the saline solution group (30 °C). In fact, preliminary results obtained by our group have indicated that the preservation of sheep ovaries in saline solution for 4 h at 38.5 °C negatively affects oocyte quality, IVM rates and cumulus cells compared to 30 °C (unpublished data). Moreover, it has been suggested that the use of low temperatures (4 °C) during long ovary storage times preserves oocyte quality possibly due to a decrease in cellular metabolism [43]. However, in our laboratory, oocytes from ovaries stored in saline solution at 4 °C for 3 h showed lower rates of MII after maturation (unpublished data), suggesting that low transport temperatures may have a detrimental effect during the transport of sheep ovaries. Differences with other authors such as Goodarzi et al. [44] could be related to the use of different transport media or even to the size of follicles from which the oocytes are obtained. More studies using a combination of different transport solutions, temperatures and storage times to retrieve the largest number of competent oocytes are necessary.

5. Conclusions

Our study has provided new insight into the complex field of oocyte survival and in vitro development throughout ovary preservation. Moreover, it has contributed to understanding the effect of ovary storage in the physiological features of immature oocytes, which had not been evaluated up to date. We have demonstrated that transport ovary times up to 5 h in saline solution are the most adequate storage conditions to maintain oocyte quality as well as developmental capacity in sheep. After that, the quality and developmental potential of oocytes and cumulus cells dramatically decreases after storage of ovaries from 7 h. Based on these results, new strategies to evaluate the possibility of saving or rescuing the developmental potential of stored oocytes will be needed for successful production of high-quality embryos.

Supplementary Materials: The following are available online at http://www.mdpi.com/2076-2615/10/12/2414/s1, Table S1: Composition of solutions used in experiments, Table S2: List of primers used in qPCR of sheep immature oocytes, Table S3: List of abbreviations, Figure S1: Measurement of caspase-3 intracellular activity in sheep oocytes collected from ovaries stored for 3, 5, 7, 9, 11 and 13 h, Figure S2: Mitochondrial membrane potential in sheep oocytes collected from ovaries stored for 3, 5, 7, 9, 11 and 13 h, Figure S3: Mitochondrial distribution patterns in sheep oocytes collected from ovaries stored for 3, 5, 7, 9, 11 and 13 h, Figure S4: Relative mRNA transcript abundance pattern of genes of interest in sheep immature oocytes collected from ovaries stored for 3, 7 and 13 h, Figure S5: Live/dead status and apoptosis, GSH and ROS levels, DNA fragmentation, caspase-3 intracellular activity, mitochondrial membrane potential and distribution, and relative mRNA transcript abundance in sheep immature oocytes collected from ovaries stored with TCM or saline solution: the results are expressed as mean ± SEM.

Author Contributions: Conceptualization, I.S.-A. and A.J.S.; methodology: A.M.-M., I.S.-A., C.M., P.P.-F., D.-A.M.-C., B.C., J.C.N. and M.R.F.-S.; formal analysis: A.M.-M., I.S.-A. and C.M.; investigation: I.S.-A. and A.J.S.; resources: A.J.S. and J.J.G.; writing: A.M.-M. and I.S.-A.; original draft preparation: I.S.-A.; writing—review and editing: A.J.S. and J.J.G.; funding acquisition: A.J.S. and J.J.G. All authors have read and agreed to the published version of the manuscript.

Funding: This research was funded by the Spanish Ministry of Economy and Competitiveness (AGL2017-89017-R). A.M.-M. was supported by a Ministry of Economy and Competitiveness scholarship. P.P.-F. and D.-A.M.-C. were supported by a University of Castilla-La Mancha scholarship. B.C. was supported by ERASMUS+−.

Conflicts of Interest: The authors declare no conflict of interest.

References

1. Amiridis, G.S.; Cseh, S. Assisted reproductive technologies in the reproductive management of small ruminants. *Anim. Reprod. Sci.* **2012**, *130*, 152–161. [CrossRef] [PubMed]
2. Garde, J.J.; Martínez-Pastor, F.; Gomendio, M.; Malo, A.F.; Soler, A.J.; Fernández-Santos, M.R.; Esteso, M.C.; García, A.J.; Anel, L.; Roldán, E.R.S. The application of reproductive technologies to natural populations of red deer. *Reprod. Domest. Anim.* **2006**, *41*, 93–102. [CrossRef] [PubMed]
3. Paramio, M.-T.; Izquierdo, D. Recent advances in in vitro embryo production in small ruminants. *Theriogenology* **2016**, *86*, 152–159. [CrossRef] [PubMed]
4. Sirard, M.A.; Desrosier, S.; Assidi, M. In vivo and in vitro effects of FSH on oocyte maturation and developmental competence. *Theriogenology* **2007**, *68*, S71–S76. [CrossRef]
5. Wang, Q.; Sun, Q.Y. Evaluation of oocyte quality: Morphological, cellular and molecular predictors. *Reprod. Fertil. Dev.* **2007**, *19*, 1–12. [CrossRef]
6. Wongsrikeao, P.; Otoi, T.; Karja, N.W.K.; Agung, B.; Nii, M.; Nagai, T. Effects of ovary storage time and temperature on DNA fragmentation and development of porcine oocytes. *J. Reprod. Dev.* **2005**, *51*, 87–97. [CrossRef]

7. Reynolds, L.P.; Grazul-Bilska, A.T.; Redmer, D.A. Angiogenesis in the female reproductive organs: Pathological implications. *Int. J. Exp. Pathol.* **2002**, *83*, 151–164. [CrossRef]
8. King, T.C. Cell Injury, Cellular Responses to Injury, and Cell Death. In *Elsevier's Integrated Pathology*; Mosby Elsevier: Maryland Heights, MO, USA, 2007; pp. 1–20, ISBN 978-0-323-04328-1.
9. Tellado, M.N.; Alvarez, G.M.; Dalvit, G.C.; Cetica, P.D. The Conditions of Ovary Storage Affect the Quality of Porcine Oocytes. *Adv. Reprod. Sci.* **2014**, *2*, 56–67. [CrossRef]
10. Sánchez-Ajofrín, I.; Iniesta-Cuerda, M.; Peris-Frau, P.; Martín-Maestro, A.; Medina-Chávez, D.; Maside, C.; Fernández-Santos, M.; Ortiz, J.; Montoro, V.; Garde, J.; et al. Beneficial Effects of Melatonin in the Ovarian Transport Medium on In Vitro Embryo Production of Iberian Red Deer (Cervus Elaphus hispanicus). *Animals* **2020**, *10*, 763. [CrossRef]
11. García-Álvarez, O.; Maroto-Morales, A.; Berlinguer, F.; Fernández-Santos, M.R.; Esteso, M.C.; Mermillod, P.; Ortiz, J.A.; Ramon, M.; Pérez-Guzmán, M.D.; Garde, J.J.; et al. Effect of storage temperature during transport of ovaries on in vitro embryo production in Iberian red deer (Cervus elaphus hispanicus). *Theriogenology* **2011**, *75*, 65–72. [CrossRef]
12. Takahashi, Y.; First, N.L. In vitro development of bovine ane-cell embryos: Influence of glucose, lactate, pyruvate, amino acids and vitamins. *Theriogelology* **1992**, *37*, 963–978. [CrossRef]
13. Sánchez-Ajofrín, I.; Iniesta-Cuerda, M.; Sánchez-Calabuig, M.J.; Peris-Frau, P.; Martín-Maestro, A.; Ortiz, J.A.; del Rocío Fernández-Santos, M.; Garde, J.J.; Gutiérrez-Adán, A.; Soler, A.J. Oxygen tension during in vitro oocyte maturation and fertilization affects embryo quality in sheep and deer. *Anim. Reprod. Sci.* **2020**, *213*, 106279. [CrossRef] [PubMed]
14. Bermejo-Álvarez, P.; Rizos, D.; Rath, D.; Lonergan, P.; Gutiérrez-Adán, A. Can Bovine In Vitro-Matured Oocytes Selectively Process X- or Y-Sorted Sperm Differentially? *Biol. Reprod.* **2008**, *79*, 594–597. [CrossRef] [PubMed]
15. Schmittgen, T.D.; Livak, K.J. Analyzing real-time PCR data by the comparative CT method. *Nat. Protoc.* **2008**, *3*, 1101–1108. [CrossRef] [PubMed]
16. Livak, K.J.; Schmittgen, T.D. Analysis of relative gene expression data using real-time quantitative PCR and the 2-ΔΔCT method. *Methods* **2001**, *25*, 402–408. [CrossRef] [PubMed]
17. Halici, Z.; Karaca, M.; Keles, O.N.; Borekci, B.; Odabasoglu, F.; Suleyman, H.; Cadirci, E.; Bayir, Y.; Unal, B. Protective effects of amlodipine on ischemia-reperfusion injury of rat ovary: Biochemical and histopathologic evaluation. *Fertil. Steril.* **2008**, *90*, 2408–2415. [CrossRef]
18. Amorim, C.A. Artificial ovary. In *Principles and Practice of Fertility Preservation*; Cambridge University Press: Cambridge, UK, 2011; pp. 448–458. ISBN 9784431559634.
19. Taylor, M.J. Biology of Cell Survival in the Cold: The Basis for Biopreservation of Tissues and Organs. In *Advances in Biopreservation*; CRC Press: New York, NY, USA, 2006; pp. 15–63, ISBN 1420004220.
20. Pedersen, H.G.; Watson, E.D.; Telfer, E.E. Effect of ovary holding temperature and time on equine granulosa cell apoptosis, oocyte chromatin configuration and cumulus morphology. *Theriogenology* **2004**, *62*, 468–480. [CrossRef]
21. Wolfe, B.A.; Wildt, D.E. Development to blastocysts of domestic cat oocytes matured and fertilized in vitro after prolonged cold storage. *J. Reprod. Fertil.* **1996**, *106*, 135–141. [CrossRef]
22. Otera, H.; Mihara, K. Mitochondrial dynamics: Functional link with apoptosis. *Int. J. Cell Biol.* **2012**, *2012*, 1–10. [CrossRef]
23. Gottlieb, E.; Armour, S.M.; Harris, M.H.; Thompson, C.B. Mitochondrial membrane potential regulates matrix configuration and cytochrome c release during apoptosis. *Cell Death Differ.* **2003**, *10*, 709–717. [CrossRef]
24. Susin, S.A.; Zamzami, N.; Kroemer, G. Mitochondria as regulators of apoptosis: Doubt no more. *Biochim. Biophys. Acta Bioenerg.* **1998**, *1366*, 151–165. [CrossRef]
25. Yang, M.; Antoine, D.; Weemhoff, J.; Jenkins, R.; Farhood, A.; Park, B.; Jaeschke, H. Biomarkers Distinguish Apoptotic and Necrotic Cell Death During Hepatic ischemia/reperfusion Injury in Mice. *Liver Transpl.* **2014**, *20*, 1372–1382. [CrossRef] [PubMed]
26. Garrity, M.M.; Burgart, L.J.; Riehle, D.L.; Hill, E.M.; Sebo, T.J.; Witzig, T. Identifying and quantifying apoptosis: Navigating technical pitfalls. *Mod. Pathol.* **2003**, *16*, 389–394. [CrossRef] [PubMed]
27. Jaeschke, H.; Lemasters, J.J. Apoptosis versus oncotic necrosis in hepatic ischemia/reperfusion injury. *Gastroenterology* **2003**, *125*, 1246–1257. [CrossRef]

28. Zhou, T.; Chuang, C.C.; Zuo, L. Molecular Characterization of Reactive Oxygen Species in Myocardial Ischemia-Reperfusion Injury. *Biomed Res. Int.* **2015**, *2015*, 1–9. [CrossRef]
29. Clanton, T.L. Hypoxia-induced reactive oxygen species formation in skeletal muscle. *J. Appl. Physiol.* **2007**, *102*, 2379–2388. [CrossRef]
30. Paradis, S.; Charles, A.L.; Meyer, A.; Lejay, A.; Scholey, J.W.; Chakfé, N.; Zoll, J.; Geny, B. Chronology of mitochondrial and cellular events during skeletal muscle ischemia-reperfusion. *Am. J. Physiol. Cell Physiol.* **2016**, *310*, C968–C982. [CrossRef]
31. Zhou, T.; Prather, E.R.; Garrison, D.E.; Zuo, L. Interplay between ROS and antioxidants during ischemia-reperfusion injuries in cardiac and skeletal muscle. *Int. J. Mol. Sci.* **2018**, *19*, 417. [CrossRef]
32. Reader, K.L.; Stanton, J.A.L.; Juengel, J.L. The role of oocyte organelles in determining developmental competence. *Biology* **2017**, *6*, 35. [CrossRef]
33. Castaneda, C.A.; Kaye, P.; Pantaleon, M.; Phillips, N.; Norman, S.; Fry, R.; D'Occhio, M.J. Lipid content, active mitochondria and brilliant cresyl blue staining in bovine oocytes. *Theriogenology* **2013**, *79*, 417–422. [CrossRef]
34. Torner, H.; Ghanem, N.; Ambros, C.; Hölker, M.; Tomek, W.; Phatsara, C.; Alm, H.; Sirard, M.A.; Kanitz, W.; Schellander, K.; et al. Molecular and subcellular characterisation of oocytes screened for their developmental competence based on glucose-6-phosphate dehydrogenase activity. *Reproduction* **2008**, *135*, 197–212. [CrossRef]
35. Gaulden, M.E. Maternal age effect: The enigma of Down syndrome and other trisomic conditions. *Mutat. Res. Genet. Toxicol.* **1992**, *296*, 69–88. [CrossRef]
36. Van Blerkom, J.; Antczak, M.; Schrader, R. The developmental potential of the human oocyte is related to the dissolved oxygen content of follicular fluid: Association with vascular endothelial growth factor levels and perifollicular blood flow characteristics. *Hum. Reprod.* **1997**, *12*, 1047–1055. [CrossRef]
37. Khosravi-Farsani, S.; Sobhani, A.; Amidi, F.; Mahmoudi, R. Mouse oocyte vitrification: The effects of two methods on maturing germinal vesicle breakdown oocytes. *J. Assist. Reprod. Genet.* **2010**, *27*, 233–238. [CrossRef]
38. Gilchrist, R.B.; Ritter, L.J.; Armstrong, D.T. Oocyte-somatic cell interactions during follicle development in mammals. *Anim. Reprod. Sci.* **2004**, *82–83*, 431–446. [CrossRef]
39. Febretrisiana, A.; Setiadi, M.A.; Karja, N.W.K. Nuclear maturation rate of sheep oocytes in vitro: Effect of storage duration and ovary temperature. *J. Indones. Trop. Anim. Agric.* **2015**, *40*, 93–99. [CrossRef]
40. Hinrichs, K.; Choi, Y.H.; Love, L.B.; Varner, D.D.; Love, C.C.; Walckenaer, B.E. Chromatin Configuration Within the Germinal Vesicle of Horse Oocytes: Changes Post Mortem and Relationship to Meiotic and Developmental Competence1. *Biol. Reprod.* **2005**, *72*, 1142–1150. [CrossRef]
41. Blondin, P.; Coenen, K.; Guilbault, L.; Sirard, M.-A. In vitro production of bovine embryos: Developmental competence is acquired before maturation. *Theriogenology* **1997**, *47*, 1061–1075. [CrossRef]
42. Ribeiro, B.I.; Love, L.B.; Choi, Y.H.; Hinrichs, K. Transport of equine ovaries for assisted reproduction. *Anim. Reprod. Sci.* **2008**, *108*, 171–179. [CrossRef]
43. Luu, V.V.; Hanatate, K.; Tanihara, F.; Sato, Y.; Do, L.T.K.; Taniguchi, M.; Otoi, T. The effect of relaxin supplementation of in vitro maturation medium on the development of cat oocytes obtained from ovaries stored at 4 °C. *Reprod. Biol.* **2013**, *13*, 122–126. [CrossRef]
44. Goodarzi, A.; Zare Shahneh, A.; Kohram, H.; Sadeghi, M.; Moazenizadeh, M.H.; Fouladi-Nashta, A.; Dadashpour Davachi, N. Effect of melatonin supplementation in the long-term preservation of the sheep ovaries at different temperatures and subsequent in vitro embryo production. *Theriogenology* **2018**, *106*, 265–270. [CrossRef]

Article

Reproductive Outcomes and Endocrine Profile in Artificially Inseminated versus Embryo Transferred Cows

Jordana S. Lopes [1,2], Estefanía Alcázar-Triviño [3], Cristina Soriano-Úbeda [1], Meriem Hamdi [4], Sebastian Cánovas [1,2], Dimitrios Rizos [4] and Pilar Coy [1,2,*]

1 Physiology of Reproduction Group, Departamento de Fisiología, Facultad de Veterinaria, Universidad de Murcia, Campus Mare Nostrum, 30100 Murcia, Spain; jordanaluisa.portugals@um.es (J.S.L.); cmsu1@um.es (C.S.-Ú.); scber@um.es (S.C.)

2 Institute for Biomedical Research of Murcia, IMIB-Arrixaca, 30120 Murcia, Spain

3 El Barranquillo S.L., Torre Pacheco, 30700 Murcia, Spain; estefania.alcazar@elbarranquillo.es

4 Department of Animal Reproduction, National Institute for Agriculture and Food Research and Technology (INIA), 28040 Madrid, Spain; mhamdi9186@hotmail.com (M.H.); drizos@inia.es (D.R.)

* Correspondence: pcoy@um.es

Received: 1 July 2020; Accepted: 3 August 2020; Published: 6 August 2020

Simple Summary: Bovine embryos are nowadays produced in laboratories, frozen and transferred to other cows. However, the percentage of pregnancies obtained after these transfers as well as difficulties found during labor, especially due to increased size of calves, are a matter of great concern. One of the possible explanations for these problems relies on the embryo being produced in in vitro conditions (laboratory settings), more specifically the culture medium (liquid) used to develop these embryos. In an attempt to better mimic what happens naturally, female reproductive liquids (from oviducts and uterus) were used as a supplement to the culture of the embryos. As controls, embryos produced using the standard protocol in the laboratory were produced, as well as embryos derived from artificial insemination of cows (in vivo). An evaluation on the pregnancy rates, how the hormonal profile of the recipients changed during pregnancy, difficulties during parturitions, and phenotype of calves were recorded. Results showed that all the groups were very similar, but many differences were noted on the hormonal profiles during pregnancy. In conclusion, all systems provided safe production of calves, but long-term analysis of these calves is necessary to understand the future impact of the laboratory protocols.

Abstract: The increasing use of in vitro embryo production (IVP) followed by embryo transfer (ET), alongside with cryopreservation of embryos, has risen concerns regarding the possible altered pregnancy rates, calving or even neonatal mortality. One of the hypotheses for these alterations is the current culture conditions of the IVP. In an attempt to better mimic the physiological milieu, embryos were produced with female reproductive fluids (RF) as supplements to culture medium, and another group of embryos were supplemented with bovine serum albumin (BSA) as in vitro control. Embryos were cryopreserved and transferred while, in parallel, an in vivo control (artificial insemination, AI) with the same bull used for IVP was included. An overview on pregnancy rates, recipients' hormonal levels, parturition, and resulting calves were recorded. Results show much similarity between groups in terms of pregnancy rates, gestation length and calves' weight. Nonetheless, several differences on hormonal levels were noted between recipients carrying AI embryos especially when compared to BSA. Some calving issues and neonatal mortality were observed in both IVP groups. In conclusion, most of the parameters studied were similar between both types of IVP derived embryos and the in vivo-derived embryos, suggesting that the IVP technology used was efficient enough for the safe production of calves.

Keywords: embryo transfer; reproductive fluids; pregnancy; vitrification; calving

1. Introduction

In 2018, more than 1 million bovine embryos were produced worldwide from which the majority were in vitro produced (IVP) [1]. This represents a continuous growth of IVP-embryos globally despite the fact that IVP embryos have lower pregnancy rate than in vivo-derived embryos (IVD) [2,3]. Its success relies on the possibility of producing more embryos than in vivo, in a shorter amount of time and that the embryonic losses after implantation are no different from IVD, therefore, compensating the potential decrease in implantation rate [3,4]. In addition, the use of IVP-embryos is a better option during season of heat-stress [5] or in cases of repeat breeders [6].

Another technique on the rise is the cryopreservation of embryos, with more than 38% of 2018's embryo transfers (ET) coming from frozen embryos [1]. The need for exchange of cattle genetics has popularized this embryo preservation technique, which has the added benefit to not have all the recipients synchronized at the time of fresh embryo production. Within these large numbers of ET of frozen embryos, almost half (>46%) were IVP-derived embryos, but only a very small percentage of these (<0.1%) were produced with oocytes from abattoir ovaries, the rest being obtained by ovum pick-up (OPU).

Although the increase in IVP is significant over the years, the quality of the embryos produced in vitro still remains inferior to IVD embryos [7–9]. This is a factor that is affected in all mammalian embryos, where culture conditions such as the type of medium, type of supplementations, gases concentration or even culture devices play a crucial role, determining the final yield and quality of the embryo [10]. Currently, research is leading us towards a more physiological approach to in vitro culture. Media supplementation is being updated, such as the addition of insulin-transferrin-selenium to culture medium [11] that increases the blastocyst yield, or the addition of reproductive fluids (RF) [12] that increases embryo quality by increasing their cryotolerance and protecting from oxydative stress. García-Martínez et al. [13] measured the levels of O_2 within the different segments of the pig oviduct and uterus in vivo, and adapted the data obtained to the IVP, showing that decreasing the O_2 concentration during IVF, though not affecting the IVF results directly, had a positive impact on blastocyst yield and quality. Also, a new device [14] named 3D oviduct-on-a-chip model was recently designed to promote a more accurate surface when culturing bovine embryos, allowing a proper fertilization of the oocytes and eliminating polyspermy and parthenogenic activation.

Several issues besides embryo quality have been pointed out regarding problems in IVP-pregnancies. Hasler [15] summarized them into four main points: (1) increased abortion rate; (2) reduced intensity of labour; (3) increased dystocia, birth weight, calf mortality and fetal abnormalities; and (4) higher percentage of male calves over female calves. However, in addition to these four issues, the fact that embryos are cryopreserved also affects their survival after warming [9,16,17], due to their high lipid content, as well as their implantation success, giving usually inferior pregnancy rates [15,18,19].

In order to further improve in vitro production and ultimately bovine-assisted reproductive technologies (ART), we designed the present study with the following specific objectives; (1) To find out whether culture media is improved by the addition of RF—acting closer to the physiological milieu—impacting the embryo implantation rate versus embryos produced in vitro with conventional culture media (BSA) or in vivo (AI); (2) To find out if hormonal levels in recipients, including P4, E2, AMH and cortisol, at day of ET and across pregnancy could help to predict ART-associated problems such as decrease implantation rate; and (3) To find out if RF-derived calves have more similarities with AI-derived calves, relative to conceptus size, gestation length, calving difficulty, and calf birth weight, than BSA-derived calves.

To fulfil our objectives, we used cow recipients allocated in three experimental groups, according to the origin of the embryo transferred: the first group received an in vitro produced embryo grown in culture media improved by supplementation with reproductive fluids (RF group); the second group received an embryo produced at the same time as those at the previous group but using conventional culture media, lacking reproductive fluids (BSA group); and the third group consisted of animals artificially inseminated with frozen-thawed semen from the same bull used in the two IVP groups (AI group).

2. Materials and Methods

All chemicals were purchased from Sigma-Aldrich Chemical Company (Madrid, Spain), unless otherwise indicated.

2.1. Ethics

The experimental work was submitted to evaluation by the CEEA (Comité Ético de Experimentación Animal) from University of Murcia. After approval, authorization from "Dirección General de Agricultura, Ganadería, Pesca y Acuicultura", Región de Murcia- nr A13170706 was given to perform the animal experiments.

2.2. Oocyte Collection and In Vitro Maturation

Ovaries from crossbred beef cycling heifers and cows were collected at the slaughterhouse. The protocol has already been described elsewhere [12]. Briefly, follicles between 2–8 mm were aspirated and intact cumulus-oocyte complexes (COC) were selected for in vitro maturation (IVM). Groups of 50 COCs were cultured in maturation medium that consisted of TCM-199 supplemented with 10% fetal calf serum and 10 ng/mL epidermal growth factor for a period of 24 h at 38.5 °C, under 5%CO_2 and high humidity.

2.3. In Vitro Fertilization

Commercially bought semen doses from one bull (Asturian Valley breed, ASEAVA, Asturias, Spain) were used for all the cycles of embryo production. Frozen semen straws were thawed in a water bath at 38 °C, and following the manufacturer's instructions from Bovipure® (Nidacon, Sweeden), the gradient with semen was centrifuged at 300× *g* for 15 min and then washed for 5 min at 300× *g*. Matured oocytes were washed in Fert-TALP [20] medium, transferred to a new dish and inseminated with 1×10^6 spermatozoa/mL. Fertilization was left to occur during 18–20 h, at 38.5 °C, under 5%CO_2 and high humidity.

2.4. Embryo Culture

Presumptive zygotes were denuded from cumulus cells by vortex for 3 min. In each replicate, putative zygotes were divided in two groups according to embryo culture medium (SOF) supplementation: bovine serum albumin (BSA group) or reproductive fluids (RF group). BSA group received supplementation of 3 mg/mL of bovine serum albumin from day 1 to day 8. RF group received supplementation of 1.25% (*v*/*v*) of oviductal fluid (from early luteal phase of the estrous cycle, NaturARTs BOF-EL—Embryocloud, Spain) from day 1 to day 4, and 1.25% (*v*/*v*) of uterine fluid (from mid-luteal phase of the estrous cycle, NaturARTs BUF-ML—Embryocloud, Spain) from day 4 to day 8, as previously described [20]. Putative zygotes were washed twice in the corresponding medium and put into culture in groups of 25 per 25 µL microdrop, covered with parafin oil (Nidoil, Nidacon, Sweden). Incubation conditions were 38.5 °C, 5% CO_2 and 5% O_2. On day 4 of culture, all embryos were washed twice in the new corresponding medium and put in a new culture dish.

2.5. Embryo Vitrification and Warming

Commercial vitrification media (Kitazato-Dibimed, Spain) and an open-system Cryotop were used following manufacturer's instructions and as previously described [21]. Embryos on day 7 or 8 of culture and on stage 6 or 7 of development [22] were submitted to vitrification and stored in liquid nitrogen until use. Commercial thawing media (Kitazato-Dibimed, Spain) was used to warm vitrified embryos. Warming was performed according to manufacturer's instructions and performed less than 4 h before embryo transfer. Embryos were loaded in 0.25 mL straws with commercial medium (BO-Transfer, IVF-Bioscience, Denmark).

2.6. Recipient Synchronization and Embryo Transfer

Holstein multiparous dairy cows from a commercial farm (El Barranquillo SL, Spain) were synchronized using double-Ovsynch protocol. Reproductive ultrasound evaluation on the previous day of the transfer was made in order to discard recipients that failed or had delayed ovulation. Recipients were on their day 6, 7 or 8 of the estrus cycle and embryo transfer (ET) was made non-surgically to the ipsilateral uterine horn from ovulation (one embryo per recipient).

2.7. Artificial Insemination

Cows were artificially inseminated with frozen-thawed semen from the same Asturian-Valley bull used for IVP. Synchronization was made the same way described earlier (double-Ovsynch protocol) and cows were inseminated on day 0 (presumptive day of estrus).

2.8. Pregnancy Detection and Follow-up until Parturition

Pregnancy was detected at day 30 ± 3 of gestation by rectal ultrasonography (Easi-Scan™, BCF Technology, Scotland, UK). Measurement of the crown-rump length (mm) was performed to the conceptuses. Confirmation of pregnancy was repeated at days 60, 90, 150, and 210 of pregnancy. If parturition had not occurred by day 283 ± 2 of gestation, labour was induced with 0,150 mg d-cloprostenol q24h. No C-sections were performed but human intervention was available when calving was difficult. Calving was considered "easy" if little or no help was necessary and "difficult" if heavy assistance was needed. Calves' weight was assessed between 0 to 4 h after birth, using a weight scale.

2.9. Blood Collection and Analysis

Blood from recipient cows was collected via puncture of the median caudal vein with plain tubes (Vacutainer, BD Spain) on the day of ET or 7 days after-estrus for AI group, from here on referred to as "day 7" in both cases. Blood from pregnant recipients was also collected on day 30, 90, 150, and 210. Samples were centrifuged at 1000× g for 20 min at room temperature and plasma collected and stored (−20 °C) until analysis. Hormone levels—anti-Müllerian (AMH, ng/mL), estradiol (E2, pg/mL), progesterone (P4, ng/mL) and cortisol (nmol/L)—were measured with ECLIA assay (electrochemiluminescence immunoassay) using a Cobas® e801 system (Roche Diagnostics GmbH, Mannheim, Germany). Five samples of estradiol were not correctly measured and gave values below the detection (<5 ng/mL; a mean value of the previous/following measurement of the individual cow was made and used as a replacement).

2.10. Statistical Analysis

For statistical analysis, we followed two different approaches: In the first one, we evaluated the three groups (AI, BSA and RF) independently; in the second one, we compared the data from AI group versus the pooled data of BSA and RF groups, with the intention of getting information about the effect of the ART procedure used (i.e., in vivo fertilized embryos (AI) versus in vitro produced embryos, where IVP = RF + BSA data).

Data were tested for normality using Shapiro–Wilk test ($p > 0.05$) followed by either one-way parametric ANOVA/t-test when approved for normality, or non-parametric Kruskal–Wallis/Mann Whitney U when not normally distributed. Multiple comparisons tests (Tukey/Dunn) were used when significant difference was found ($p < 0.05$). Hormonal parameters were analysed using ANOVA for repeated measures, with Geisser–Greenchouse's correction applied when data did not follow sphericity. Multiple comparisons tests were performed using Tukey for group analysis and Sidak for type of ART analysis, and $p < 0.05$ was considered significant.

Data presented are mean ± SEM, unless otherwise indicated. The software used was GraphPad Prism version 8.4.0 for Windows (GraphPad Software, San Diego, CA, USA).

3. Results

3.1. Rate of In Vitro Produced Embryos, Pregnancy Maintenance and Conceptus Size after Embryo Transfer/Artificial Insemination, Were Similar between Groups

Cleavage and blastocyst rate with BSA or RF as supplements to the embryo culture medium was not significantly different (Table 1).

Table 1. Total presumptive zygotes, cleavage and blastocyst rate of in vitro produced embryos bovine serum albumin (BSA) or reproductive fluids (RF).

Group	Total Presumptive Zygotes	Cleavage Rate		Blastocyst Yield	
	n	%	*n*	%	*n*
BSA	360	85.6 ± 1.9	308	26.7 ± 2.3	96
RF	429	85.6 ± 1.7	367	25.9 ± 2.1	111

Percentages are shown as mean ± SEM.

Pregnancy rates were also similar between groups. Table 2 describes the confirmation of pregnancy at different timelines and no statistical difference was found between groups, both in terms of pregnancy rates as well as pregnancy loss.

Table 2. Pregnancy confirmation during gestation and parturitions of artificial insemination (AI), bovine serum albumin (BSA), reproductive fluids (RF) and IVP (BSA and RF groups data combined) groups.

Group	Recipients	Day 30		Day 60		Day 90		Day 150		Day 210		Parturition	
	n	%	*n*	%	*n*	%	*n*	%	*n*	%	*n*	%	*n*
AI	35	22.9	8	22.9	8	22.9	8	22.9	8	22.9	8	22.9	8
BSA	45	22.2	10	17.8	8	17.8	8	17.8	8	17.8	8	17.8	8
RF	54	22.2	12	18.5	10	18.5	10	16.7	9	16.7	9	16.7	9
IVP (BSA + RF)	99	22.2	22	18.2	18	18.2	18	17.3	17	17.3	17	17.3	17

Percentages are means and n are number of animals.

Conceptus size was also not significantly different between groups (Figure 1).

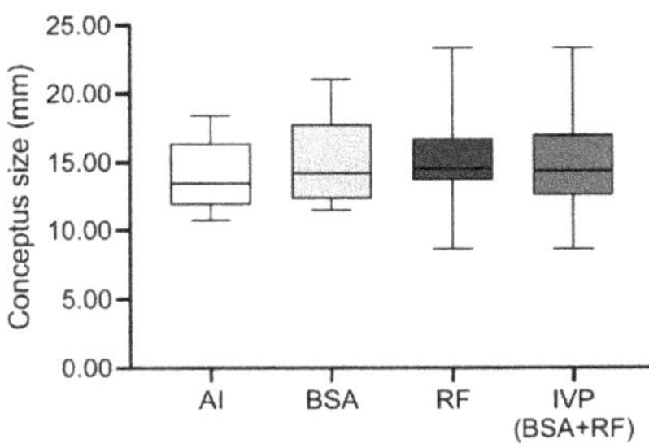

Figure 1. Distribution of conceptus size at day 30 of gestation of embryos produced by artificial insemination (AI), bovine serum albumin (BSA), reproductive fluids (RF), or in vitro-produced (IVP, corresponding of BSA + RF). The boxplot is represented by the first quartile, the median and the third quartile, with minimum and maximum as whiskers.

3.2. Recipient's Hormonal Levels at Day 7, 30, 90, and 210 of Gestation Showed Differences at Specific Time-Points

Figure 2 represents the evolution of hormonal levels through day 7, 30, 90, and 210 of pregnancy; their statistical differences and influence of the day of collection; the group; and day x group interaction. Cortisol levels were significantly affected by the day of collection, and particularly at day 7, showed a tendency to be higher in BSA recipients than RF recipients (Figure 2A, $p < 0.09$). AMH levels of AI recipients showed a significant decrease on day 30 when compared to IVP (Figure 2B, $p < 0.05$). P4 concentrations were influenced by the day of collection, being significantly lower on AI recipients when facing RF or IVP ($p < 0.01$) but just a tendency when compared to BSA (Figure 2C, $p < 0.09$). E2 had a tendency to be influenced by the group or day x group interaction, but no other significant difference was shown (Figure 2D). E2/P4 ratio was highly variable due to the day, group and day x group interaction, having AI recipients higher mean levels on day 7 vs. BSA/RF/IVP ($p < 0.01$) and a tendency on day 210 vs. BSA (Figure 2E, $p < 0.09$).

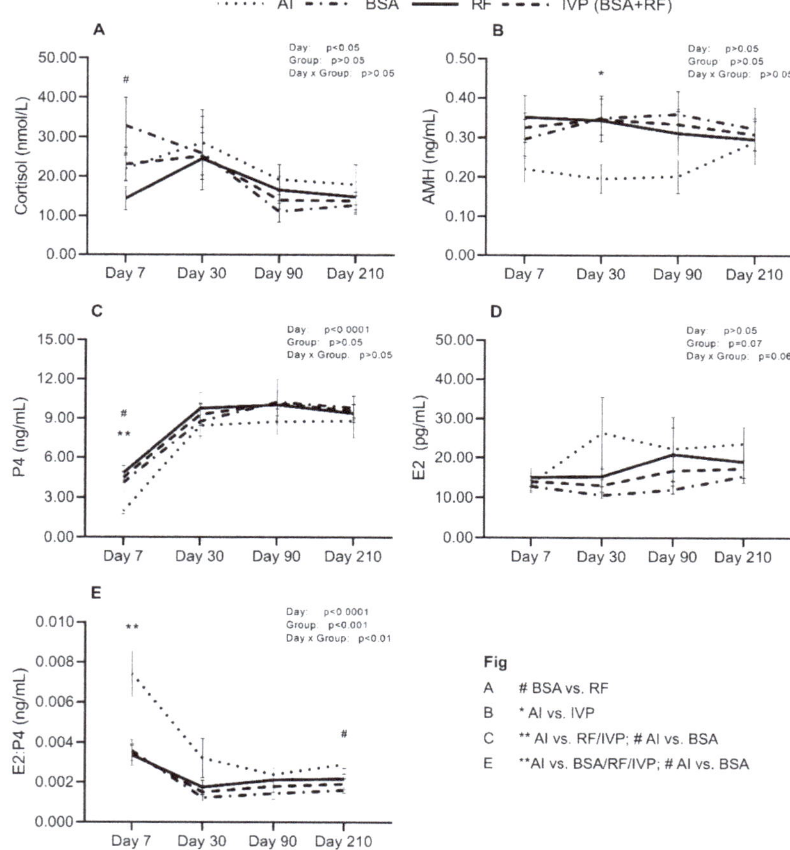

Figure 2. Distribution of mean ± SEM values, of (**A**). Cortisol, (**B**). Anti-Müllerian hormone (AMH), (**C**). Progesterone (P4), (**D**). Estrogen (E2) and (**E**). E2:P4 ratio, in pregnant recipients from artificial insemination (AI) group, bovine serum albumin group (BSA), reproductive fluids group (RF) or in vitro produced (IVP, corresponding of BSA + RF) across gestation time. * $p < 0.05$, ** $p < 0.01$ and # $p < 0.09$.

3.3. Gestation Length, Parturitions and Neonatal Period

Mean gestation length, minimum and maximum length for the non-induced parturitions is shown in Table 3.

Table 3. Gestation length of non-induced parturitions from pregnant cows of artificial insemination (AI), bovine serum albumin (BSA), reproductive fluids (RF) or in vitro-produced (IVP, corresponding of BSA + RF).

Group	*n*	Gestation Length	Minimum	Maximum
AI	8	281.1 ± 0.7	277	284
BSA	5	280.4 ± 1.4	275	282
RF	5	273.6 ± 9.7	240	298 *
IVP (BSA + RF)	10	277.0 ± 4.8	240	298 *

Gestation length is represented as mean ± SEM; * this parturition was not induced by decision of the farmers.

One premature calf was delivered naturally with 240 days and another recipient, due to productivity reasons, was not induced and the gestation lasted 298 days. Induced parturitions were necessary in some cases for BSA and RF groups (Table 4), mostly due to the fear of increased birth weight and consequently calving difficulties, and resulted in a higher tendency of IVP vs. AI of induced parturitions ($p = 0.0573$). Calving ease showed no statistical differences between groups. Percentage of male calves was high in absolute values, but without statistical differences between groups.

Table 4. Induced parturitions, calving ease score, male calves, and neonatal mortality within groups of artificial insemination (AI), bovine serum albumin (BSA), reproductive fluids (RF) or in vitro produced (IVP, corresponding of BSA + RF).

Group	Parturitions	Induced Parturition		Calving Ease Easy		Calving Ease Difficult		Male Calves		Neonatal Mortality	
	n	%	*n*	%	*n*	%	*n*	%	*n*	%	*n*
AI	8	0 †	0	100	8	0	0	75.0	6	0 *	0
BSA	8	37.5	3	87.5	7	12.5	1	62.5	5	12.5	1
RF	9	44.4	4	77.8	7	22.2	2	55.6	5	44.4 *	4
IVP (BSA+RF)	17	41.2 †	7	82.4	14	17.6	3	58.8	10	29.4	5

Calving ease was scored as easy if it required little to no assistance or difficult if it needed moderate to heavy assistance (i.e., surgery/veterinarian intervention). † AI vs. IVP $p = 0.0573$; * AI vs. RF $p = 0.0752$.

Weight at birth was also not different between groups (Figure 3).

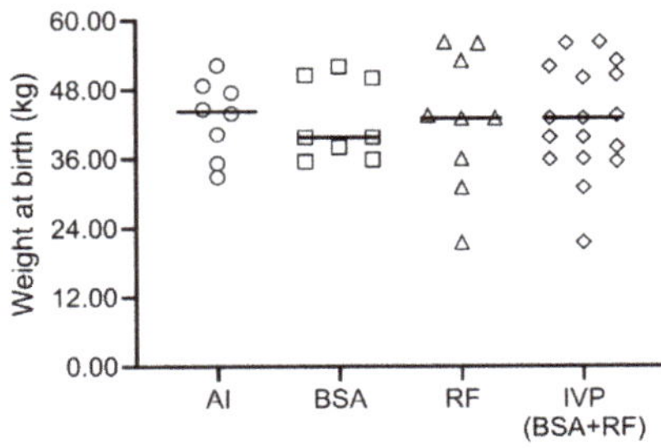

Figure 3. Weight at birth of calves born through artificial Insemination (AI) group, bovine serum albumin group (BSA), reproductive fluids group (RF) or in vitro-produced (IVP, corresponding of BSA + RF). The line represents the median and each symbol represents one animal.

Neonatal mortality happened in one calf for BSA group and in four calves for RF group and statistically did not show any difference between groups (Table 4), but a tendency of higher mortality in RF vs. AI happened. Details on age at death, sex, birth weight, and cause of death are shown in Table 5.

Table 5. Cases of neonatal mortality from calves born by embryo transfer of vitrified-warmed embryos produced with reproductive fluids (RF) or bovine serum albumin (BSA) as supplements to culture medium.

Characteristic	Case 1	Case 2	Case 3	Case 4	Case 5
Age at death (days)	12	0	2	13	0
Sex	Female	Male	Female	Male	Female
Birth weight (kg)	21.5	56.0	43.0	52.0	56.2
Group	RF	RF	RF	BSA	RF
Cause of death	Premature calf [ND]	Dystocia	Septicaemia	Diarrhoea	Dystocia, spine fracture

[ND] Stands for not determined.

4. Discussion

The in vitro embryo production in cattle industry has been drastically increasing over the past few years. Today, it is necessary not only to optimize its final yield, but also to assure that we are producing high quality and healthy animals and that calving problems are minimum. To this end, culture conditions and recipient selection and monitoring appear as main points to keep under control. Culture conditions of IVP are still in need of improvement, not only in cattle but in mammals in general [10,23,24]. The pre-implantation embryo represents a critical stage where all environmental conditions might reach higher relevance than at any other stage. As reviewed by Vajta et al. [10], current culture conditions, ranging from media composition to temperature of incubation, are not a unanimity, suggesting that there is still a lot of unknown factors in the biological environment that need to be studied and addressed. The addition of RF, such as oviductal and uterine fluids, has been proposed before [12,25] as a potential tool to improve the embryo quality. Hamdi et al. [12], using both fluids as supplements to embryo culture media, obtained similar blastocyst rates to those reached with fetal calf serum or even BSA supplementation. Despite not showing any improvement of blastocyst yield, embryos produced with RF supplementation showed higher survivability after vitrification-warming than serum-derived embryos, a downregulation of genes related with oxidative stress in comparison with BSA-derived embryos as well as significantly lower reactive oxygen species when compared to both serum and BSA groups. However, no data related to the ability of these embryos to implant or to develop to term after being transferred into recipient cows had been published until now, thus our work is the first producing live birth calves from IVP embryos cultured with reproductive fluids as supplements.

Pregnancy rates from vitrified-warmed IVP embryos are known to be inferior when compared to fresh IVP embryo transfers [5,26,27], with very few studies showing equal values [28]. Our results showed similar values at the first diagnosis for both BSA and RF groups, and these rates were slightly lower when compared to some studies that used abattoir-derived oocytes in IVP plus vitrification [5,26,29], but within the range or even higher than others that used slow freezing method [6,30]. Our AI pregnancy rates were lower than expected, probably explained by the high temperatures during the season in southern Spain, which induces heat-stress to the cows. Pregnancy loss was not significantly different between groups, and the percentage of lost pregnancies is in accordance to another study [5] that used vitrified embryos and had 25.9% pregnancy loss between diagnoses and calving.

Progesterone, being a key hormone associated with uterus preparation to receive the embryo, is one of the most studied hormones in cattle. According to Stronge et al. [31], low levels of P4 during the first 5 days after estrus are associated with low fertility. In our data, we found the normal physiological response that is the increase of P4 between day 7 and day 30. However, between the pregnant animals of our experimental groups, lower levels of P4 in AI recipients were found when compared to RF group or even IVP. This difference might be related to the fact that IVP recipients were chosen carefully, with evaluation of corpus luteum on the day prior to ET, and non-conforming recipients were discarded. On the contrary, AI recipients were inseminated after synchronization, without later confirmation of ovulation. It could be that AI recipients, even though pregnancy was achieved, may have ovulated later than expected in comparison to IVP recipients and consequently, the P4 levels were lower. Nevertheless, both our AI and IVP groups had P4 values that are within the range of normality for day 7 [32–36]. After this difference on day 7, by day 30 and onwards, P4 concentrations in all groups were similar. This could also help to explain why we did not find any differences regarding the conceptus size. Higher levels of circulating P4 have been associated with conceptus elongation as early as day 14 [36,37]. Supplementation of progesterone in early days post-estrus has also been responsible for higher conceptus length [38,39]. Although day 7 P4 concentrations from our groups differ significantly between AI recipients and RF/IVP recipients, the fact that by day 30 P4 levels were similar, might have led to non-existing differences in size of conceptus. Whether the rise of P4 concentrations on AI group between day 7 and day 30 was sufficient to catch up with IVP groups or if the possible higher embryo quality (and interferon-τ secretion) from AI group were responsible for this lack of difference, remains to be explained.

The role of cortisol during pregnancy is still not totally understood. It is known that glucocorticoids have roles in many physiological processes (reviewed by [40,41]), but its importance on modulating the uterine environment and consequently, influencing embryo implantation, has been the subject of attention. In cattle, interferon-τ is part of the maternal-recognition system and it is secreted by the trophoblast into the uterus especially between day 13–21 of pregnancy, and as so, Majewska et al. [40] correlated the relationship between cortisol and interferon-τ. This study, after using epithelial and stromal cells from pregnant and non-pregnant cows, concluded that interferon-τ modulates the conversion of cortisone to cortisol just after 12 h of incubation, but that response is also dose-dependent, meaning that there is a necessary dose of interferon-τ to achieve this modulation. This is an important step, as by increasing cortisol levels in vivo, the prostaglandins will be down-regulated and the cow will be able to maintain the corpus luteum active. In our study, the day of collection showed a high influence over the results. Mean levels of cortisol by day 30 were very similar between groups, and then, in general, showed a decrease during the rest of the pregnancy. The only non-conforming group was BSA recipients that showed really high levels at day 7, then a decrease on day 30 continuing until day 90, and finishing with a slight increase at day 210. However, this response did not influence, as far as we understand, the outcomes of the pregnancies. Whether the vitrified-warmed IVP embryos differ from those produced in vivo in the expression of interferon-τ is not yet clear (reviewed by [42]).

Anti-Müllerian hormone is a growth factor produced by granulosa cells that has been positively correlated with pregnancy rate and maintenance, among other traits [43,44]. This hormone might be used as a biomarker for fertility, and even though its value varies much between individuals and breeds [43], it is quite stable within one individual and does not vary much within the estrus cycle. In general, our AI recipients showed lower levels of AMH when compared to the other groups, but this difference was only significant when comparing to IVP data on day 30. All the obtained values are in the mean average range of AMH concentration in Holstein cows [43], which is around 0.250 ng/mL, and could be considered an intermediate value (not too high nor too low). This is important because the same study [43] unveiled that cows with low AMH had greater risk for early pregnancy loss, which was not the case of our AI recipients. Nevertheless, these studies need to be taken into careful consideration, since they only use AI as a pregnancy method, and we are not fully confident that the embryo does not itself have an influence over their mother's hormonal values, as it happens with foetal sex [45].

Oestrogen levels were quite similar within groups of pregnant cows. However, when evaluating levels of E2, it is important to compare with P4 levels, thus the necessity of E2:P4 ratio. This ratio showed a high influence of the day, group and day x group interaction, and much higher levels for AI recipients vs. BSA/RF/IVP on day 7, as expected after the differences found on P4 levels. Later on, at day 210 there was again a tendency of difference between AI and BSA recipients. With the exception of day 7 differences, where the reason for the higher E2:P4 was the low P4 concentrations, the day 210 tendency was due to high levels of E2. Fuchs et al. [46] described that concentrations of E2 tend to decrease after day 20 of estrus until day 250 where they start to rise again to prepare for parturition. Nonetheless, they also describe E2 concentrations at day 20 around 24 pg/mL, which is within the range for our AI group, meaning that it is not that AI recipients have high levels of E2, but rather that BSA cows expressed lower levels than expected. However, all these eight BSA recipients gave birth to live calves.

Gestation lengths are reported to be longer for IVP pregnancies in some studies [15,47–52], but similar to AI in others [5,53]. It should be noted that the percentage of induced parturitions was higher in IVP (regardless of the RF or BSA supplementation) pregnancies than AI pregnancies (with a tendency to be significantly different), and those pregnancies lengths (n = 7) were not accounted for mean gestation length, which if it did, would probably contribute to an increase in the mean gestation length. Moreover, one of the RF-pregnancies was not induced and was led to term without induction, lasting 298 days. Although gestation length is considered a highly heritable trait (reviewed by [54]), it is important to refer to the fact that the same bull was used in both AI and IVP groups. The mean value for gestation length for calves from Asturian Valley breed is 286.6 days for female and 287.5 days for male calves [55], but since AI cows were Holstein and there is no traceability on the female donors for IVP embryos (crossbred beef), it is not possible to attribute these differences to any of the estimated breed values.

Parturitions in IVP-derived pregnancies are known to have a tendency to be more difficult than AI pregnancies [29,51,52,56]. In our study, three IVP pregnancies were considered difficult while all AI calving were evaluated as easy, not being statistically relevant. The proportion of male calves being higher in IVP vs. AI has also been pointed as a distinct characteristic [15]. In fact, some studies reported higher percentage of male calves [3,50,56,57], while on the contrary others reported a similar proportion [58–60]. Nevertheless, our results were that AI group had the highest proportion of males vs. female, followed by BSA group, and the lowest proportion was given by the RF group.

Weight at birth was not statistically different between groups. This is in disagreement with previous studies that reported higher birth weight for IVP calves [30,47–50,52,53,57,59,61,62], but in agreement with others [63]. Interestingly, our heaviest calf was IVP-derived (56.2 kg) as well as our lightest (21.5 kg). The AI *Asturian x Holstein* combination brought a high proportion of calves with weights within 40–50 kg.

Mortality during the neonatal period happened only with IVP calves. Numabe et al. [64], Behboodi et al. [57] and Van Wagtendonk-De Leeuw et al. [53] reported higher incidence of perinatal mortality in IVP-derived calves than AI-derived calves. Both calves that died during parturition were heavy calves (>50 kg), which is an attribute given to the IVP-origin. The calf that died 2 days after birth was in agreement with Jenkins et al. [54] that associated longer gestation to higher perinatal mortality. Another calf died after surviving 12 days of premature deliver (240 days). Finally, the last calf that died at 13 days old was due to one of the most common causes of death in neonatal calves, diarrhoea.

5. Conclusions

In this study we applied new strategies in ART towards the more physiological approach of IVP, by including natural reproductive fluids in the culture media. With our limited sample size, we were not able to find worthy differences at the first stages of embryo development, implantation and parturitions, derived from the addition of such fluids. By contrast, we found that most of the parameters studied were similar between both types of IVP derived embryos and the in vivo-derived embryos, suggesting the IVP technology used was efficient enough for the safe production of calves. Since there are changes that may only be detected phenotypically later in life, as one may understand from other works [53], it is imperative to study the development and growth of these animals in order to reach more consistent conclusions. Moreover, it is important to assess if the changes that have been previously found in pre-implantation stages are in any form present in the live animals, particularly regarding large offspring syndrome. The fact that these ART are being used globally to improve genetics of the herd or even to overcome heat stress in affected areas, make them essential tools that need to be studied thoroughly.

Author Contributions: Conceptualization, P.C.; methodology, J.S.L., D.R. and P.C.; formal analysis, J.S.L.; investigation, J.S.L., E.A.-T., C.S.-Ú., M.H. and S.C.; writing—original draft preparation, J.S.L. and P.C.; writing—review and editing, J.S.L., E.A.-T., C.S.-Ú., M.H., S.C., D.R. and P.C.; funding acquisition, D.R. and P.C. All authors have read and agreed to the published version of the manuscript.

Funding: This research was funded by European Union, Horizon 2020 Marie Sklodowska-Curie Action, grant number REPBIOTECH675526 and as well as by the Ministry of Economy and Competitiveness (Spain), grants number AGL2015-66341-R & AGL2015-70140-R MINECO-FEDER and Fundación Séneca, grant number 20040/GERM/16.

Acknowledgments: The authors are grateful to Transformación Ganadera De Leganés SA; Matadero Madrid Norte, San Agustin de Guadalix; and Carnica Colmenar SC, in Madrid, Spain for providing access to biological material (ovaries); also to Roberth and rest of the staff from El Barranquillo SL for their precious assistance with the animals. Extended gratitude towards Francisco Javier Ibáñez López for statistical support and Mikhael Poirier for English editing.

Conflicts of Interest: The authors declare no conflict of interest.

References

1. Viana, J. IETS Data Retrieval Committee. 2018 statistics of embryo production and transfer in domestic farm animals. *Embryo Technol. Newsl.* **2019**, *37*, 7–25.
2. Drost, M.; Ambrose, J.D.; Thatcher, M.-J.; Cantrell, C.K.; Wolfsdorf, K.E.; Hasler, J.F.; Thatcher, W.W. Conception rates after artificial insemination or embryo transfer in lactating dairy cows during summer in florida. *Theriogenology* **1999**, *52*, 1161–1167. [CrossRef]
3. Pontes, J.H.F.; Nonato-Junior, I.; Sanches, B.V.; Ereno-Junior, J.C.; Uvo, S.; Barreiros, T.R.R.; Oliveira, J.A.; Hasler, J.F.; Seneda, M.M. Comparison of embryo yield and pregnancy rate between in vivo and in vitro methods in the same Nelore (*Bos indicus*) donor cows. *Theriogenology* **2009**, *71*, 690–697. [CrossRef] [PubMed]
4. Taverne, M.; Breukelman, S.; Perényi, Z.; Dieleman, S.; Vos, P.; Jonker, H.; de Ruigh, L.; van Wagtendonk-de Leeuw, J.M.; Beckers, J.-F. The monitoring of bovine pregnancies derived from transfer of in vitro produced embryos. *Reprod. Nutr. Dev.* **2002**, *42*, 613–624. [CrossRef] [PubMed]
5. Stewart, B.M.; Block, J.; Morelli, P.; Navarette, A.E.; Amstalden, M.; Bonilla, L.; Hansen, P.J.; Bilby, T.R. Efficacy of embryo transfer in lactating dairy cows during summer using fresh or vitrified embryos produced in vitro with sex-sorted semen. *J. Dairy Sci.* **2011**, *94*, 3437–3445. [CrossRef] [PubMed]
6. Dochi, O.; Takahashi, K.; Hirai, T.; Hayakawa, H.; Tanisawa, M.; Yamamoto, Y.; Koyama, H. The use of embryo transfer to produce pregnancies in repeat-breeding dairy cattle. *Theriogenology* **2008**, *69*, 124–128. [CrossRef]
7. Holm, P.; Booth, P.J.; Callesen, H. Kinetics of early in vitro development of bovine in vivo-and in vitro-derived zygotes produced and/or cultured in chemically defined or serum-containing media. *Reproduction* **2002**, *123*, 553–565. [CrossRef]
8. Rizos, D.; Clemente, M.; Bermejo-Alvarez, P.; De La Fuente, J.; Lonergan, P.; Gutiérrez-Adán, A. Consequences of in vitro culture conditions on embryo development and quality. *Reprod. Domest. Anim.* **2008**, *43*, 44–50. [CrossRef]
9. Sudano, M.J.; Paschoal, D.M.; da Silva Rascado, T.; Magalhães, L.C.O.; Crocomo, L.F.; de Lima-Neto, J.F.; da Cruz Landim-Alvarenga, F. Lipid content and apoptosis of in vitro-produced bovine embryos as determinants of susceptibility to vitrification. *Theriogenology* **2011**, *75*, 1211–1220. [CrossRef]
10. Vajta, G.; Rienzi, L.; Cobo, A.; Yovich, J. Embryo culture: Can we perform better than nature? *Reprod. Biomed. Online* **2010**, *20*, 453–469. [CrossRef]
11. Wydooghe, E.; Heras, S.; Dewulf, J.; Piepers, S.; Van Den Abbeel, E.; De Sutter, P.; Vandaele, L.; Van Soom, A. Replacing serum in culture medium with albumin and insulin, transferrin and selenium is the key to successful bovine embryo development in individual culture. *Reprod. Fertil. Dev.* **2014**, *26*, 717–724. [CrossRef] [PubMed]

12. Hamdi, M.; Lopera-vasquez, R.; Maillo, V.; Sanchez-Calabuig, M.J.; Núnez, C.; Gutierrez-Adan, A.; Rizos, D. Bovine oviductal and uterine fluid support in vitro embryo development. *Reprod. Fertil. Dev.* **2018**, *30*, 935–9345. [CrossRef] [PubMed]
13. García-Martínez, S.; Sánchez Hurtado, M.A.; Gutiérrez, H.; Sánchez Margallo, F.M.; Romar, R.; Latorre, R.; Coy, P.; López Albors, O. Mimicking physiological O_2 tension in the female reproductive tract improves assisted reproduction outcomes in pig. *Mol. Hum. Reprod.* **2018**, *24*, 260–270. [CrossRef] [PubMed]
14. Ferraz, M.; Henning, H.; Costa, P.F.; Malda, J.; Melchels, F.P.; Wubbolts, R.; Stout, T.A.E.; Vos, P.L.A.M.; Gadella, B.M. Improved bovine embryo production in an oviduct-on-a-chip system: Prevention of poly-spermic fertilization and parthenogenic activation. *Lab Chip* **2017**, *17*, 905–916. [CrossRef]
15. Hasler, J.F. In-vitro production of cattle embryos: Problems with pregnancies and parturition. *Hum. Reprod.* **2000**, *15*, 47–58. [CrossRef] [PubMed]
16. Paschoal, D.M.; Sudano, M.J.; Schwarz, K.R.L.; Maziero, R.R.D.; Guastali, M.D.; Crocomo, L.F.; Magalhães, L.C.O.; Martins, A.; Leal, C.L.V.; da Cruz Landim-Alvarenga, F. Cell apoptosis and lipid content of in vitro–produced, vitrified bovine embryos treated with forskolin. *Theriogenology* **2017**, *87*, 108–114. [CrossRef] [PubMed]
17. Rizos, D.; Ward, F.; Duffy, P.; Boland, M.P.; Lonergan, P. Consequences of bovine oocyte maturation, fertilization or early embryo development in vitro versus in vivo: Implications for blastocyst yield and blastocyst quality. *Mol. Reprod. Dev.* **2002**, *61*, 234–248. [CrossRef]
18. Ferraz, P.A.; Burnley, C.; Karanja, J.; Viera-Neto, A.; Santos, J.E.P.; Chebel, R.C.; Galvão, K.N. Factors affecting the success of a large embryo transfer program in Holstein cattle in a commercial herd in the southeast region of the United States. *Theriogenology* **2016**, *86*, 1834–1841. [CrossRef]
19. Sanches, B.V.; Lunardelli, P.A.; Tannura, J.H.; Cardoso, B.L.; Colombo Pereira, M.H.; Gaitkoski, D.; Basso, A.C.; Arnold, D.R.; Seneda, M.M. A new direct transfer protocol for cryopreserved IVF embryos. *Theriogenology* **2016**, *85*, 1147–1151. [CrossRef]
20. Parrish, J.J. Bovine in vitro fertilization: In vitro oocyte maturation and sperm capacitation with heparin. *Theriogenology* **2014**, *81*, 67–73. [CrossRef]
21. Lopes, J.S.; Canha-Gouveia, A.; París-Oller, E.; Coy, P. Supplementation of bovine follicular fluid during in vitro maturation increases oocyte cumulus expansion, blastocyst developmental kinetics, and blastocyst cell number. *Theriogenology* **2019**, *126*, 222–229. [CrossRef] [PubMed]
22. Bó, G.A.; Mapletoft, R.J. Evaluation and classification of bovine embryos. *Anim. Reprod.* **2013**, *10*, 344–348.
23. Lonergan, P.; Fair, T. The ART of studying early embryo development: Progress and challenges in ruminant embryo culture. *Theriogenology* **2014**, *81*, 49–55. [CrossRef] [PubMed]
24. Sunde, A.; Brison, D.; Dumoulin, J.; Harper, J.; Lundin, K.; Magli, M.C.; Van den Abbeel, E.; Veiga, A. Time to take human embryo culture seriously. *Hum. Reprod.* **2016**, *31*, 2174–2182. [CrossRef] [PubMed]
25. Canovas, S.; Ivanova, E.; Romar, R.; García-Martínez, S.; Soriano-Úbeda, C.; García-Vázquez, F.A.; Saadeh, H.; Andrews, S.; Kelsey, G.; Coy, P. DNA methylation and gene expression changes derived from assisted reproductive technologies can be decreased by reproductive fluids. *eLife* **2017**, *6*, 1–24. [CrossRef] [PubMed]
26. Block, J.; Bonilla, L.; Hansen, P.J. Efficacy of in vitro embryo transfer in lactating dairy cows using fresh or vitrified embryos produced in a novel embryo culture medium. *J. Dairy Sci.* **2010**, *93*, 5234–5242. [CrossRef]
27. Chebel, R.C.; Demétrio, D.G.B.; Metzger, J. Factors affecting success of embryo collection and transfer in large dairy herds. *Theriogenology* **2008**, *69*, 98–106. [CrossRef]
28. Do, V.H.; Catt, S.; Amaya, G.; Batsiokis, M.; Walton, S.; Taylor-Robinson, A.W. Comparison of pregnancy in cattle when non-vitrified and vitrified in vitro-derived embryos are transferred into recipients. *Theriogenology* **2018**, *120*, 105–110. [CrossRef]
29. Bonilla, L.; Block, J.; Denicol, A.C.; Hansen, P.J. Consequences of transfer of an in vitro-produced embryo for the dam and resultant calf. *J. Dairy Sci.* **2014**, *97*, 229–239. [CrossRef]
30. Lazzari, G.; Wrenzycki, C.; Herrmann, D.; Duchi, R.; Kruip, T.; Niemann, H.; Galli, C. Cellular and molecular deviations in bovine in vitro-produced embryos are related to the large offspring syndrome. *Biol. Reprod.* **2002**, *67*, 767–775. [CrossRef]
31. Stronge, A.J.H.; Sreenan, J.M.; Diskin, M.G.; Mee, J.F.; Kenny, D.A.; Morris, D.G. Post-insemination milk progesterone concentration and embryo survival in dairy cows. *Theriogenology* **2005**, *64*, 1212–1224. [CrossRef]
32. Schrick, F.N.; Inskeep, E.K.; Butcher, R.L. Pregnancy rates for embryos transferred from early postpartum beef cows into recipients with normal estrous cycles. *Biol. Reprod.* **1993**, *49*, 617–621. [CrossRef] [PubMed]

33. Silva, J.C.; Costa, L.L.; Silva, J.R. Plasma progesterone profiles and factors affecting embryo-fetal mortality following embryo transfer in dairy cattle. *Theriogenology* **2002**, *58*, 51–59. [CrossRef]
34. Breukelman, S.P.; Perényi, Z.; Taverne, M.A.M.; Jonker, H.; van der Weijden, G.C.; Vos, P.L.A.M.; de Ruigh, L.; Dieleman, S.J.; Beckers, J.F.; Szenci, O. Characterisation of pregnancy losses after embryo transfer by measuring plasma progesterone and bovine pregnancy-associated glycoprotein-1 concentrations. *Vet. J.* **2012**, *194*, 71–76. [CrossRef] [PubMed]
35. Shorten, P.R.; Ledgard, A.M.; Donnison, M.; Pfeffer, P.L.; McDonald, R.M.; Berg, D.K. A mathematical model of the interaction between bovine blastocyst developmental stage and progesterone-stimulated uterine factors on differential embryonic development observed on Day 15 of gestation. *J. Dairy Sci.* **2018**, *101*, 736–751. [CrossRef]
36. O'Hara, L.; Scully, S.; Maillo, V.; Kelly, A.K.; Duffy, P.; Carter, F.; Forde, N.; Rizos, D.; Lonergan, P. Effect of follicular aspiration just before ovulation on corpus luteum characteristics, circulating progesterone concentrations and uterine receptivity in single-ovulating and superstimulated heifers. *Reproduction* **2012**, *143*, 673–682. [CrossRef]
37. Spencer, T.E.; Forde, N.; Lonergan, P. The role of progesterone and conceptus-derived factors in uterine biology during early pregnancy in ruminants. *J. Dairy Sci.* **2016**, *99*, 5941–5950. [CrossRef] [PubMed]
38. O'Hara, L.; Forde, N.; Kelly, A.K.; Lonergan, P. Effect of bovine blastocyst size at embryo transfer on day 7 on conceptus length on day 14: Can supplementary progesterone rescue small embryos? *Theriogenology* **2014**, *81*, 1123–1128. [CrossRef]
39. Clemente, M.; de La Fuente, J.; Fair, T.; Al Naib, A.; Gutierrez-Adan, A.; Roche, J.F.; Rizos, D.; Lonergan, P. Progesterone and conceptus elongation in cattle: A direct effect on the embryo or an indirect effect via the endometrium? *Reproduction* **2009**, *138*, 507–517. [CrossRef]
40. Majewska, M.; Lee, H.Y.; Tasaki, Y.; Acosta, T.J.; Szostek, A.Z.; Siemieniuch, M.; Okuda, K.; Skarzynski, D.J. Is cortisol a modulator of interferon tau action in the endometrium during early pregnancy in cattle? *J. Reprod. Immunol.* **2012**, *93*, 82–93. [CrossRef]
41. Michael, A.E.; Papageorghiou, A.T. Potential significance of physiological and pharmacological glucocorticoids in early pregnancy. *Hum. Reprod. Update* **2008**, *14*, 497–517. [CrossRef] [PubMed]
42. Ealy, A.D.; Wooldridge, L.K.; Mccoski, S.R. Post-transfer consequences of in vitro-produced embryos in cattle. *J. Anim. Sci.* **2019**, *30*, 2555–2568. [CrossRef] [PubMed]
43. Ribeiro, E.S.; Bisinotto, R.S.; Lima, F.S.; Greco, L.F.; Morrison, A.; Kumar, A.; Thatcher, W.W.; Santos, J.E.P. Plasma anti-Müllerian hormone in adult dairy cows and associations with fertility. *J. Dairy Sci.* **2014**, *97*, 6888–6900. [CrossRef] [PubMed]
44. Nawaz, M.Y.; Jimenez-Krassel, F.; Steibel, J.P.; Lu, Y.; Baktula, A.; Vukasinovic, N.; Neuder, L.; Ireland, J.L.H.; Ireland, J.J.; Tempelman, R.J. Genomic heritability and genome-wide association analysis of anti-Müllerian hormone in Holstein dairy heifers. *J. Dairy Sci.* **2018**, 1–13. [CrossRef] [PubMed]
45. Stojsin-Carter, A.; Costa, N.N.; De Morais, R.; De Bem, T.H.; Costa, M.P.; Carter, T.F.; Gillis, D.J.; Neal, M.S.; Ohashi, O.M.; Miranda, M.S.; et al. Fetal sex alters maternal anti-Mullerian hormone during pregnancy in cattle. *Anim. Reprod. Sci.* **2017**, *186*, 85–92. [CrossRef] [PubMed]
46. Fuchs, A.-R.; Helmer, H.; Behrens, O.; Liu, H.-C.; Antonian, L.; Chang, S.M.; Fields, M.J. Oxytocin and Bovine Parturition: A Steep Rise in Endometrial Oxytocin Receptors Precedes Onset of Labor1. *Biol. Reprod.* **2005**, *47*, 937–944. [CrossRef]
47. Kruip, T.A.M.; Den Daas, J.H.G. In vitro produced and cloned embryos: Effects on pregnancy, parturition and offspring. *Theriogenology* **1997**, *47*, 43–52. [CrossRef]
48. Sinclair, K.D.; Broadbent, P.J.; Dolman, D.F. In vitro produced embryos as a means of achieving pregnancy and improving productivity in beef cows. *Anim. Sci.* **1995**, *60*, 55–64. [CrossRef]
49. Yang, B.; Im, G.; Park, S. Characteristics of Korean native, Hanwoo, calves produced by transfer of in vitro produced embryos. *Anim. Reprod. Sci.* **2001**, *67*, 153–158. [CrossRef]
50. Park, Y.-S.; Kim, S.-S.; Kim, J.-M.; Park, H.-D.; Byun, M.-D. The effects of duration of in vitro maturation of bovine oocytes on subsequent development, quality and transfer of embryos. *Theriogenology* **2005**, *64*, 123–134. [CrossRef]
51. Pimenta-Oliveira, A.; Oliveira-Filho, J.P.; Dias, A.; Gonçalves, R.C. Morbidity-mortality and performance evaluation of Brahman calves from in vitro embryo production. *BMC Vet. Res.* **2011**, *7*, 79. [CrossRef] [PubMed]

52. Van Wagtendonk-De Leeuw, A.M.; Aerts, B.J.G.; Den Daas, J.H.G. Abnormal offspring following in vitro production of bovine preimplantation embryos: A field study. *Theriogenology* **1998**, *49*, 883–894. [CrossRef]
53. Siqueira, L.G.B.; Dikmen, S.; Ortega, M.S.; Hansen, P.J. Postnatal phenotype of dairy cows is altered by in vitro embryo production using reverse X-sorted semen. *J. Dairy Sci.* **2017**, *100*, 5899–5908. [CrossRef] [PubMed]
54. Jenkins, G.M.; Amer, P.; Stachowicz, K.; Meier, S. Phenotypic associations between gestation length and production, fertility, survival, and calf traits. *J. Dairy Sci.* **2016**, *99*, 418–426. [CrossRef]
55. Goyache, F.; Fernandez, I.; Alvarez, I.; Royo, L.J.; Gutierrez, J.P. Gestation length in the asturiana de los valles beef cattle breed and its relationship with birth weight and calving ease. *Arch. Zootec.* **2002**, *51*, 431–439.
56. Van Wagtendonk-de Leeuw, A.M.; Mullaart, E.; de Roos, A.P.W.; Merton, J.S.; den Daas, J.H.G.; Kemp, B.; de Ruigh, L. Effects of different reproduction techniques: AI, moet or IVP, on health and welfare of bovine offspring. *Theriogenology* **2000**, *53*, 575–597. [CrossRef]
57. Behboodi, E.; Anderson, G.B.; BonDurant, R.H.; Cargill, S.L.; Kreuscher, B.R.; Medrano, J.F.; Murray, J.D. Birth of large calves that developed from in vitro-derived bovine embryos. *Theriogenology* **1995**, *44*, 227–232. [CrossRef]
58. Schmidt, M.; Greve, T.; Avery, B.; Beckers, J.F.; Sulon, J.; Hansen, H.B. Pregnancies, calves and calf viability after transfer of in vitro produced bovine embryos. *Theriogenology* **1996**, *46*, 527–539. [CrossRef]
59. Jacobsen, H.; Schmidt, M.; Holm, P.; Sangild, P.T.; Vajta, G.; Greve, T.; Callesen, H. Body dimensions and birth and organ weights of calves derived from in vitro produced embryos cultured with or without serum and oviduct epithelium cells. *Theriogenology* **2000**, *53*, 1761–1769. [CrossRef]
60. Merton, J.S.; Knijn, H.M.; Flapper, H.; Dotinga, F.; Roelen, B.A.J.; Vos, P.L.A.M.; Mullaart, E. Cysteamine supplementation during invitro maturation of slaughterhouse-and opu-derived bovine oocytes improves embryonic development without affecting cryotolerance, pregnancy rate, and calf characteristics. *Theriogenology* **2013**, *80*, 365–371. [CrossRef]
61. Hasler, J.F.; Henderson, W.B.; Hurtgen, P.J.; Jin, Z.Q.; McCauley, A.D.; Mower, S.A.; Neely, B.; Shuey, L.S.; Stokes, J.E.; Trimmer, S.A. Production, freezing and transfer of bovine IVF embryos and subsequent calving results. *Theriogenology* **1995**, *43*, 141–152. [CrossRef]
62. McEvoy, T.G.; Sinclair, K.D.; Broadbent, P.J.; Goodhand, K.L.; Robinson, J.J. Post-natal growth and development of Simmental calves derived from in vivo or in vitro embryos. *Reprod. Fertil. Dev.* **1998**, *10*, 459–464. [CrossRef] [PubMed]
63. Chavatte-Palmer, P.; Heyman, Y.; Richard, C.; Monget, P.; LeBourhis, D.; Kann, G.; Chilliard, Y.; Vignon, X.; Renard, J.P. Clinical, hormonal, and hematologic characteristics of bovine calves derived FROM nuclei from somatic cells. *Biol. Reprod.* **2002**, *66*, 1596–1603. [CrossRef] [PubMed]
64. Numabe, T.; Oikawa, T.; Kikuchi, T.; Horiuchi, T. Production efficiency of Japanese black calves by transfer of bovine embryos produced in vitro. *Theriogenology* **2000**, *54*, 1409–1420. [CrossRef]

Article

Study of the Metabolomics of Equine Preovulatory Follicular Fluid: A Way to Improve Current In Vitro Maturation Media

Pablo Fernández-Hernández [1,2], María Jesús Sánchez-Calabuig [3,4], Luis Jesús García-Marín [1,5], María J. Bragado [1,6], Alfonso Gutiérrez-Adán [3], Óscar Millet [7], Chiara Bruzzone [7], Lauro González-Fernández [1,6,†] and Beatriz Macías-García [1,2,*,†]

1 Research Group of Intracellular Signaling and Technology of Reproduction, Research Institute INBIO G – C, University of Extremadura, 10003 Cáceres, Spain; pablofh@unex.es (P.F.-H.); ljgarcia@unex.es (L.J.G.-M.); jbragado@unex.es (M.J.B.); lgonfer@unex.es (L.G.-F.)
2 Department of Animal Medicine, Faculty of Veterinary Sciences, University of Extremadura, 10003 Cáceres, Spain
3 Department of Animal Reproduction, INIA, 28040 Madrid, Spain; mariasanchezcalabuig@gmail.com (M.J.S.-C.); agutierr@inia.es (A.G.-A.)
4 Department of Animal Medicine and Surgery, Faculty of Veterinary Sciences, University Complutense of Madrid, 28040 Madrid, Spain
5 Department of Physiology, Faculty of Veterinary Sciences, University of Extremadura, 10003 Cáceres, Spain
6 Department of Biochemistry and Molecular Biology and Genetics, Faculty of Veterinary Sciences, University of Extremadura, 10003 Cáceres, Spain
7 Precision Medicine and Metabolism Lab., CICbioGUNE, 48160 Vizcaya, Spain; omillet@cicbiogune.es (Ó.M.); cbruzzone@cicbiogune.es (C.B.)
* Correspondence: bemaciasg@unex.es
† These authors contributed equally to the present work as senior investigators.

Received: 20 April 2020; Accepted: 14 May 2020; Published: 19 May 2020

Simple Summary: Commercial in vitro embryo production in horses by ICSI (intracytoplasmic sperm injection) is currently used to produce embryos clinically. However, the successful pregnancy and foaling rates obtained after ICSI are only 10% of the oocytes matured in vitro. Conditions used for oocyte in vitro maturation are not optimized for equine oocytes. Hence, in the present work, we aimed to elucidate the major metabolites present in equine preovulatory follicular fluid obtained from postmortem mares using proton nuclear magnetic resonance spectroscopy (^{1}H-NMR). Twenty-two metabolites were identified; among these, nine of them are not included in the composition of in vitro maturation media. Hence, our data suggest that the currently used media for equine oocyte maturation can be further improved.

Abstract: Production of equine embryos in vitro is currently a commercial technique and a reliable way of obtaining offspring. In order to produce those embryos, immature oocytes are retrieved from postmortem ovaries or live mares by ovum pick-up (OPU), matured in vitro (IVM), fertilized by intracytoplasmic sperm injection (ICSI), and cultured until day 8–10 of development. However, at best, roughly 10% of the oocytes matured in vitro and followed by ICSI end up in successful pregnancy and foaling, and this could be due to suboptimal IVM conditions. Hence, in the present work, we aimed to elucidate the major metabolites present in equine preovulatory follicular fluid (FF) obtained from postmortem mares using proton nuclear magnetic resonance spectroscopy (^{1}H-NMR). The results were contrasted against the composition of the most commonly used media for equine oocyte IVM: tissue culture medium 199 (TCM-199) and Dulbecco's modified eagle medium/nutrient mixture F-12 Ham (DMEM/F-12). Twenty-two metabolites were identified in equine FF; among these, nine of them are not included in the composition of DMEM/F-12 or TCM-199 media, including (mean ± SEM): acetylcarnitine (0.37 ± 0.2 mM), carnitine (0.09 ± 0.01 mM), citrate (0.4 ± 0.04

mM), creatine (0.36 ± 0.14 mM), creatine phosphate (0.36 ± 0.05 mM), fumarate (0.05 ± 0.007 mM), glucose-1-phosphate (6.9 ± 0.4 mM), histamine (0.25 ± 0.01 mM), or lactate (27.3 ± 2.2 mM). Besides, the mean concentration of core metabolites such as glucose varied (4.3 mM in FF vs. 5.55 mM in TCM-199 vs. 17.5 mM in DMEM/F-12). Hence, our data suggest that the currently used media for equine oocyte IVM can be further improved.

Keywords: IVM; oocytes; equine; metabolomic

1. Introduction

Production of equine embryos in vitro is currently a commercial technique and a reliable way of producing embryos for vitrification or uterine/oviductal transfer [1]. In order to produce those embryos, immature oocytes are retrieved from postmortem ovaries or live mares by ovum pick-up (OPU) [2], matured in vitro (IVM), fertilized by intracytoplasmic sperm injection (ICSI), and cultured until day 8–10 of development [3]. However, among all the oocytes used for ICSI, in the best of the scenarios, roughly 10% of them end up in successful pregnancy and foaling [1,4]. Surprisingly, when in vivo matured equine oocytes are transferred to oviducts of live mares to produce equine offspring, the likelihood of pregnancy rises to 75%, contrasting with the 40% obtained when the oocytes transferred are matured in vitro [1]. These results highlight the fact that the media and conditions used for oocytes matured in vitro (IVM) largely differ from the physiological conditions required for correct nuclear and cytoplasmic maturation in horses, therefore decreasing the oocyte's developmental competence.

It has to be noted that the base media more commonly used for equine IVM are tissue culture medium 199 (TCM-199) or Dulbecco's modified eagle medium/nutrient mixture F-12 Ham (DMEM/F-12), which are generally chosen depending on the preferences of the laboratory where IVM is performed, and core differences exist among them [5]. Furthermore, none of these media have been developed specifically for equine IVM; instead, they were developed for cell culture, albeit equine cumulus–oocyte complexes (COCs) are capable of maturing with similar efficiency in either media [3,6,7].

To try to better understand the physiological conditions that equine COCs require and improve current IVM conditions, several reports have tried to address the metabolic requirements of equine COCs in vitro [5,6,8], the differences between the proteomic profiles of equine COCs maturated in vivo or in vitro [8], or the differential expression and localization of glycosidic residues in equine COCs matured in vitro vs. in vivo [9], among other approaches. All these reports have revealed a specific metabolic profile of equine COCs matured in vitro and important differences between equine COCs matured in vitro vs. in vivo. However, no research has been conducted to develop a defined oocyte equine-specific maturation medium. Hence, in the present work, we aimed to elucidate the metabolomic composition of equine preovulatory follicular fluid (FF). To do this, the major metabolites present in equine preovulatory follicular fluid were analyzed by high-field proton nuclear magnetic resonance spectroscopy (^{1}H-NMR) and the results were contrasted against the composition of the formerly mentioned media according to the manufacturer's specifications.

2. Materials and Methods

2.1. Collection of Equine Follicular Fluid

Follicular fluid was obtained immediately postmortem at a commercial slaughterhouse, on four separate days. At the time of evisceration, the entire mare reproductive tract was extracted and carefully inspected. The ovaries were examined and those tracts having a preovulatory follicle ≥35 mm in diameter, associated with uterine edema on examination of the opened endometrial surface (vivid endometrial folds with a gelatinous appearance), were sampled as preovulatory. The fluid was collected using a 10 mL plastic syringe attached to a 20 g hypodermic needle. The fluid obtained was separated into 1.5 mL Eppendorf tubes and centrifuged for 2 min in a microcentrifuge at room temperature (RT) to remove large cellular masses. The supernatant was retrieved, transferred to a clean tube, and placed in dry ice until its arrival at the laboratory (4–5 h). Once at the laboratory, the fluid was thawed and centrifuged at 16,000× *g* at 4 °C for 20 min, and the supernatant was transferred to a clean tube. The samples were then kept at −80 °C until analysis.

2.2. Sample Preparation

Samples of follicular fluid from six different mares (one sample per mare) at the preovulatory stage (PRE) were used ($n = 6$). For the preparation of the nuclear magnetic resonance samples (NMR), the follicular fluids were pretreated. A methanol extraction was performed with the following protocol: samples were defrosted at room temperature for 30 min slowly on ice and 170 µL of follicular fluid of each sample were placed in a 1.5 mL Eppendorf tube and 1.3 mL of a mixture of methanol:deuterated water in a ratio 2:1 was added to the follicular fluid. The Eppendorf tube with the extraction was placed at 4 °C with agitation for 4 h. The mixture was centrifuged at 4 °C at 25,000× *g* for 30 min. The supernatant was transferred to a new 2 mL Eppendorf tube and the samples were plunged in liquid N_2. Once the mixtures were frozen, samples were subjected to lyophilization. For sample analysis, the lyophilized product was resuspended with 500 µL of 0.2 M potassium phosphate buffer in deuterium oxide (D_2O) with a pH of 7.4 ± 0.5 and 1.11 µL of TSP (3-(Trimethylsilyl) propanoic acid), to reach a final volume of 500 µL. Samples were briefly vortexed and 500 µL of the follicular fluid/buffer mixture were finally pipetted into a 5 mm NMR tube. In all cases, sample preparation was manually done at RT.

2.3. NMR Measurements

Samples were measured at 298 K in an 800 MHz Bruker spectrometer (AVANCE III, Bruker Biospin GmbH, Reinsthetten, Germany) equipped with a ^{1}H detected cryoprobe with z-gradient and automatic tuning and matching unit. Optimization of experimental conditions included automated tuning and matching, automated locking, and automated shimming using TopShim. The 90° hard pulse was optimized to be sample-specific and the presaturation field strength was adjusted to 25 Hz. To minimize the interference of the water content in the NMR spectrum, solvent suppression techniques were applied.

For each sample, one-dimensional (1D) ^{1}H-NMR spectra were collected using a Carr–Purcell–Meiboom–Gill (CPMG) pulse sequence; 2D *J*-resolved included 800 × 40 points. Data analysis was done using the TopSpin 3.5 software (Bruker Biospin GmbH, Reinsthetten, Germany). Free induction decays were multiplied by an exponential function equivalent to 0.3 Hz line-broadening before applying a Fourier transformation. All transformed spectra were automatically corrected for phase and baseline distortions and referenced to the DSS singlet at 0 ppm for further analysis.

The 2D-*J*res experiment was also routinely included in the acquisition package, along with the 1D ^{1}H-NOESY. This experiment separates J-couplings and chemical shifts in the 2D plane and provides a useful and simplified proton-decoupled projection spectrum. A standard pulse sequence with a water peak suppression was used. After 16 dummy scans, 2 free induction decays (FIDs) were accumulated into 8 k × 40 data points at a spectral width of 16 ppm.

For assignment purposes, a battery of experiments including 2D-^{1}H, ^{13}C-HSQC, 2D-^{1}H, ^{1}H-TOCSY, and 2D ^{1}H-^{1}H-NOESY were recorded in a Bruker Avance III 800 MHz spectrometer (Figure S1). The chemical shift, multiplet type, and number of contributing nuclei to each metabolite are provided in Table 1 and were determined following previously validated methods [10–12].

Table 1. Chemical shift assignment, multiplicity, and number of contributing protons for the identified metabolites.

Metabolite Name	Chemical Shift (ppm)	Multiplet Type	Protons Contributing
Acetylcarnitine	3.8 3.2 2.6	dd s dd	9
Acetate	1.91	s	3
Alanine	1.46 3.76	d q	3
Aspartate	3.89 2.66	dd dd	2
Carnitine	3.41 3.21 2.42	m s m	9
Choline	3.18	s	9
Citrate	2.52	d	2
Creatine	3.02	s	3
Creatine phosphate	3.03	s	3
Fumarate	6.51	s	2
Glucose	5.22 3.88 3.72 3.52 3.45 3.23	d dd m dd m dd	1
Glucose-1-phosphate	3.75 3.39	m t	1
Glycine	3.54	s	2
Histamine	7.99 7.14	s s	1
Histidine	7.9 7.09	d d	1
Isoleucine	0.99	d	3
Lactate	4.1 1.3	q d	3
Leucine	3.72 1.70 0.94	m m t	6
Pyruvate	2.4	s	3
Succinate	2.39	s	
Threonine	4.24 1.31	m	1
Valine	1.02 0.97	m d	3

Multyplet type: s—singlet; d—doublet; t—triplet; dd—doublet of doublets; q—quadruplet; m—multiplet.

2.4. Commercial Media Composition

The composition of commercial media routinely used in our laboratory: TCM-199 with Earle´s salts (Ref. 31150022; Thermo Fisher Scientific (Waltham, MA, USA)) and DMEM/F-12 (Ref. 11320033; Thermo Fisher Scientific (Waltham, MA, USA)) were directly extracted from the manufacturer´s website.

2.5. Statistical Analysis

Data were analyzed using descriptive statistics to establish the mean, standard error of the mean, minimum, and maximum for each metabolite using the software Sigma Plot (ver. 12.0) for windows (Systat Software, Chicago, IL, USA).

3. Results

3.1. Metabolite Identification

The chemical shift assignment for each metabolite was performed using a random follicular fluid sample; the spectra of the FF (Figure S1) was contrasted against the identified metabolites that were chosen based on our previous study in horses [13]. The list of the chemical shift for each proton nucleus of each metabolite is provided in Table 1. The measured concentrations of 22 metabolites, which were selected based on the report of González-Fernández et al. (2020) [13], and the known identification capabilities of the NMR facility for preovulatory follicular fluid samples are presented in Table 2. Pyruvate and succinate are characterized by one single peak with the same chemical shift and cannot be discriminated; therefore, they are presented as the sum of both metabolites. All metabolites were detected in all the samples except for acetylcarnitine, which could not be detected in one sample submitted to NMR analysis.

Table 2. Concentrations of metabolites detected in equine preovulatory follicular fluid.

Metabolite	Follicular Fluid (mM)
Acetlylcarnitine	0.37 ± 0.2 (0.02–1.148)
Acetate	1.5 ± 0.6 (0.5–4.1)
Alanine	1.1 ± 0.14 (0.74–1.6)
Aspartate	2.7 ± 0.3 (2.1–3.7)
Carnitine	0.09 ± 0.01 (0.04–0.14)
Choline	0.03 ± 0.01 (0.01–0.09)
Citrate	0.4 ± 0.04 (0.3–0.5)
Creatine	0.36 ± 0.14 (0.15–0.9)
Creatine phosphate	0.36 ± 0.05 (0.2–0.6)

Table 2. *Cont.*

Metabolite	Follicular Fluid (mM)
Fumarate	0.05 ± 0.007 (0.03–0.07)
Glucose	4.3 ± 0.4 (3.1–5.5)
Glucose-1-phosphate	6.9 ± 0.4 (5.75–8.9)
Glycine	3.2 ± 0.5 (1.26–4.8)
Histamine	0.25 ± 0.01 (0.2–0.27)
Histidine	0.05 ± 0.009 (0.04–0.07)
Isoleucine	0.6 ± 0.07 (0.4–0.9)
Lactate	27.3 ± 2.2 (19.3–35.02)
Leucine	0.5 ± 0.05 (0.4–0.8)
Pyruvate + Succinate	0.16 ± 0.03 (0.08–0.3)
Threonine	0.35 ± 0.03 (0.14–0.8)
Valine	0.13 ± 0.02 (0.09–0.2)

The results are presented as mean ± SEM (minimum value−maximum value); the values correspond to 6 different mares (n = 6).

3.2. Comparison of the Metabolites Present in Commercial Media and Equine Preovulatory Follicular Fluid

Among the 22 metabolites identified in native preovulatory FF, nine of them are not present in TCM-199 or DMEM/F-12 according to the manufacturer's specifications (Table 3). These metabolites were: acetylcarnitine, carnitine, citrate, creatine, creatine phosphate, fumarate, glucose-1-phosphate, histamine, and lactate. Other metabolites such as acetate is present in FF and TCM-199 but not in DMEM/F-12, while pyruvate is included in the composition of DMEM/F-12 and possibly is present in FF (as it cannot be discriminated from succinate) but not in TCM-199. Vivid differences exist in the concentration of core metabolites such as lactate (27.3 mM in FF vs. 0 mM in TCM-199 and DMEM/F-12), glucose (4.3 mM in FF vs. 5.55 mM in TCM-199 vs. 17.5 mM in DMEM/F-12), alanine (1.1 mM in FF vs. 0.28 mM in TCM-199 vs. 0.05 mM in DMEM/F-12), aspartate (2.7 mM in FF vs. 0.22 mM in TCM-199 vs. 0.05 mM in DMEM/F-12), or glycine (3.2 mM in FF vs. 0.67 mM in TCM-199 vs. 0.25 mM in DMEM/F-12).

Table 3. Presence and concentration of the metabolites found in equine preovulatory follicular fluid, and in TCM-199 and DMEM/F-12 media (manufacturer's specifications).

Metabolite	Follicular Fluid (mM)	TCM-199 (mM)	DMEM/F-12 (mM)
Acetate	1.5	0.61	-
Acetlylcarnitine	0.37	-	-
Alanine	1.1	0.28	0.05
Aspartate	2.7	0.22	0.05
Carnitine	0.09	-	-
Citrate	0.4	-	-
Creatine	0.36	-	-
Creatine phosphate	0.36	-	-
Choline	0.03	0.003	0.064
Fumarate	0.05	-	-
Glucose (d-Glucose)	4.3	5.55	17.5
Glucose-1-phosphate	6.9	-	-
Glycine	3.2	0.67	0.25
Histamine	0.25	-	-
Histidine	0.05	0.1	0.15
Isoleucine	0.6	0.3	0.41
Lactate	27.3	-	-
Leucine	0.5	0.46	0.45
Pyruvate + Succinate	0.16	-	0.5
Threonine	0.35	0.25	0.45
Valine	0.13	0.21	0.45

4. Discussion

In the present work, the metabolome of equine preovulatory FF was investigated using ^{1}H-NMR. Our work revealed the presence of at least 22 metabolites including carbohydrates, amino acids, and intermediate metabolites (Table 2). The metabolome of equine FF at different dominant follicular stages (early dominant, late dominant, and healthy preovulatory stage) has previously been described by Gérard et al. in 2002 [14]. In their work, they did not detect apparent differences in the pattern or concentration of the metabolites detected among the studied stages. These authors described eight peaks corresponding to chemical groups of sugar chains and N-acetyl groups of glycoconjugates, CH3 groups of lipoproteins, trimethylamines, acetate, alanine, creatine/creatinine, and polyamines, plus a non-identified peak at 3.1 ppm, but quantitative identification of the peaks is not provided [14]. In our work, we detected 21 peaks (as succinate and pyruvate were overlapped; Table 1); an explanation for the differences found in the number of peaks (8 vs. 21) can be easily explained as Gérard et al. (2002) used a 200 MHz Bruker spectrometer (in our work we used an 800 MHz spectrometer and a cryoprobe) and the samples were directly diluted in deuterated water (instead of being previously subjected to a methanol extraction and lyophilization as in the present work), likely resulting in higher water interferences and lower spectra resolution [14]. Hence, in the formerly mentioned work, some peaks as citrate are suspected, while in our work, it was detected in all samples due to a better resolution of current NMR spectrometers (Tables 1 and 2). It must be mentioned that citrate and fumarate are not routinely added to base equine IVM media (Table 3) as both are intermediate metabolites produced by the metabolism of glucose in the tricarboxylic acid cycle (TCA); specifically, citrate comes from acetyl-CoA and oxaloacetate in the tricarboxylic acid cycle (TCA). Citrate acts as a key substrate for epigenetic modifiers in the oocyte [15,16] and is a link between TCA, β-hydroxybutyrate, and lipid metabolism in FF [17,18], so adequate supplementation could be important during equine IVM and should be added to TCM-199 (Table 3). Regarding pyruvate, this molecule has been previously reported to range between 0.03 and 0.13 mM in FF from early and late dominant equine follicles, respectively [19]. This metabolite has been demonstrated to be crucial for adequate oocyte metabolism [5] and is also involved in active reactive oxygen species scavenging [20]. In equine oocytes, it has been demonstrated that when DMEM/F-12 is supplemented with pyruvate at 0.15 mM, this induces an increase in glycolytic activity, without affecting mitochondrial oxidative phosphorylation [5]. The concentration of pyruvate above reported for equine FF [19] coincides with our work in which 0.16 mM ± 0.03 was observed. However, in our experiments, succinate could not be discriminated from pyruvate, and thus, exact values cannot be provided; nevertheless, as per previous reports, pyruvate addition to equine IVM media and the concentration at which it is supplemented needs to be seriously considered.

The research group of Gérard et al. also published in 2015 another report in which they performed a comparative metabolomic study of porcine, equine, and bovine FF [21]. In this report, even when the extraction method remained the same as in their previous work, a 500 mHz ^{1}H-NMR spectrometer was used and the resolution of the analysis was improved. In this work, they described the presence and concentration of some metabolites that coincide with the ones reported in the present work such as acetate, alanine, citrate, and glucose (alpha and beta, while in our work, glucose and glucose-1-phosphate were detected) [21]. However, the concentration of other metabolites reported in their work such as histidine (0.05 ± 0.009 mM in our work vs. 0.27 mM [21]) and valine (0.13 ± 0.02 mM in our work vs. 0.37 mM [21]) do not agree with our findings. These differences can be attributed to the different equipment used, the sample extraction method, or the fact that Gérard et al. (2015) recovered the FF by transvaginal aspiration of follicles around 33 mm and in our work postmortem FF was obtained from bigger follicles (35–45 mm). Interestingly, lactate concentration largely differs in our report compared with that obtained by Gerard et al. [21] (27.3 mM ± 2.2 vs. 0.70 mM ± 0.18, respectively), the time between mare decease and sample obtention being around 30 min. This time lapse could lead to lactate postmortem accumulation [22] as previously observed for equine oviductal fluid [13]. Nevertheless, the high lactate concentration observed in the present work (19.3–35.02 mM) cannot be solely explained as the metabolism of the typical amount of available glucose (2 molecules of lactate per 1 of glucose consumed [17]), as the concentration of glucose observed in our work matches previous reports [14,19,21,23], and thus appears to reflect a high level of lactate in the equine FF. However, it has to be noticed that the equine FF analyzed by Gérard et al. (2015) [21] was retrieved when the dominant follicle reached 33 mm, while it has been demonstrated that in equine follicles reaching 39 mm 24 h after equine crude gonadotropin administration, the lactate values reach 4 mM, this concentration of lactate being closely related to adequate meiosis resumption in equine oocytes [19]. Similar findings among lactate production during IVM and oocyte competence in the horse have been recently reported in vitro [5], highlighting the important relationship existing between anaerobic glycolysis and oocyte meiotic competency in the horse, strongly suggesting that FF may be providing energy to the oocyte in the form of lactate as postulated in humans [17]. Furthermore, high lactate concentrations in the FF have been linked to improved pregnancy rates in humans [24], and this metabolite could be an important additive contributing to osmolarity adjustment, as reported in human oviductal fluid [25]; hence, lactate would need to be supplemented to equine IVM media (Table 3).

Notably in our NMR spectra, different intracellular metabolites such as glucose-1-phosphate, creatine, creatine phosphate, and carnitine were found. These metabolites cannot be incorporated into the oocytes in their molecular form and may have been released from the granulosa cells due to cellular damage. However, recent investigations have demonstrated that a wide variety of intracellular metabolites are also present in the oviductal fluid of cows and horses [13,26], embedded in extracellular vesicles [27]. The compounds included in these vesicles could enter the oocytes/embryos via fusion as previously demonstrated in mouse embryos [28], contributing to the oocyte's developmental competence and metabolism as observed in cows [29]. It is known that carnitine plays a major role in the catabolism of lipids, allowing the transport of fatty acids from the cytosol to the mitochondria, where they are metabolized through beta-oxidation, which has also been identified in human FF [30]. Surprisingly, this work demonstrated the presence of carnitine in the FF but did not find expression of the enzymes involved in the carnitine synthesis either in oocytes or in cumulus cells. In contrast, the enzymes related to beta-oxidation were highly expressed in the oocyte and cumulus cells as also demonstrated in horses [8]. Therefore, this compound may also need to be somehow included in equine IVM media to ensure adequate oocyte lipid metabolism [30], as low concentration of carnitine in the FF has been associated with low reproductive performance outcome in sows [31]. The presence of creatine in the equine FF has been previously reported at similar concentrations to the ones reported in the present work [14]. Creatine and creatine phosphate are produced as a result of arginine and glycine metabolism. Even when arginine was not found in equine FF (Tables 1 and 2), arginine was present in TCM-199 (0.33 mM) and DMEM/F-12 (0.7 mM). Interestingly, arginine depletion during the final 6 h of IVM of human oocytes was associated with a higher maturation potential [32,33]. Hence, even when arginine was not detected in our experiments (Table 1), our data suggest that an active metabolism of arginine could be occurring during in vivo maturation of equine oocytes explaining the depletion of this amino acid in the FF and thus, the arginine present in the equine IVM media (TCM-199 and DMEM/F-12) could be needed. Glycine is known to be an organic osmolyte that regulates osmolarity in cells and embryos [34] and is one of the most abundant amino acids in follicular and uterine fluids [35]. Interestingly, its concentration in the FF has been demonstrated to predict the cleavage rate of oocytes after insemination (being a stronger marker of cleavage capacity in lower grade oocytes) as well as to be a good marker of the blastocyst rate in bovine [36]. Hence, considering the vivid difference existing in glycine concentration among TCM-199, DMEM/F-12, and native equine preovulatory FF (Table 3), the concentration at which this amino acid is added to equine IVM media should be carefully evaluated.

Interestingly, glucose-1-phosphate was detected in equine preovulatory FF (Table 1). This molecule is derived from glycogen, which is generally metabolized in the liver but is also a metabolic source used by granulosa cells in humans, pigs, bulls, and mice [15,37,38]. Thus, in view of our data, the equine oocyte also relies on glycogen metabolism as is also proposed in previous reports [5]. This is an interesting finding as, in our experiments, the amount of glucose-1-phosphate found (6.9 mM ± 0.4) surpassed the quantity of glucose (4.3 ± 0.4 mM), indicating that glycogen metabolism could support equine oocyte maturation in vivo, and more research is needed in this field, as this energy source is not generally considered in equine oocytes.

Other metabolites such as histamine, which has recently been found in equine oviductal fluid [13], are also present in horse FF and could be involved in ovulation induction, as observed in rabbits and rats [39,40] and as also postulated in horses [41].

Our results demonstrate that the base media used for equine IVM and the composition of preovulatory equine FF greatly differ. One limitation of the present study is that the composition of fetal bovine serum (FBS) that is usually added at concentrations ranging from 10% to 20% to equine IVM media has not been considered [4,42]. It is well known that FBS composition greatly varies among batches, and thus, a significant error could be introduced if the composition of a single batch was considered; this is why FBS composition is not considered in the present report.

5. Conclusions

In conclusion, our data provide new insights into equine preovulatory follicular fluid composition comparing it with the most widely used commercial media available for equine IVM (TCM-199 and DMEM/F-12). Our data provide new metabolite information that should be considered to design specific equine IVM media and help to improve current in vitro fertilization outcomes in the equine species. More research is warranted to better understand the metabolic requirements of equine oocytes and the relationship among metabolism, oocyte meiotic competence, and developmental competence.

Supplementary Materials: The following are available online at http://www.mdpi.com/2076-2615/10/5/883/s1, Figure S1. Metabolite assignment on 1D ^{1}H NMR spectra of an FF sample. (**A**) Chemical shift region from 6.4 to 8.0 ppm, (**B**) chemical shift region from 3 to 4.2 ppm, (**C**) chemical shift region from 1.8 to 2.5 ppm, (**D**) chemical shift region from 0.8 to 1.5 ppm.

Author Contributions: Conceptualization, L.G.-F. and B.M.-G.; Data curation, P.F.-H. and Ó.M.; Formal analysis, P.F.-H., Ó.M., C.B., and B.M.-G.; Funding acquisition, M.J.S.-C., L.J.G.-M., M.J.B., A.G.-A., and B.M.-G.; Investigation, M.J.S.-C., A.G.-A., C.B., and L.G.-F.; Methodology, M.J.S.-C. and Chiara Bruzzone; Project administration, B.M.-G.; Resources, Ó.M.; Software, C.B.; Supervision, Ó.M.; Validation, C.B.; Writing—original draft, P.F.-H., M.J.B.,

A.G.-A., and B.M.-G.; Writing—review & editing, M.J.S.-C., L.J.G.-M., Ó.M., and L.G.-F. All authors have read and agreed to the published version of the manuscript.

Funding: This study was supported by Spanish Ministry of Economy, Industry and Competitiveness and Fondo Europeo de Desarrollo Regional (FEDER) (AEI/FEDER/UE); References: RTI2018-093548-B-I00, AGL2017-84681-R and RYC-2017-21545 (this last awarded to B. Macías-García). L. González-Fernández was supported by regional grant "Atracción y retorno de talento investigador a Centros de I+D+i pertenecientes al Sistema Extremeño de Ciencia, Tecnología e Innovación" from "Junta de Extremadura" (Spain); Reference: TA18008. P. Fernández-Hernández was supported by a grant "Acción II" from the University of Extremadura (Ref. Beca RC4).

Acknowledgments: The help of the veterinary team (Gerardo, Carmen, and Jesús) at the slaughterhouse (Incarsa, Burgos, Spain) is warmly appreciated. This work is dedicated to all the victims of Covid-19 who will not directly benefit from our results but would possibly be relieved to know that science never stops, not even in difficult scenarios. To Daniela and Bruno González-Macías and all the kids who are valiantly coping with confinement.

Conflicts of Interest: The authors declare no conflict of interest.

References

1. Hinrichs, K. Assisted reproductive techniques in mares. *Reprod. Domest. Anim.* **2018**, *53*, 4–13. [CrossRef] [PubMed]
2. Velez, I.C.; Arnold, C.; Jacobson, C.C.; Norris, J.D.; Choi, Y.H.; Edwards, J.F.; Hayden, S.S.; Hinrichs, K. Effects of repeated transvaginal aspiration of immature follicles on mare health and ovarian status: Effect of transvaginal aspiration on health. *Equine Vet. J.* **2012**, *44*, 78–83. [CrossRef] [PubMed]
3. Choi, Y.-H.; Velez, I.C.; Macías-García, B.; Riera, F.L.; Ballard, C.S.; Hinrichs, K. Effect of clinically-related factors on in vitro blastocyst development after equine ICSI. *Theriogenology* **2016**, *85*, 1289–1296. [CrossRef]
4. Galli, C.; Duchi, R.; Colleoni, S.; Lagutina, I.; Lazzari, G. Ovum pick up, intracytoplasmic sperm injection and somatic cell nuclear transfer in cattle, buffalo and horses: From the research laboratory to clinical practice. *Theriogenology* **2014**, *81*, 138–151. [CrossRef] [PubMed]
5. Lewis, N.; Hinrichs, K.; Leese, H.J.; Argo, C.M.; Brison, D.R.; Sturmey, R. Energy metabolism of the equine cumulus oocyte complex during in vitro maturation. *Sci. Rep.* **2020**, *10*, 3493. [CrossRef] [PubMed]
6. González-Fernández, L.; Sánchez-Calabuig, M.J.; Alves, M.G.; Oliveira, P.F.; Macedo, S.; Gutiérrez-Adán, A.; Rocha, A.; Macías-García, B. Expanded equine cumulus-oocyte complexes exhibit higher meiotic competence and lower glucose consumption than compact cumulus-oocyte complexes. *Reprod. Fertil. Dev.* **2018**, *30*, 297–306. [CrossRef] [PubMed]
7. Galli, C.; Lazzari, G. The manipulation of gametes and embryos in farm animals. *Reprod. Domest. Anim. Zuchthyg.* **2008**, *43* (Suppl. 2), 1–7. [CrossRef]
8. Walter, J.; Huwiler, F.; Fortes, C.; Grossmann, J.; Roschitzki, B.; Hu, J.; Naegeli, H.; Laczko, E.; Bleul, U. Analysis of the equine "cumulome" reveals major metabolic aberrations after maturation in vitro. *BMC Genom.* **2019**, *20*, 588. [CrossRef]
9. Accogli, G.; Douet, C.; Ambruosi, B.; Martino, N.A.; Uranio, M.F.; Deleuze, S.; Dell'Aquila, M.E.; Desantis, S.; Goudet, G. Differential expression and localization of glycosidic residues in in vitro- and in vivo-matured cumulus-oocyte complexes in equine and porcine species. *Mol. Reprod. Dev.* **2014**, *81*, 1115–1135. [CrossRef]
10. Embade, N.; Cannet, C.; Diercks, T.; Gil-Redondo, R.; Bruzzone, C.; Ansó, S.; Echevarría, L.R.; Ayucar, M.M.M.; Collazos, L.; Lodoso, B.; et al. NMR-based newborn urine screening for optimized detection of inherited errors of metabolism. *Sci. Rep.* **2019**, *9*, 13067. [CrossRef]
11. Beckonert, O.; Keun, H.C.; Ebbels, T.M.D.; Bundy, J.; Holmes, E.; Lindon, J.C.; Nicholson, J.K. Metabolic profiling, metabolomic and metabonomic procedures for NMR spectroscopy of urine, plasma, serum and tissue extracts. *Nat. Protoc.* **2007**, *2*, 2692–2703. [CrossRef]
12. Wishart, D.S. Quantitative metabolomics using NMR. *TrAC Trends Anal. Chem.* **2008**, *27*, 228–237. [CrossRef]
13. González-Fernández, L.; Sánchez-Calabuig, M.J.; Calle-Guisado, V.; García-Marín, L.J.; Bragado, M.J.; Fernández-Hernández, P.; Gutiérrez-Adán, A.; Macías-García, B. Stage-specific metabolomic changes in equine oviductal fluid: New insights into the equine fertilization environment. *Theriogenology* **2020**, *143*, 35–43. [CrossRef] [PubMed]
14. Gérard, N.; Loiseau, S.; Duchamp, G.; Seguin, F. Analysis of the variations of follicular fluid composition during follicular growth and maturation in the mare using proton nuclear magnetic resonance (1H NMR). *Reprod. Camb. Engl.* **2002**, *124*, 241–248. [CrossRef] [PubMed]

15. Gu, L.; Liu, H.; Gu, X.; Boots, C.; Moley, K.H.; Wang, Q. Metabolic control of oocyte development: Linking maternal nutrition and reproductive outcomes. *Cell. Mol. Life Sci.* **2015**, *72*, 251–271. [CrossRef] [PubMed]
16. Harvey, A.J. Mitochondria in early development: Linking the microenvironment, metabolism and the epigenome. *Reproduction* **2019**, *157*, R159–R179. [CrossRef]
17. Piñero-Sagredo, E.; Nunes, S.; de los Santos, M.J.; Celda, B.; Esteve, V. NMR metabolic profile of human follicular fluid. *NMR Biomed.* **2010**, *23*, 485–495. [CrossRef]
18. Guo, X.; Wang, X.; Di, R.; Liu, Q.; Hu, W.; He, X.; Yu, J.; Zhang, X.; Zhang, J.; Broniowska, K.; et al. Metabolic Effects of FecB Gene on Follicular Fluid and Ovarian Vein Serum in Sheep (Ovis aries). *Int. J. Mol. Sci.* **2018**, *19*, 539. [CrossRef]
19. Gérard, N.; Prades, I.; Couty, M.; Labberté, M.; Daels, P.; Duchamp, G. Concentrations of glucose, pyruvate and lactate in relation to follicular growth, preovulatory maturation and oocyte nuclear maturation stage in the mare. *Theriogenology* **2000**, *53*, 372.
20. Guarino, V.A.; Oldham, W.M.; Loscalzo, J.; Zhang, Y.-Y. Reaction rate of pyruvate and hydrogen peroxide: Assessing antioxidant capacity of pyruvate under biological conditions. *Sci. Rep.* **2019**, *9*, 19568. [CrossRef]
21. Gérard, N.; Fahiminiya, S.; Grupen, C.G.; Nadal-Desbarats, L. Reproductive Physiology and Ovarian Folliculogenesis Examined via ^{1}H-NMR Metabolomics Signatures: A Comparative Study of Large and Small Follicles in Three Mammalian Species (*Bos taurus*, *Sus scrofa domesticus* and *Equus ferus caballus*). *OMICS J. Integr. Biol.* **2015**, *19*, 31–40. [CrossRef] [PubMed]
22. Tews, J.K.; Carter, S.H.; Roa, P.D.; Stone, W.E. Free amino acids and related compounds in dog brain: Post-mortem and anoxic changes, effects of ammonium chloride infusion, and levels during seizures induced by picrotoxin and by pentylenetetrazol. *J. Neurochem.* **1963**, *10*, 641–653. [CrossRef] [PubMed]
23. Engle, C.C.; Foley, C.W. Certain physiochemical properties of uterine tubal fluid, follicular fluid, and blood plasma in the mare. *Am. J. Vet. Res.* **1975**, *36*, 149–154. [PubMed]
24. Wallace, M.; Cottell, E.; Gibney, M.J.; McAuliffe, F.M.; Wingfield, M.; Brennan, L. An investigation into the relationship between the metabolic profile of follicular fluid, oocyte developmental potential, and implantation outcome. *Fertil. Steril.* **2012**, *97*, 1078–1084.e8. [CrossRef] [PubMed]
25. Ménézo, Y.; Guérin, P.; Elder, K. The oviduct: A neglected organ due for re-assessment in IVF. *Reprod. Biomed. Online* **2015**, *30*, 233–240. [CrossRef] [PubMed]
26. Lamy, J.; Gatien, J.; Dubuisson, F.; Nadal-Desbarats, L.; Salvetti, P.; Mermillod, P.; Saint-Dizier, M. Metabolomic profiling of bovine oviductal fluid across the oestrous cycle using proton nuclear magnetic resonance spectroscopy. *Reprod. Fertil. Dev.* **2018**, *30*, 1021. [CrossRef] [PubMed]
27. Gatien, J.; Mermillod, P.; Tsikis, G.; Bernardi, O.; Janati Idrissi, S.; Uzbekov, R.; Le Bourhis, D.; Salvetti, P.; Almiñana, C.; Saint-Dizier, M. Metabolomic Profile of Oviductal Extracellular Vesicles across the Estrous Cycle in Cattle. *Int. J. Mol. Sci.* **2019**, *20*, 6339. [CrossRef]
28. Marinaro, F.; Macías-García, B.; Sánchez-Margallo, F.M.; Blázquez, R.; Álvarez, V.; Matilla, E.; Hernández, N.; Gómez-Serrano, M.; Jorge, I.; Vázquez, J.; et al. Extracellular vesicles derived from endometrial human mesenchymal stem cells enhance embryo yield and quality in an aged murine model. *Biol. Reprod.* **2018**. [CrossRef]
29. Hung, W.-T.; Hong, X.; Christenson, L.K.; McGinnis, L.K. Extracellular Vesicles from Bovine Follicular Fluid Support Cumulus Expansion1. *Biol. Reprod.* **2015**, *93*. [CrossRef]
30. Montjean, D.; Entezami, F.; Lichtblau, I.; Belloc, S.; Gurgan, T.; Menezo, Y. Carnitine content in the follicular fluid and expression of the enzymes involved in beta oxidation in oocytes and cumulus cells. *J. Assist. Reprod. Genet.* **2012**, *29*, 1221–1225. [CrossRef]
31. Chen, M.; Zhang, B.; Cai, S.; Zeng, X.; Ye, Q.; Mao, X.; Zhang, S.; Zeng, X.; Ye, C.; Qiao, S. Metabolic disorder of amino acids, fatty acids and purines reflects the decreases in oocyte quality and potential in sows. *J. Proteom.* **2019**, *200*, 134–143. [CrossRef] [PubMed]
32. Hemmings, K.E.; Maruthini, D.; Vyjayanthi, S.; Hogg, J.E.; Balen, A.H.; Campbell, B.K.; Leese, H.J.; Picton, H.M. Amino acid turnover by human oocytes is influenced by gamete developmental competence, patient characteristics and gonadotrophin treatment. *Hum. Reprod.* **2013**, *28*, 1031–1044. [CrossRef] [PubMed]
33. Houghton1, F.D.; Hawkhead, J.A.; Humpherson, P.G.; Hogg, J.E.; Balen, A.H.; Rutherford, A.J.; Leese, H.J. Non-invasive amino acid turnover predicts human embryo developmental capacity. *Hum. Reprod.* **2002**, *17*, 999–1005. [CrossRef] [PubMed]

34. Steeves, C.L.; Baltz, J.M. Regulation of intracellular glycine as an organic osmolyte in early preimplantation mouse embryos. *J. Cell. Physiol.* **2005**, *204*, 273–279. [CrossRef]
35. Hugentobler, S.A.; Diskin, M.G.; Leese, H.J.; Humpherson, P.G.; Watson, T.; Sreenan, J.M.; Morris, D.G. Amino acids in oviduct and uterine fluid and blood plasma during the estrous cycle in the bovine. *Mol. Reprod. Dev.* **2007**, *74*, 445–454. [CrossRef] [PubMed]
36. Sinclair, K.; Lunn, L.; Kwong, W.; Wonnacott, K.; Linforth, R.; Craigon, J. Amino acid and fatty acid composition of follicular fluid as predictors of in-vitro embryo development. *Reprod. Biomed. Online* **2008**, *16*, 859–868. [CrossRef]
37. Uzbekova, S.; Salhab, M.; Perreau, C.; Mermillod, P.; Dupont, J. Glycogen synthase kinase 3B in bovine oocytes and granulosa cells: Possible involvement in meiosis during in vitro maturation. *Reproduction* **2009**, *138*, 235–246. [CrossRef]
38. Otani, T.; Maruo, T.; Yukimura, N.; Mochizuki, M. Effect of insulin on porcine granulosa cells: Implications of a possible receptor mediated action. *Acta Endocrinol.* **1985**, *108*, 104–110. [CrossRef]
39. Schmidt, G.; Owman, C.; Sjoberg, N.-O. Histamine induces ovulation in the isolated perfused rat ovary. *Reproduction* **1986**, *78*, 159–166. [CrossRef]
40. Kobayashi, Y.; Wright, K.H.; Santulli, R.; Kitai, H.; Wallach, E.E. Effect of Histamine and Histamine Blockers on the Ovulatory Process in the in Vitro Perfused Rabbit Ovary. *Biol. Reprod.* **1983**, *28*, 385–392. [CrossRef]
41. Watson, E.D.; Sertich, P.L. Concentrations of arachidonate metabolites, steroids and histamine in preovulatory horse follicles after administration of human chorionic gonadotrophin and the effect of intrafollicular injection of indomethacin. *J. Endocrinol.* **1991**, *129*, 131–139. [CrossRef] [PubMed]
42. González-Fernández, L.; Macedo, S.; Lopes, J.S.; Rocha, A.; Macías-García, B. Effect of Different Media and Protein Source on Equine Gametes: Potential Impact During In Vitro Fertilization. *Reprod. Domest. Anim. Zuchthyg.* **2015**, *50*, 1039–1046. [CrossRef]

MDPI
St. Alban-Anlage 66
4052 Basel
Switzerland
Tel. +41 61 683 77 34
Fax +41 61 302 89 18
www.mdpi.com

Animals Editorial Office
E-mail: animals@mdpi.com
www.mdpi.com/journal/animals

www.ingramcontent.com/pod-product-compliance
Lightning Source LLC
LaVergne TN
LVHW070128170726
843515LV00007B/1765
9783036508283